T0254778

# Water Resources Development and Management

More information about this series at http://www.springer.com/series/7009

Alexandra Nauditt · Lars Ribbe
Editors

# Land Use and Climate Change Interactions in Central Vietnam

## LUCCi

*Editors*
Alexandra Nauditt
Institute for Technology and Resources Management in the Tropics and Subtropics
Cologne (Deutz)
Germany

Lars Ribbe
Institute for Technology and Resources Management in the Tropics and Subtropics
Cologne (Deutz)
Germany

ISSN 1614-810X ISSN 2198-316X (electronic)
Water Resources Development and Management
ISBN 978-981-10-9669-3 ISBN 978-981-10-2624-9 (eBook)
DOI 10.1007/978-981-10-2624-9


Softcover reprint of the hardcover 1st edition 2017


Printed on acid-free paper

This Springer imprint is published by Springer Nature
The registered company is Springer Nature Singapore Pte Ltd.
The registered company address is: 152 Beach Road, #21-01/04 Gateway East, Singapore 189721, Singapore

# Contents

# Editors and Contributors

## About the Editors

**Alexandra Nauditt** is an environmental scientist with a focus on hydrology and river basin management (www.basin-info.net) at the Institute for Technology and Resources Management in the Tropics and Subtropics (ITT). Her research focusses on water management and hydrological drought assessment on catchment scale. Since 1999, she has been involved in educational and research projects at ITT dealing with river basin assessment and management, drought risk analyses, water allocation and stakeholder involvement as well as interactions between climate variability, hydrology and land uses in semi-arid and tropical environments with a regional focus on South East Asia and Latin America. Recently she has been coordinating the BMBF funded research projects "Land use and Climate change Interactions in Central Vietnam—LUCCi" (www.lucci-vietnam.info) and "Water Use Efficiency in Semi Arid Central Chile" (www.hidro-limari.info).

**Lars Ribbe** is Professor of Integrated Land and Water Resources Management and director of the Institute for Technology and Resources Management in the Tropics and Subtropics (ITT), TH Köln. His work areas include river basin assessment, modelling and management and he is specifically interested in developing knowledge systems that help decision-makers to cope with prevailing water resources related challenges such as water scarcity and drought, floods and pollution. Recently he has been involved in bridging scientific and sectoral approaches in the context of the water, energy and food security nexus by forming an interdisciplinary group on that matter at Cologne University of Applied Sciences.

## Contributors

**Le Van An** is an associate professor and the Rector of the Hue University of Agriculture and Forestry (HUAF), Hue University, Vietnam. He obtained his MSc and Ph.D. in Agriculture at the Swedish University of Agricultural Sciences (SLU), Sweden. He is also a Professor at Okayama University, Japan for Global Partnership and Education overseas. He is coordinator for a number of research projects on community development, climate change response and natural resources management.

**Valerio Avitabile** is a postdoc researcher at the Laboratory of Geo-information Science and Remote Sensing of Wageningen University (the Netherlands) and has received the Ph.D. degree from the Friedrich-Schiller University (Jena, Germany) in 2012 working on thesis entitled "Optical Remote Sensing for Biomass Estimation in the Tropics: the case study of Uganda". He has expertise on environment monitoring and assessment from local to regional scale. His research activities focus on forest monitoring from remote sensing with specialisation on the assessment of forest biomass in the tropics and the integration of field and remote sensing data streams.

**Astrid Bos** is a Ph.D. candidate at the Laboratory of Geo-information Science and Remote Sensing of Wageningen University (The Netherlands) where she is working in collaboration with the Center for International Forestry Research (CIFOR) on the Global Comparative Study on REDD+. She has expertise in impact evaluation and land use and land cover change monitoring and modelling. Her interests lie in interdisciplinary research, in which remote sensing and GIS are combined with social science techniques. She holds two M.Sc. degrees in Geographical Sciences (2013) and Environmental Sciences (2012) of the Utrecht University (NL) and a B.Sc. degree in International Development Studies (2008) of the Wageningen University (NL).

**Johannes Cullmann** is leading the World Meteorological Organisation's climate and water related activities since October 2015. Prior, he coordinated international water affairs within the German government and with international partners. He was the President of UNESCO's water programme from 2012–2014. In his function as a department head in the German Federal Institute for Hydrology, he was the German representative in the Commission for the hydrology of the Rhine river. He was responsible for the German support for UNEP's water quality related activities and he is familiar with data sharing arrangements in support of creating political trust in disputed situations. Mr. Cullmann worked for the Max Planck Institute in Brazil and for the German Development Cooperation in Chile. He was a senior researcher and teacher at German Universities for 5 years with a focus on hydrological modelling, flood forecasting and artificial intelligence. Mr. Cullmann holds a Master of Hydrology, a Ph.D. in flood forecasting as well as a Habilitation in Hydrology from Technical University in Dresden, Germany. He also completed a Master of Public Administration at the Hertie School of Governance in Berlin.

**Pham Manh Cuong** is the former national REDD+ focal point and Director of Vietnam REDD+ Office of Vietnam and received a Ph.D. degree from the University of Goettingen, Germany in 2005. He worked for more than twelve years at the National Forest Inventory and Planning Institute (FIPI) before becoming a policy-maker at the Ministry of Agriculture and Rural Development (MARD) of Vietnam. Currently, he is on a two-year assignment at the Department of

Agriculture and Rural Development (DARD) of Tuyen Quang Province. He has great experience in implementing practical management, policy formulation and international cooperation in the forestry sector, and working with multicultural and international organisations.

**Dang Quang Thinh** is a Ph.D. student at Karlsruhe Institute of Technology, Institute of Meteorology and Climate Research (KIT/IMK-IFU), Germany. He holds a Master degree in Hydrology and Water Resources at Institute for Water Education, UNESCO-IHE, The Netherlands. His research fields focus on catchment hydrology and water resources, and impact of climate change on water balance and agricultural sector. He holds expertise in hydrological–hydraulic modelling and hydrological extreme analysis. Another focus of his work is the optimisation modelling. He is a hydrological modeller in the LUCCi project.

**Manfred Fink** is senior researcher and lecturer at the Chair of Geographic Information Science, Department of Geography, Friedrich-Schiller-University Jena, Germany. He received a Diploma in Geography, Soil Science, Water management and Zoology from Kiel University and a Ph.D. in Geography from University of Jena. The focus of his research emphasizes on the development, extension and application of eco-hydrological models as well as the assimilation of primary data in the field by to designing and operation monitoring programs. The research field comprises the development and application of geographic information processing methods to analyze and prepare data for model input. He led and participated several international projects with partners from Turkey, Poland and China. Recently he is working on the question, how to communicate the implications of climate change to people working in water management.

**A.B.M. Firoz** is a researcher at the Institute for Technology and Resources Management in the Tropics and Subtropics (ITT) at TH Köln. He is a specialist in river basin management modelling and hydrogeological msodels. He has worked on a wide variety of hydrological and hydrogeological issues around the world. For more than five years he worked in water management projects in Vietnam and Bangladesh for ITT. His special interests include the development and management of groundwater resources, basin scale hydrological and hydrogeological modelling and the management of water resources at river basin scale including reservoir modelling and irrigation water management.

**Duong Van Hau** is a lecturer at the department of chemistry of the Hue University of Agriculture and Forestry. He obtained a BSc from the Hue College of Education in 2005 and MSc in Hue Univeristy of Science in 2009. He has worked on developing methods for measurement of greenhouse gas emissions (GHG). He is currently pursuing a Ph.D. at the Hue University of Science. His research focuses on nano materials with potential widespread agricultural application such as chitosan nano, chitosan hydrogels, carbon nano…in relation to climate change adaptation and mitigation.

**Martin Herold** is Professor in remote sensing with Wageningen University (the Netherlands) and has received the Ph.D. degree from the University of California at Santa Barbara (CA, USA) in 2004, and the Habilitation (second Ph.D.) degree in 2009. He has worldwide experiences in global land cover observation and change monitoring. He leads the Global Land Cover Panel of the UN Global Terrestrial Observing System and has been coordinating the ESA GOFC GOLD Land Cover Project Office. He is a leading expert on REDD MRV and has worked for UNFCCC secretariat, FAO, World bank, and several national governments including Norway, Germany, New Zealand, Guyana, Indonesia, Vietnam, and Fiji.

**Nadine Herold** is a soil scientist at Alterra, Wageningen University and Research Centre (the Netherlands) and has received the Ph.D. degree from the Friedrich-Schiller University (Jena, Germany) in 2013. Her expertise is in the field of nutrient and natural resource management, smart and sustainable soil management in the context of climate change adaptation and mitigation, land use change and management. She has ample overseas experience and is frequently involved in consultations on policy level regarding sustainable soil and land management and natural-resource-based activities (particularly agriculture). She has work experience in Uganda, Vanuatu, Vietnam, New Zealand and Germany.

**Vu Quang Hien** completed in 2013 a Master degree the Institute of Ecology and Biological Resources (IEBR) of Vietnam in the botanic section with a thesis named "Flora assessment in Ben En national park", and from 2004 to 2008 studied silviculture at the Vietnam Forestry University. Currently, he is a researcher and officer at the Division of Training and International Cooperation under Forest Inventory and Planning Institute (FIPI) of Vietnam. His present activities focus on the issues of Climate Change, Forest Sustainable Management, Forestry Community, and he is also the focal point and coordinator of several FIPI projects and activities such as those related to the Global Forest Observation Initiative (GFOI), SilvaCarbon, and the Institutional Partnership For Improving Forecast Capacity in Vietnam's Forest Sector.

**Trong Nghia Hoang** is a Crop Scientist/Researcher at Hue University of Agriculture and Forestry. He holds an MSc in Crop Science from Hue University of Agriculture and Forestry. His fields of expertise include greenhouse gas analysis by gas chromatography, crop science and environmental science. He has worked with local and international partners on evaluating mitigation options for rice systems in Central Vietnam.

**Hoang Ho Dac Thai** is the Director of the Institute of Natural Resources and Environmental Science at Hue University. He holds a doctorate degree in Forestry from the Albert-Ludwigs-Universität Freiburg and has a strong track record in advisory services for the fields of forest landscape ecology, silvicultural treatments in tropical forests and coastal dunes system restoration in Vietnam. His research focuses on ecosystem-based disaster risk reduction and ecological assessment of particularly coastal environments.

**Harald Kunstmann** is deputy director and head of the department "Regional Climate Systems" at Karlsruhe Institute of Technology (KIT), Institute of Meteorology and Climate Research (IMK-IFU), at Campus Alpin, Garmisch-Partenkirchen, Germany, and holds the Chair for Regional Climate and Hydrology at University of Augsburg. His expertise comprises regional climate and hydrology modeling, the development of coupled atmosphere-hydrology model systems, the integration of remote sensing information in hydrometerological analyses and the establishment and operation of terrestrial observatories. He is member of the Scientific Steering Committee of TERENO (http://www.tereno.net), vice president of the *International Commission on the Coupled Land-Atmosphere System* (IAHS-ICCLAS) and was associate editor of *Journal of Hydrology* from 2010–2015.

**Patrick Laux** is a senior researcher at Karlsruhe Institute of Technology, Institute of Meteorology and Climate Research (KIT/IMK-IFU). He holds a Ph.D. degree in climatology from the Institute of Hydraulic Engineering, University of Stuttgart, and a diploma degree in Applied Environmental Sciences, University of Trier. His research centres around climate change and climate variability, and their impacts on water balance and agricultural productivity. He holds expertise in regional climate modelling based on dynamical as well as statistical downscaling approaches for various regions worldwide, mostly semiarid regions, such as sub-Saharan Africa. Another focus of his work is seasonal climate prediction. He was the leading climate scientist in the LUCCi project.

**Markus Meinhardt** is research assistant at the Chair of Geographic Information Science at the Friedrich-Schiller-University Jena. Germany. With his Bachelor degree in Geography and Master in Geoinformatics he is able to understand, analyze and solve environmental problems from the ecological but also from the technical point of view. Concerning the research presented here, he collected landslide data during a three month research stay in Vietnam, subsequently using the data for landslide susceptibility modeling. He also gathered a wealth of experience in the implementation and monitoring of measurement networks concerning climate, soil, and water quality and -quantity. By mid-2017 he will obtain his Ph.D. in hydrological modelling and climate change analysis.

**Ngo Duc Minh** is working for the CGIAR Program of Climate Change, Agriculture and Food Security (CCAFS) as an Associate Scientist at IRRI Vietnam/CCAFS SEA Office in Hanoi. He obtained a Ph.D. in Environmental Sciences in 2013 from the University of the Philippines at Los Baños. He has been an Environmental/Soil Scientist at the Soils and Fertilizers Research Institute of Vietnam from 2002–2011 and also served as a visiting researcher in the Soils Department of The James Hutton Institute, Scotland in 2009 and 2010. He has worked on a number of international collaborative research projects with international partners in the research fields relevant to sustainable management of water-land resources and nutrient, mitigation of environment pollution in irrigated rice-based systems, land-use and climate change… in Southeast Asia.

**Udo Nehren** is a senior researcher and lecturer in Physical Geography and Ecosystem Management at the Institute for Technology and Resources Management in the Tropics and Subtropics, TH Köln, Germany. He received a diploma in Geography and Geosciences from Trier University, an M.Eng. in Technology in the Tropics from TH Köln, and a Ph.D. in Geography from University of Leipzig. His research emphasises on landscape evolution, ecosystem-based disaster risk reduction and adaptation, and ecosystem management with regional foci in Latin America and SE Asia. He published several books and articles on these topics and is one of the developers of the Massive Open Online Course "Disasters and Ecosystems: Resilience in a Changing Climate"

**Phuong Ngoc Bich Nguyen** is a researcher at Vietnam Institute of Meteorology, Hydrology and Climate Change (IMHEN). She holds a Master degree in climatology from the Hanoi University of Science (HUS). Her research centers on climate variability, climate prediction and climate change. She holds expertise in regional climate modelling based on dynamical as well as statistical downscaling approaches for Vietnam and South East Asia regions. She was hired as a climate scientist in the LUCCi project.

**Rui Pedroso** is a lecturer and senior researcher at the Institute for Technology and Resources Management in the Tropics and Subtropics at the TH-Köln. He has a Ph.D. in Agricultural Economics and an M.Sc. in Economics (both from the University of Bonn), his research interests are irrigation in general and design of pressurised irrigation systems in particular, as well as economic analysis of farming systems. His methodological focus is on econometrics and optimisation modelling. Dr. Pedroso was coordinating the agricultural component of the BMBF funded research project "Land use and Climate change Interactions in Central Vietnam—LUCCi" (www.lucci-vietnam.info)

**Arun Kumar Pratihast** is a postdoc researcher at the Laboratory of Geo-information Science and Remote Sensing of Wageningen University (the

Netherlands). He received his Ph.D. in Wageningen University and at the Institute for Technology and Resources Management in the Tropics and Subtropics (ITT), Cologne University of Applied Sciences (CUAS), Germany. His primary expertise lies in citizen science, geo-information technologies and mobile application development for environmental monitoring. He has explored innovative ways to combining satellite data and community-based observations for near real-time forest monitoring. He has involved in various projects as the USAID community-based interactive monitoring system project in Peru and the Google project near real-time global deforestation monitoring using Google Earth Engine.

**Claudia Raedig** (Ph.D.) is a senior researcher and lecturer in the field of biodiversity and connectivity conservation at the Institute for Technology and Resources Management in the Tropics and Subtropics at TH Köln, University of Applied Sciences, Germany. Her research focuses on connectivity conservation management, particularly on the development of strategies for enhancing connectivity for biodiversity conservation in tropical and coastal ecosystems in Southeast Asia and Latin America.

**Giulia Salvini** is a Ph.D. candidate at the Laboratory of Geo-information Science and Remote Sensing of Wageningen University (the Netherlands). Her research interests span a wide range of topics including policy analysis, landscape planning, participatory approaches and environmental governance in the context of sustainable development and climate change adaptation and mitigation. In particular, she is interested in agriculture innovation, forest conservation and sustainable landscape management via multi-stakeholder platforms. She holds an M.Sc. degree in Natural resources management from the University of Bologna, Italy (2007) and she worked as a researcher at Vrije Universiteit, Amsterdam (2007–2009) and at Ca Foscari University, Venice (2009–2011).

**Bjoern Ole Sander** is a scientist in IRRI's climate change research group with a focus on greenhouse gas (GHG) mitigation technologies. He holds a Ph.D. in Chemistry from the University of Kiel (Germany) and joined IRRI in 2010. As climate change specialist, Ole analyzes the GHG balance of various cropping systems in Vietnam, Bangladesh and the Philippines. He also identifies suitable environmental and policy conditions to support dissemination of mitigation technologies, particularly the water saving method alternate wetting and drying (AWD).

**Michael Schultz** is a Ph.D. candidate at the Laboratory of Geo-information Science and Remote Sensing of Wageningen University (the Netherlands), where he is contributing to the REDD+ research activities of the chair group. He received undergraduate degree and M.Sc. degree in Geoinformatics and remote sensing from the Friedrich-Schiller-University (Jena, Germany), where he was involved in different projects such as the HyEurope 2010 campaign, the ILMS project and contributed to GOFC-GOLD activities. His research interests include land cover and land cover change mapping, remote sensing time series analysis, and forest monitoring.

**Harro Stolpe** is Professor of Water and Groundwater Management at the Ruhr University Bochum and Director of the Institute "Environmental engineering and ecology, EEE". He has more than 30 years of experience in the areas of environmental engineering, ecology, hydrogeology and water management. He has led a large number of research projects focusing on groundwater, ecology, integrated water resources management, environmental planning, human affected ecosystems, and application of GIS-based modelling in resources management in Vietnam, Germany and South Africa.

**Agnes Tirol-Padre** is a project scientist at the Crop and Environmental Sciences Division (CESD) of the International Rice Research Institute (IRRI). She holds a Ph.D. in Soil Science from Chiba University, Japan and MSc in Agricultural Chemistry from the University of the Philippines at Los Baños. She has worked on developing protocols for measurement, reporting and verification of greenhouse gas emissions and evaluating mitigation options for rice systems in Asia. She has been coordinating climate change mitigation research in partnership with local institutes in Vietnam, Indonesia, Thailand and Philippines. Her fields of expertise include environmental science, conservation agriculture, soil fertility and nutrient cycling.

**Dang Hoa Tran** is an associate professor and the Dean of the Faculty of Agronomy, at the Hue University of Agriculture and Forestry. He received his Ph.D. from Kyushu University, Japan in 2007. His research focus is on climate change adaptation and mitigation in agriculture and safe agriculture. He is currently at work on some international collaborative projects on Climate Smart Agriculture.

**Viet Quoc Trinh** is a senior researcher in water resources management at the Institute for Technology and Resources Management in the Tropics and Subtropics, TH Köln, Germany. He received a bachelor degree in environmental geography from Vietnam National University-Hochiminh City, an M.Sc. degree in Technology and Resources Management from TH Köln, and a Ph.D. degree in Environmental Engineering from Ruhr University Bochum. His expertise focuses on hydrodynamic modelling, GIS application, irrigation performance, risk assessment of saltwater intrusion, water shortages and flood.

**Hannes Tünschel** is M.Sc. student at the Chair of Geographic Information Science Department of Geography, Friedrich-Schiller-University Jena, Germany. He has a Bachelor degree in Geography with focus on geoinformatics and ecology which allows a holistic solution of spatial questions. He has longtime experiences in implementation and monitoring of measurement networks concerning climate, soil, and water quality and -quantity as well as spectrometric measurements on the one hand and in communicating and teaching GIS knowledge on the other hand. During a two month internship in central Vietnam he collected the landslide data which are used in this research.

**Moussa Waongo** is a lecturer and the coordinator of Agrometeorological training at AGRHYMET Regional Centre, a specialized institution of the Permanent Interstates Committee for Drought Control in the Sahel (CILSS). He holds a Ph.D. degree in agroclimatology from the Institute of Geography, faculty of Applied Computer Science, University of Augsburg, Germany. He also obtained an Engineer degree in Agrometeorology at AGRHYMET Regional Centre, Niger and a master degree in applied physics at the faculty of Natural Science, University of Ouagadougou, Burkina Faso. His research centres are meteorology, climate and agriculture with a focus on the impact of climate variability and climate change in crop production. He holds expertise in meteorology, climate, crop modelling and seasonal climate prediction. He performed studies aiming to the improvement of crop water management in the Sahel, a water-scarse region in West Africa.

**Reiner Wassmann** is the Coordinator of Climate Change Research at IRRI. He is also affiliated with the Karlsruhe Institute of Technology (Germany) where he holds a permanent position as Senior Scientist with several delegations to work at IRRI. Reiner Wassmann holds a Ph.D. in Biology from Göttingen University (Germany) and works on climate change research since 1987 and focuses on rice production systems since 1991. While his initial research addressed GHG emissions and mitigation, his current portfolio covers a wide range of aspects related to rice systems including adaptation. Geographically his ongoing research is concentrated in Southeast Asia.

**Franziska Zander** is a Geographer who made her minor subject computer science especially database management systems and applications to her main field of work. Since 2008 she works as a researcher at the Chair of Geographic Information Science at the Friedrich-Schiller-University in Jena and takes care about the environmental data of her colleagues and project partner. As one of the main developers of the web-based environmental Information RBIS (River Basin Information System) she is not only responsible for the administration and maintenance of the system but also always looking forward to integrated new functions and data types to fit current needs and challenges.

**S. Alfonso** Institute for Technology and Resources Management in the Tropics and Subtropics (ITT), TH Köln, University of Applied Sciences, Cologne, Germany

**Christian Fischer** Department of Geography, Friedrich-Schiller University of Jena, Jena, Germany

**Wolfgang-Albert Flügel** Department of Geography, University of Jena, Jena, Germany

**Tran Thi Ngan** College of Agriculture, Hue University, Hue, Vietnam

**Anh Thu Nguyen** Institute for Technology and Resources Management in the Tropics and Subtropics, Technische Hochschule Köln - University of Applied Sciences, Köln, Germany

**Uyen Nguyen** Institute for Technology and Resources Management in the Tropics and Subtropics, Technische Hochschule Köln - University of Applied Sciences, Köln, Germany

**Khac Phuc Le** Faculty of Agronomy, College of Agriculture and Forestry, Hue University, Hue, Vietnam

**N.D. Trung** Institute for Technology and Resources Management in the Tropics and Subtropics (ITT), TH Köln, University of Applied Sciences, Cologne, Germany

# Abbreviations

| | |
|---|---|
| ABM | Agent-based model |
| AUC | Area under the curve |
| AWD | Alternate wetting and drying |
| BAU | Business as usual |
| BMBF | German Ministry of Research and Education |
| CBD | Convention on biological diversity |
| CC | Climate change |
| CDS | Coastal dune system |
| CF | Continuous flooding |
| CSA | Climate-smart agriculture |
| CV | Coefficient of variation |
| DARD | Department of Agriculture and Rural Development |
| DAWACO | Danang Water Supply Company |
| DBH | Diameter at breast height |
| DEM | Digital elevation model |
| DONRE | Department of Natural Resources and Environment |
| DSO | Danang Statistics Office |
| EbA | Ecosystem-based adaptation |
| Eco-DRR | Mangroves, ecosystem-based disaster risk reduction |
| EVN | Electricity Vietnam |
| FAO | Food and Agricultural Organization |
| GDP | Gross domestic product |
| GHG | Greenhouse gas |
| GIS | Geographic information system |
| GPS | Global positioning system |
| GSO | General Statistics Office of Vietnam |
| HRU | Hydrological response unit |
| IMC | Irrigation Management Company |
| IPCC | Intergovernmental Panel on Climate Change |

| | |
|---|---|
| ITT | Institute for Technology and Resources Management in the Tropics and Subtropics |
| IWRM | Integrated water resources management |
| LUCCi | Land Use and Climate Change Interactions in Central Vietnam |
| LULC | Land use land cover |
| MARD | Ministry of Agriculture and Rural Development of Vietnam |
| MONRE | Ministry of Natural Resources and Environment of Vietnam |
| MOST | Ministry of Science and Technology of Vietnam |
| NFI | National forest inventory |
| NGO | Non-governmental organisation |
| PES | Payment for ecosystem services |
| PPC | Provincial People's Committee |
| PRA | Participator rural appraisal |
| QSO | Quangnam Statistics Office |
| RBIC | Vu Gia Thu Bon River Basin Information Centre |
| RBIS | River Basin Information System |
| RCHM | Regional Centre for Hydrology and Meteorology |
| REDD+ | Reducing emissions from deforestation and forest degradation |
| RS | Remote sensing |
| SA | Summer-autumn crop |
| SDG | Sustainable development goal |
| SLR | Sea level rise |
| SOC | Soil organic carbon |
| SWI | Saltwater intrusion |
| TE | Technical efficiency |
| TFP | Total factor productivity |
| UN | United Nations |
| UNFCCC | United Nations Framework Convention on Climate Change |
| VAWR | Vietnam Academy of Water Resources |
| VGTB | Vu Gia Thu Bon River Basin |
| VND | Vietnamese currency |
| WEF | Water-energy and food security nexus |
| WS | Winter-spring crop |
| WTP | Water treatment plant |
| WUO | Water user organisation |
| WWTP | Wastewater treatment plant |

# List of Figures

**Biophysical and Socio-economic Features of the LUCCi—Project Region: The Vu Gia Thu Bon River Basin**



**Vu Gia Thu Bon River Basin Information System (VGTB RBIS)—Managing Data for Assessing Land Use and ClimateChange Interactions in Central Vietnam**

**Forest Change and REDD+ Strategies**

**Measuring GHG Emissions from Rice Production in Quang Nam Province (Central Vietnam): Emission Factors for Different Landscapes and Water Management Practices**

**Hydrological and Agricultural Impacts of Climate Change in the Vu Gia-Thu Bon River Basin in Central Vietnam**

**Impacts of Land-Use/Land-Cover Change and Climate Change on the Regional Climate in the Central Vietnam**

## Integrated River Basin Management in the Vu Gia Thu Bon Basin



## Land Use Adaption to Climate Change in the Vu Gia–Thu Bon Lowlands: Dry Season and Rainy Season

**Distributed Assessment of Sediment Dynamics in Central Vietnam**



**Sand Dunes and Mangroves for Disaster Risk Reduction and Climate Change Adaptation in the Coastal Zone of Quang Nam Province, Vietnam**

## Hydrological Drought Risk Assessment in an Anthropogenically Impacted Tropical Catchment, Central Vietnam



## Conclusion

# List of Tables

## Measuring GHG Emissions from Rice Production in Quang Nam Province (Central Vietnam): Emission Factors for Different Landscapes and Water Management Practices



## Hydrological and Agricultural Impacts of Climate Change in the Vu Gia-Thu Bon River Basin in Central Vietnam



## Integrated River Basin Management in the Vu Gia Thu Bon Basin

## Land Use Adaption to Climate Change in the Vu Gia–Thu Bon Lowlands: Dry Season and Rainy Season



## Distributed Assessment of Sediment Dynamics in Central Vietnam



## Hydrological Drought Risk Assessment in an Anthropogenically Impacted Tropical Catchment, Central Vietnam



## Conclusion

# Summary LUCCi

In the scope of the German-Vietnamese research project "Land Use and Climate Change Interactions in Central Vietnam" (LUCCi) funded by the German and Vietnamese research ministries BMBF and MOST, a variety of feedbacks in a dynamically developing socio-ecological system were assessed during the period of 2010–2016. The study region, the Vu Gia Thu Bon river basin, provides manifold ecosystem services for the provinces Da Nang and Quang Nam and supplies the population with water, food and energy. It is exposed to numerous challenges as rapid hydropower development, urbanisation, industrialisation, and expanding tourism along the coasts of both provinces.

Strategies for sustainable resource use were developed in the framework of the LUCCi project that can cope with the increasing pressure placed on the land and water resources in the region. Crucial local and national stakeholders who are entrusted with the management of the land and water resources were included from the start to ensure demand-oriented research and simultaneously to contribute to capacity development. The research results were made available to the province governments as well as relevant national and regional institutions.

This book presents selected application-oriented results of the multifarious research project LUCCi, in which numerous German and Vietnamese scientists as well as international organisations participated. Especially, the usability of the results for resource planning by the involved regional and national stakeholders was taken into consideration in selecting the contributions.

# Introduction

**Alexandra Nauditt and Lars Ribbe**

Interdisciplinary research projects which address sustainable regional development or river basin management are scarce. However, such research projects offer a relevant potential to obtain a holistic understanding of dynamically developing regions equally considering the resilience of local ecosystems, socio-economic development and climate change related hazards. The German–Vietnamese co-funded research project LUCCi—Land Use and Climate Change Interactions in Central Vietnam—gave us the opportunity to analyse a variety of feedbacks between people and the environment at different scales. Our LUCCi consortium consists of five German university departments, two Vietnamese universities as well as a number of international research institutions and NGOs working in the region. Up to date, 13 Vietnamese and German scholarship funded Ph.D. theses are being realized within the project and around 20 M.Sc. theses covering a wide range of disciplines. Thanks to the close cooperation among Vietnamese and German partners, researchers and students the information only available in Vietnamese could also be evaluated.

In 2009, the Vietnamese government and its Ministry of Science and Technology (MOST) prioritized funding for research and development activities in the provinces of Quang Nam and Da Nang in order to deal with the challenges related to socio-economic development such as hydropower development, tourism, urbanization and related migration processes. At the same time, the research programme "Sustainable land management" of the German Ministry of Research and Education (BMBF) encouraged us to apply for such an interdisciplinary project together with our Vietnamese partner at the Vietnam Academy of Water Resources

A. Nauditt (✉) · L. Ribbe
Institute for Technology and Resources Management in the Tropics and Subtropics, Technical University of Applied Sciences, Cologne, Germany
e-mail: alexandra.nauditt@th-koeln.de

L. Ribbe
e-mail: lars.ribbe@th-koeln.de



A. Nauditt and L. Ribbe (eds.), *Land Use and Climate Change Interactions in Central Vietnam*, Water Resources Development and Management,
DOI 10.1007/978-981-10-2624-9_1

in Hanoi to jointly work in this strongly dynamic study region in Central Vietnam. Since then, we—the LUCCi consortium—have had the opportunity to closely accompany local and national development processes and get insight in planning and provincial decision making.

The local research and capacity development demand for the region has been discussed since the beginning of the application process. Stakeholders dealing with agricultural land use, water management, hydropower, forestry and biodiversity from all planning levels were involved including the national level, as well as the provincial, district and even commune levels. The active support from the national level to work in this prioritized region was extremely helpful to identify stakeholders responsible for the research topics addressed.

We structured the project in three phases: assessment and modelling of individual subsystems, model integration and scenario development and an implementation and transfer period. Due to the low data availability, quality and access, data collection and data generation took much longer than expected. Spatial information, for instance, needed to be converted to international standards or be newly generated while hydrometeorological data were costly and delivered in inadequate formats; socioeconomic data were only available in the form of statistical yearbooks. At the same time, the different models were set up and calibrated. The third year and second phase focused on the integration of the different modelling results interlinking modelling approaches in their spatial and temporal dimensions. Key system indicators and related units were determined in collaboration with the stakeholders serving as subjects to be investigated and simulated. Scenarios were developed which then served to test the sensitivity of the different system components addressed and as a knowledge basis for the elaboration of land use strategies. The implementation phase is still ongoing and started with the development of stakeholder friendly information products and the establishment of the VGTB river basin information center. During the whole period, a strong involvement of all stakeholders was indispensable in order to become aware of the research demand to obtain data and information about planning processes and local challenges. Results developed in disciplinary and interdisciplinary working groups were discussed during several consortial workshops per year in Germany and Vietnam. Furthermore, stakeholder workshops and meetings were organized several times a year to present the latest findings and be updated on the research demand, boundary conditions of the different systems, land and water use planning issues as well as the institutional setup.

This book presents selected application-oriented results of the multifarious research project LUCCi, in which numerous German and Vietnamese scientists as well as international organizations participated (www.lucci-vietnam.info/partners/). Especially, the usability of the results for resource planning by the involved regional and national stakeholders was taken into consideration in selecting the contributions. Thus, the book starts with an introduction to the particular features of the project concept and management, followed by the Chapter "Biophysical and Socioeconomic Features of the LUCCi—Project Region: The Vu Gia Thu Bon River Basin", a description of the study region, considering basic biophysical and socioeconomic characteristics, and the Chapter "Vu Gia Thu Bon River Basin Information System (VGTB RBIS)—Managing Data for Assessing Land Use and Climate Change Interactions in Central Vietnam", an introduction to the Vu Gia Thu Bon River basin information system and how data are managed within the LUCCi project and beyond. A remote sensing-based investigation of the land cover used as basis for estimating the above ground biomass and thus its carbon sink is then detailed and based on this, the potential implementation of REDD mechanisms (UN Programme: Reducing Emissions from Deforestation and Forest Degradation) in the project region are illustrated in Chapter "Forest Change and REDD+ Strategies". Tree species biodiversity as well as the related potential for conservation and developing protection corridors in forest stands are dealt with in Chapter "Connectivity Conservation Management: A Biodiversity Corridor for Central Vietnam". Agricultural land use is addressed in Chapters "Rice-Based Cropping Systems in the Delta of the Vu Gia Thu Bon River Basin in Central Vietnam" describing the particular characteristics of rice farming systems and their socioeconomic relevance and "Measuring GHG Emissions from Rice Production in Quang Nam Province (Central Vietnam): Emission Factors for Different Landscapes and Water Management Practices" which demonstrates the carbon footprint of rice production from field observations and model simulations of Greenhouse Gas emissions. The Chapter "Hydrological and Agricultural Impacts of Climate Change in the Vu Gia-Thu Bon River Basin in Central Vietnam", summarizes some of the climate (change) impact modelling results. Water allocation as a controversial issue is discussed in Chapter "Integrated River Basin Management in the Vu Gia Thu Bon Basin" which analyes the water supply and demand perspectives and deals with conflicts related to the water–food–energy nexus. Analyses of impacts of land use changes on regional climate are shown in Chapter "Impacts of Land-Use/Land-Cover Change and Climate Change on the Regional Climate in the Central Vietnam". Chapter "Land Use Adaption to Climate Change in the Vu Gia–Thu Bon Lowlands: Dry Season and Rainy Season" deals with risks in the dry and wet season such as saltwater intrusion and flood in the densely populated and vulnerable delta region and presents water management options. The Chapter "Distributed Assessment of Sediment Dynamics in Central Vietnam" presents results about the spatial distribution of erosion risk in the region and climate change adaptation through ecosystem-based coastal zone management is addressed in Chapter "Sand Dunes and Mangroves for Disaster Risk Reduction and Climate Change Adaptation in the Coastal Zone of Quang Nam Province, Vietnam". "Hydrological Drought Risk Assessment in an Anthropogenically Impacted Tropical Catchment, Central Vietnam" deals with typical drought characteristics and the implications of anthropogenic alterations of the hydrological system. Issues related to the project sustainability and current follow-up activities are described in the Conclusion.

These contributions are based on numerous analyses and model results as well as their integration, which are available to readers in technical and expert publications or project reports. Further information products such as maps, analyses and overview documents were presented to the stakeholders in English and Vietnamese and can be accessed via the Vu Gia Thu Bon River Basin Information Centre (RBIC) that was founded by the Vietnam Academy of Water Resources and the Institute for Technology and Resources Management in Da Nang in 2014 (http://www.basin-info.net/river-basins/vu-gia-thu-bon-information-centre-vietnam).

The project results have been presented twice each year to the stakeholders of the two provinces Da Nang and Quang Nam as well as to National agencies such as the Ministries of Agriculture (MARD) and Environment (MONRE). The Provincial governments represented by the vice chairmen of their People's Committees (PPC) supported the project activities in the region and issued the necessary permissions for fieldwork. They were enthusiastic about the project results which were presented in a stakeholder friendly format of maps, figures, flyers and posters in Vietnamese language. They confirmed that the outcoming knowledge and recommendations regarding land and water strategies would feed into the next planning schedule of the provinces.

Such a close cooperation with the stakeholders as well as the difficult access to data and fieldwork permissions requires the permanent presence of project representatives in the region. Therefore, during the first four project years, we rented a LUCCi project house for researchers, Ph.D. and M.Sc. students in Da Nang. The "VuGia-ThuBon River Basin Information Center" (VGTB RBIC) in Da Nang at the Central Department of the Vietnam Academy of Water Resources has the objective to support resources management on basin scale and the communication among the water-related stakeholders in the River Basin. It offers a cross sector neutral space to discuss fair water allocation, land use changes and strategies among all involved stakeholders and provides access to the database on the Vu Gia Thu Bon river basin (Chapter "Vu Gia Thu Bon River Basin Information System (VGTB RBIS)—Managing Data for Assessing Land Use and Climate Change Interactions in Central Vietnam") and information products elaborated in the context of the LUCCI project.

The transfer of the results to other provinces is ongoing. The VuGia-ThuBon Information Center will serve as an example for an information center which can be built up in any other river basin in Vietnam. The National Agricultural and Environmental Ministries (MARD and MONRE) in Hanoi are frequently informed and interviewed regarding their interest to transfer the work and results.

The research project LUCCi was able to fill essential knowledge gaps in sustainable management of land and water resources in the Vu Gia Thu Bon river basin. However, the research team also faced large challenges and there are still quite a few open research issues that are closely connected to deficits in the existing monitoring systems. Important environmental and societal basis data are not systematically recorded and are only spatially and temporally available incompletely. Since the data infrastructure in the project region is still in a rudimentary state, scientifically founded analyses and coordinated resource management is difficult.

We hope that the LUCCi project and this book have been able to and will continue to raise awareness for these high-value knowledge products and were able to contribute to the growing understanding of a very dynamic region.

# Biophysical and Socio-economic Features of the LUCCi—Project Region: The Vu Gia Thu Bon River Basin

Viet Quoc Trinh, Alexandra Nauditt, Lars Ribbe and A.B.M. Firoz

## Introduction to the Study Region

The project area is located in Central Vietnam and comprises the entire Vu Gia Thu Bon river basin with a total area of 10,350 km$^2$ extending from 14°54′N to 16°13′N and from 107°12′E to 108°44′E. The basin borders on the Huong River Basin to the north and the East Sea (South-China Sea) to the east. It shares borders with the Mekong River Basin to the west and southwest, and with the Tra Khuc River Basin to the south. The administrative boundaries include the Vietnamese provinces Quang Nam and Da Nang as well as very small parts of Kon Tum and Quang Ngai. Although Da Nang City does not belong to the river basin, it was included in the study area due to its relevant socio-economic features (Fig. 1).

V.Q. Trinh (✉) · A. Nauditt · L. Ribbe · A.B.M. Firoz
TH Köln, Institute for Technology and Resources Management in the Tropics and Subtropics, Cologne, Germany
e-mail: trinhquocviet1981@gmail.com

A. Nauditt
e-mail: alexandra.nauditt@th-koeln.de

L. Ribbe
e-mail: lars.ribbe@th-koeln.de

A.B.M. Firoz
e-mail: abm.firoz@th-koeln.de



A. Nauditt and L. Ribbe (eds.), *Land Use and Climate Change Interactions in Central Vietnam*, Water Resources Development and Management,
DOI 10.1007/978-981-10-2624-9_2

**Fig. 1** Location of LUCCi project region in South East Asia

The project area covers the entire Quang Nam Province, as well as seven inland districts of the Da Nang Municipality. Together, the 9450 $km^2$ of the Quang Nam and Da Nang areas account for 91.8 % of the total project area of 10,350 $km^2$. The remaining 850 $km^2$ or 8.2 % of the basin are part of six mountainous communes of the Dak Glei District in Kon Tum Province. Figure 2 shows the administrative structure relevant to the project region indicating the dominance of rural districts in Quang Nam province.

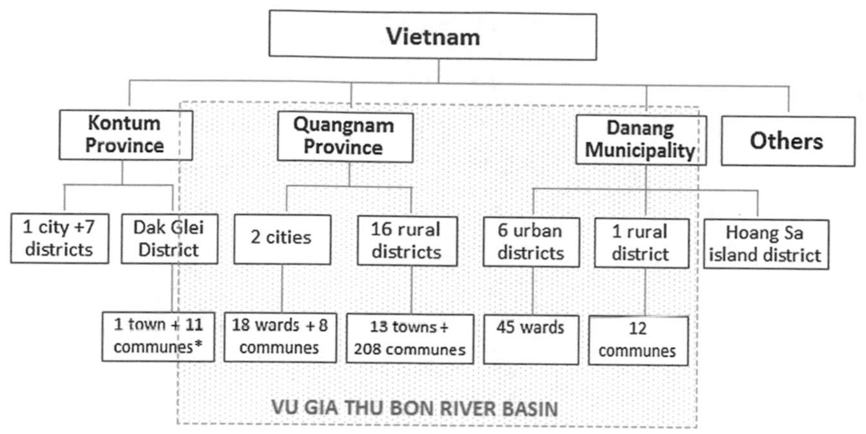


**Fig. 2** Administrative organisation in the project area

## Biophysical Characteristics

The Vu Gia Thu Bon (VGTB) basin is located at the eastern side of the Truong Son Mountain Range. The Thu Bon (205 km) and the Vu Gia (145 km) are the two major rivers that receive water from 19 first-order streams. Khang, Ngon Thu Bon, Ly Ly are among the major first-order streams of the Thu Bon and Giang, Bung, Con and Tuy Loan are major first-order streams of the Vu Gia. The river basin belongs to the South Central Coast Region of Vietnam within the tropical monsoon climate zone. Compared to the rest of Vietnam, this area has the longest dry season of eight months (from January to August), which is characterised by low rainfall frequently leading to droughts, and very high rainfall in wet seasons causing large-scale floods.

Topography is characterised by the Bach Ma Mountain Range in the north with peaks over 1000 m, by the South Truong Son mountains in the west with peaks over 2000 m, and by the Kon Tum mountain mass in the south, which runs out to the sea with Mount Ngoc Linh as the highest peak in the whole river basin (2598 m). The topography divides the area into three major landscapes from the west to the east. Based on the topographical features and major economic activities, the landscapes of the basin can be classified into highlands, midlands and lowlands.

According to the 2010 land use inventory of Ministry of Natural Resources and Environment (MONRE), the largest proportion of land is forested (64 %) in the highlands and midlands (compare Chapter "Forest Change and REDD+ Strategies"). The land used for cultivating annual crops makes up about 5 % of the total area and is mainly concentrated in the coastal plains as well as along the major rivers. The rural and urban settlements account for 5 % of the total area. Due to urbanisation and

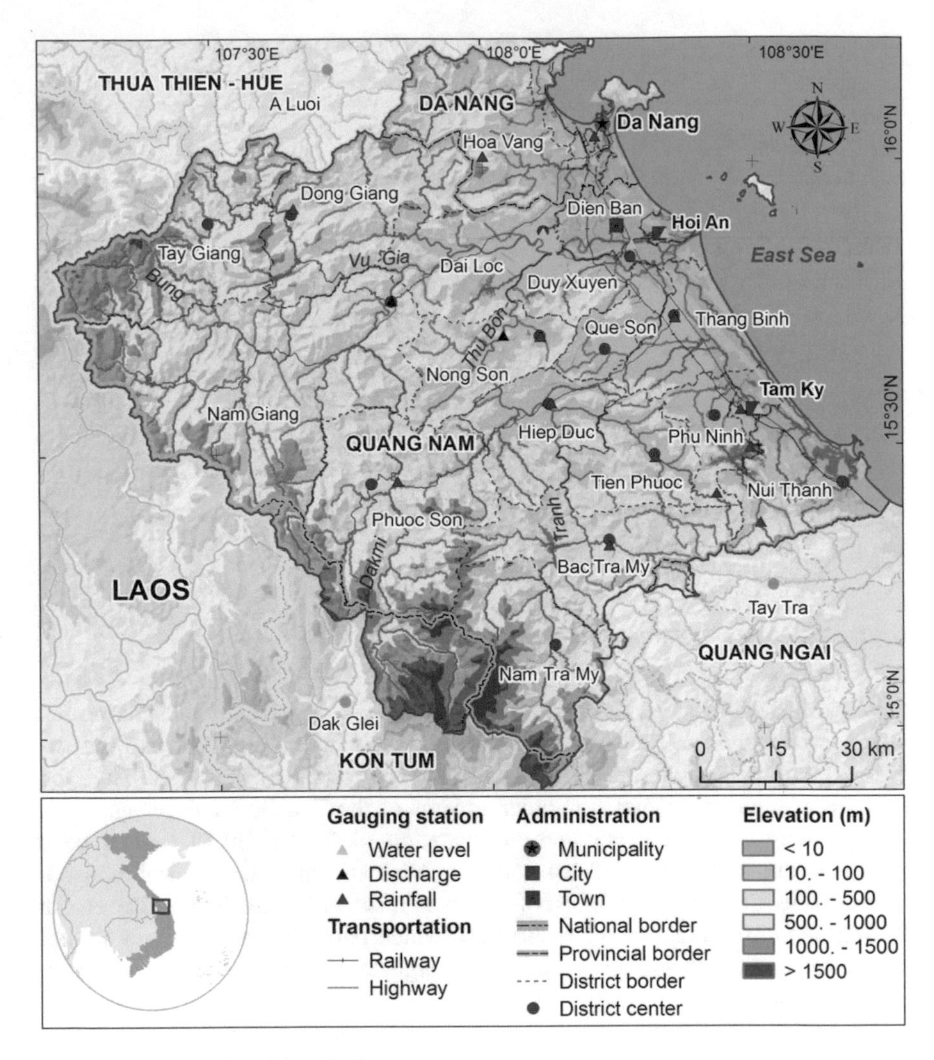


**Fig. 3** Vu Gia Thu Bon River Basin

industrialisation, large parts of land along the coastal zone have been repurposed from agricultural usage to other non-agricultural usage (Fig. 3).

*Major features of the basin are described in Table* 1.

## Climate

Due to its location below the Bach Ma hill range and the eastern direction of the West Truong Son mountain range, the climate in the VGTB river basin shows

**Table 1** Features of the LUCCi project area

| Location | Extended VGTB in Central Vietnam |
|---|---|
| Coordinate system | 14°54′N–16°13′N; 107°12′E–108°44′E |
| Area | 12,577 $km^2$ (of which VGTB primary is 10,350 $km^2$) |
| Primary stream | Thu Bon (205 km) and Vu Gia (145 km) |
| Origin | Mount Ngoc Linh (2598 m); Mount Mang (1600 m) |
| Outlet | East Sea (through three mouths of Cua Han, Cua Dai and Truong Giang) |
| Highest point | Mount Ngoc Linh (2598 m) |
| Lowest point | Cua Dai Estuary (0 m) |
| Tributaries | Thu Bon: Tranh, Khang, Ngon Thu Bon and Ly Ly<br>Vu Gia: Dakmi, Bung, Con and Tuy Loan |
| Geological features | Granite, Conglomerate, Limestone, Sandstone |
| Soil types | Ferrosol, Fluvisol and Acrenosol |
| Population | Approx. 2.5 million inhabitants |
| City | Da Nang, Hoi An |
| Land uses (2010) | Forest lands (64.0 %); paddy rice (5.0 %); and Settlement lands (4.5 %) |

*Source* Adopted from RCHM (2013), MONRE (2002, 2009, 2014a, b, c, d, e, f) and Trinh (2014)

similar characteristics as the Southern Central Vietnamese climate with relatively warm winters, dry summers affected by dry westerly winds, and a strong rainy season with typhoons lasting from September to December.

Rainfall during the wet season accounts for 65–80 % of the total annual rainfall, with 40–50 % of the annual rainfall in October and November causing severe floods regularly (RCHM 2013).

The extended dry season lasts from January to August and is regularly accompanied by droughts, which usually occur from February to April and account for 3–5 % of the total annual rainfall. In May and June there is a second rainfall peak, which is more pronounced in the north-western part of the area.

Rainfall in the basin is unevenly spatially distributed and has a strong orographic component ranging from about 2000 mm in the flood plain to more than 4000 mm in the upper parts of Song Vu Gia, with the highest rainfall measured in Tra My. However, there are only twelve rainfall stations in the river basin, none of which are above 400 m of elevation (Fig. 4).

The annual average temperature varies between 24.5 and 25.5 °C in the mountains and between 25.5 and 26.0 °C at the coast and tends to increase gradually from North to South. Temperature data is only available for the climate stations in Da Nang and Tra My for the period 1977–2010. Maximum temperature occurs in June and July with maximum temperatures above 40 °C on days with dry westerly winds, and minimum temperatures in December and January. The lowest recorded daily temperatures at both stations have been as low as 10 °C, but the monthly mean temperatures are above 20 °C for all months. According to RCHM (2013), the following climate values were observed and calculated:

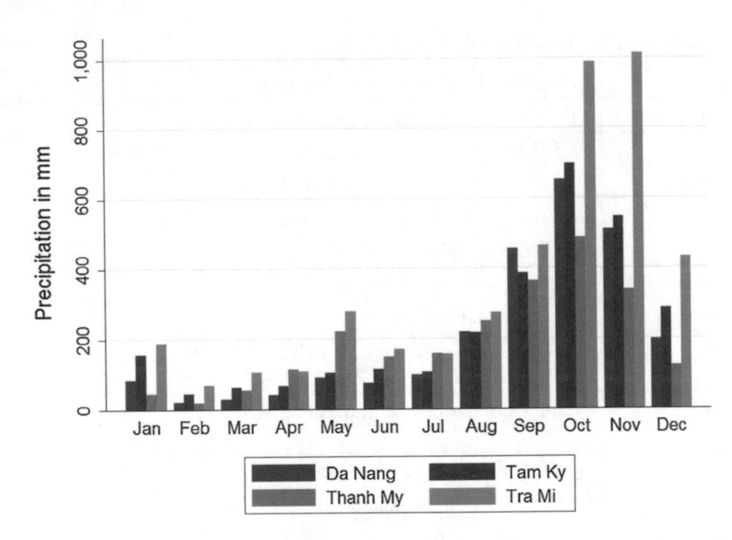


**Fig. 4** Monthly precipitation at four selected stations whereas Tam Ky (South) and Da Nang (North) are located at the coast and Tra My (wetter South) and Thanh My (drier North) more inland

- The relative humidity is around 77 % in the dry season (April to September) and can reach up to 93 % in the wet season (October to March).
- The average total annual radiation ranges from about 140–150 kcal/cm$^2$. The average annual radiation balance ranges from 75 to 100 kcal/cm$^2$.
- The average annual sunshine hours vary from 1800 h in the high mountains to more than 2000 h in the coastal plain.
- The average annual wind speed is 0.7–1.3 m/s in the mountains, whereas it is 1.3–1.6 m/s in the coastal plain. The maximum observed wind speed is 34 m/s in summer, and 25 m/s in the rainy season at the Tra My station. In the coastal plain area, the wind speed reaches 40 m/s.

## *Hydrology*

The VGTB river basin is one of the nine largest basins in Vietnam. With a drainage area of 10,350 km$^2$, the two main rivers of the basin are Vu Gia and Thu Bon, which originate in the Truong Son Mountain Range near the border to Lao P.D.R. Below the mountain range, the river and its tributary streams flow through the medium flat terrain of the coastal plain and are cross linked approximately 36 km upstream from the coast.

During the flood season, the two rivers interact through the Quang Hue and Vinh Dien rivers forming a braided river delta system. This connection is often broken in the dry season (Trinh 2014). The basin is some 100 km wide and 120 km long, and has a wide, flood-prone coastal plain backed by the steep Truong Son Mountain

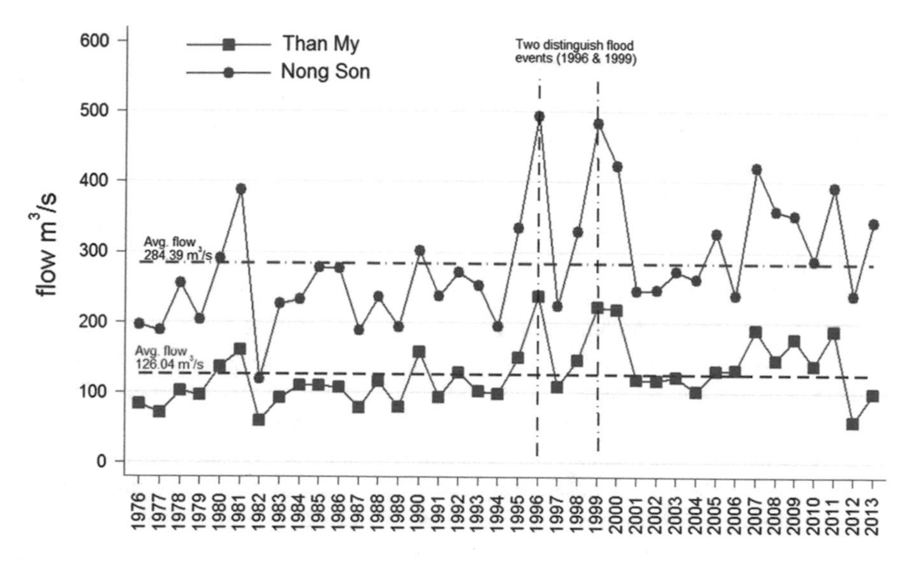


**Fig. 5** Comparison of annual discharge rates at Nong Son (Thu Bon) and Thanh My (Vu Gia)

Range that rises to 2600 m. The amount of surface water flowing in the basin is directly related to precipitation, which varies geographically and seasonally. The majority of rainfall occurs in the wet season (September to December), with 70 % of runoff generated during these 3 months. In contrast, during the height of the dry season (February to March) only 12 % of runoff is generated (ICEM 2008). The basin has one of the longest dry seasons in Vietnam with 9 months of the year receiving only 30 % of the rainfall, resulting in severe water shortages and problems with saline intrusion at the coast (Trinh 2014).

Hydrological stations at Thanh My on the Vu Gia, and Nong Son on the Thu Bon have continuously gauged daily discharge since 1976. Observed data since 1976 show that the variability of river discharge is high and that the Southern Thu Bon receives much more rainfall with higher discharge. The highest flow at Nong Son was 10,600 $m^3/s$ (04-11-1999), while the lowest flow was only 4.63 $m^3/s$ (17-08-1977) (Fig. 5).

Hydro-meteorological data were collected at the Centres for Hydrology and Meteorology in each Province and IMHEN.

## Socio-economical and Demographical Change

The economy of the region is rapidly changing from agriculture-based to industry and service-based, with an annual growth rate of over 10 % in the last decade. The percentage of the population working in the agricultural sector is decreasing.

In 2014, more than half of the population in the region worked in industry and service sectors, and the movement of labour forces from the agricultural sector to industry and services sectors has been increasing.

The economy of the VGTB basin is diverse, including a changing primary sector consisting of agriculture, forestry, fishery, and handicrafts, and strongly growing secondary and tertiary sectors. The basin has experienced strong economic growth in the secondary and tertiary sectors but limits to growth are projected if infrastructure development does not keep up with the pace of change. Local economies are still relatively underdeveloped compared to North and South Vietnam. A spatial distribution of economic activities in the region is apparent. Major economic centres are located in the lowlands, where dense population, high concentration of industry, services and intensive agricultural activities support dynamic growth. The economic activities in the midlands are limited to forestry, perennial crops and small-scale annual crops. Hydropower generation and forestry are two major economic activities in the midlands and uplands. Sparse population and limited access restrict economic development of the uplands. Table 2 summarises the major socio-economic features of the basin.

After separating the Quang Nam-Da Nang Province into the two provinces Quang Nam and Da Nang in 1997, the new provincial-level administrative units have been experiencing rapid economic growth since then. In the Quang Nam Province, the annual mean economic growth rate was over 8.1 % in 1997–2000, 11.9 % in 2001–2010 and 11.5 % in 2011–2014 (QSO 2005, 2014). The Da Nang municipality has also experienced high economic growth over the period from 1997 to 2014, with an annual mean growth rate of 10.3 % in 1997–2000, 11.7 % in 2001–2010 and 9.9 % in 2011–2014 (DSO 2010). High growth rates helped to narrow the development gap between the region and other economic centres in South and North Vietnam. In 2014, the GDP per capita in Quang Nam was around 1635 USD, which was approximately 79 % of the GDP per capita of Vietnam, and

**Table 2** Major socio-economic features of the VGTB basin

| Landscape | Area ($km^2$) | Population (person) | Density (per./ $km^2$) | Land use | Economic activities |
|---|---|---|---|---|---|
| Lowlands | 3370 | 2,024,357 | 600 | Paddy rice, annual crop, settlement, Industrial use, special use | Processing and light industries service, tourism, intensive agriculture |
| Midlands | 4120 | 364,016 | 88 | Forest, perennial crops, annual crop, paddy rice | Hydropower generation, forest product exploitation, agriculture |
| Uplands | 7220<br>5585 | 142,508<br>84,666 | 15 | Forest, perennial crops, annual crop | Hydropower generation, forest product exploitation |

*Source* Synthesised from QSO (2014) and DSO (2014)

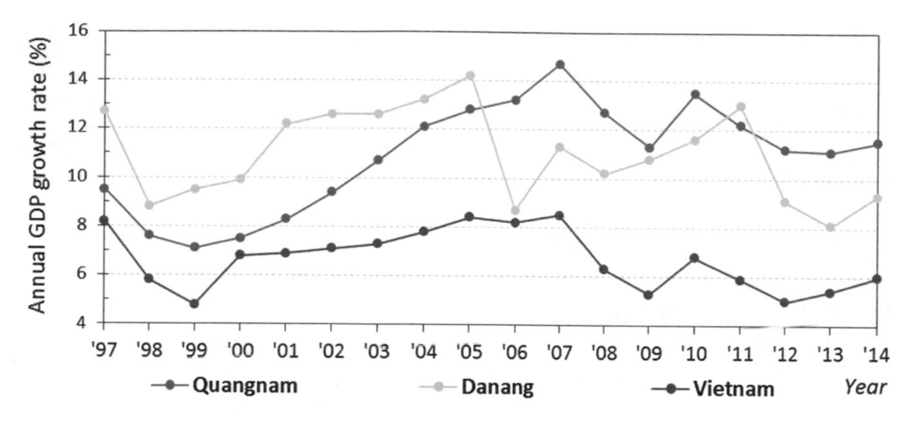


**Fig. 6** Annual GDP growth rate of Quang Nam and Da Nang in comparison to Vietnam in 1997–2014 (*Data* GSO, QSO, DSO in 1997–2014)

in Da Nang it was around 2800 USD, which was approximately 140 % of the GDP per capita of Vietnam (QSO 2014; DSO 2014) (Fig. 6).

The enormous growth in Quang Nam Province is mainly driven by the development of the secondary (industry and construction) and tertiary (service) sectors. The development of these sectors has induced a structural change in the economy. The contribution of the agricultural sector to the GDP in Quang Nam decreased from 48 % in 1997 to 17 % in 2014, whereas the contribution of the industrial and services sectors increased from 52 % in 1997 to 83 % in 2014. Consequently, although a high proportion of the population still depends on agriculture, forestry and fishing, the number of people working in this sector decreased remarkably from 78 % in 1997 to 51 % in 2013 (QSO 2005, 2014).

In contrast to Quang Nam, the economic development of the Da Nang municipality was dominantly fuelled by the development of industry and services. The proportion of agriculture in the economy decreased from 7.9 % in 2000 to 3 % in 2011, while the share of industry and construction increased from 40.3 to 46 % and the share of services sector was relatively stable at 51 % GDP in the same period (DSO 2014).

The VGTB is home to approximately 2.5 million inhabitants (2013), 80 % of which live in the coastal lowlands, 45 % of which live in the urban areas. The mean population density in the basin was 197 persons per km$^2$ but distributed unevenly with high density in the lowlands, low in the midlands and very low in the uplands, with a density lower than 20 persons per km$^2$, except for some concentrations in small district centres. The density of population in midlands, which are transactional areas between uplands and lowlands, varies from 20 to 100 persons per km$^2$. The population is densely concentrated in the lowlands, particularly in the cities, where the density is over 5000 persons per km$^2$.

The annual mean growth rate of population during 1995–2013 was 0.58 % in the Quang Nam Province and 2.52 % in the Da Nang municipality (GSO 2014). Migration plays an important role in population growth of the region. The large flux

**Table 3** Net migration rate of Quang Nam and Da Nang in 2005–2013

| Year | | 2005 | 2007 | 2008 | 2009 | 2010 | 2011 | 2012 | 2013 |
|---|---|---|---|---|---|---|---|---|---|
| Net migration rate (‰) | Da Nang | 3.8 | 7.6 | 6.5 | 15.3 | 26.4 | 14.9 | 11.2 | 5.8 |
| | Quang Nam | −4.3 | −4.5 | −4.3 | −8.0 | −9.7 | −2.3 | −3.6 | −1.5 |

*Data* GSO (2014)

of migration caused low growth rates in the Quang Nam Province, while a high immigration rate supplied more labour force for the Da Nang municipality. The high growth rate in Da Nang was fuelled by rapid urbanisation. The low population growth in Quang Nam will pose a challenge for providing an adequate labour force for the local economy in the future.

The population of Quang Nam in 2013 was 1,449,000, growing on average by 7722 persons each year. During 1995–2014, the population of Da Nang increased to 355,500 persons, growing on average by 19,750 persons each year. The average annual growth rate of the population in Quang Nam between 2006 and 2014 was 0.5 %, a decrease compared to 0.7 % in 1995–2005 (QSO 2005). Unlike Quang Nam, the population in Da Nang increased with a higher rate in 2006–2014 (2.6 %) in comparison to 2.4 % in 1995–2005 (DSO 2010, 2014). The differences in population growth between Quang Nam and Da Nang resulted from migration (Table 3).

Quang Nam has a young population: 20.7 % of the population is under 15 years old, while only 9 % of the population is over 65 years old (GSO 2014). The urbanisation rate in Quang Nam is low with only 19.2 % of the population living in urban areas, compared to 32.2 % in Vietnam and 37.0 % in the Southern Central Coast (PPC Quang Nam) (GSO 2014). However, due to rapid industrialisation, urbanisation will speed up in the coming years. According to the Strategy for Urban Development to 2020 prepared by PPC Quang Nam in cooperation with UN Habitat, approximately 41.7 % of the population in the province will live in urban areas by 2020 (PPC Quang Nam 2014). The urban centres are concentrated in the lowlands, along the 1A National Highway.

The total population of Da Nang City in 2013 was 887,435 persons with an urbanisation rate of the city is the highest in Vietnam, with 86.9 % of the population living in urban areas in 2009 compared with 32.2 % in Vietnam (GSO 2010). For the whole VGTB River Basin, the urbanisation rate is 44.8 %, a high rate compared to other regions in Vietnam such as 29.3 % in the Red Delta, 57.2 % in the South East Region and 22.8 % in the Mekong Delta (GSO 2014). The region also has a small family size with, on average, 3.76 persons per family in Quang Nam and 3.91 persons per family in Da Nang compared to 3.82 persons per family in Vietnam.

## *Agricultural Sector*

Paddy cultivation and livestock farming are the two main agricultural activities in the basin. Paddy cultivation is the most important agricultural activity. Two crops of paddy are planted per year in the lowlands and areas along major rivers, where ample water resources exist for irrigation. Even in the valleys of the mountainous areas, at least one crop of paddy rice is planted per year. Paddy rice is planted intensively in the Dai Loc, Dien Ban, Duy Xuyen, Que Son, Thang Binh and Phu Ninh districts in the Quang Nam Province and the Hoa Vang District in the Da Nang Municipality. Other annual crops such as cassava, maize, sweet potato, peanut and sesame are also planted in the lowlands, particularly in the coastal areas. Pepper, rubber and cinnamon are the main perennial crops in the region and are planted in the upstream parts of the Thu Bon sub-basin.

From 1995 to 2013, planted areas of paddy in the region decreased by 21 % from 106,200 ha in 1995 to 87,900 ha in 2013. The planted area in Da Nang, in particular, decreased by 60 % from 13,400 ha (1995) to 5400 ha (2013) due to the expansion of settlement areas (GSO 2014). However, paddy production during this period increased by 30 % from 362,000 tonnes (1995) to 471,000 tonnes (2013) due to a significant increase in paddy yield (GSO 2014). The paddy yield in the region has increased by 67 % from 3.02 tonnes/ha (1995) to 5.05 tonnes/ha (2013) (QSO 2005, 2014) (Fig. 7).

According to land use planning, the paddy planting areas in the region will continuously decrease to 42,711 ha in 2020 from 43,979 ha in 2015 (MONRE 2014a, b, c, d, e and f). The reduction will mainly take place in the area surrounding Da Nang, where the urbanisation and industrialisation processes are occurring rapidly.

The typical livestock reared in the basin are buffaloes, cattle, pigs, poultry and goats. Although its contribution to the agricultural sector is not as high as crop cultivation, livestock farming is very important given the local consumption habits and culture. Like in other regions of Vietnam, animal farming is typically undertaken by small-scale farm households. The sizes of herds depend on the type of animals, ranging from 3 to 20 heads per farm household. Pig and poultry raising activities are mainly undertaken in the lowland districts such as the Dai Loc, Dien Ban, Que Son, Duy Xuyen, Thang Binh, Nui Thanh and Phu Ninh districts, where cropping is more developed. This is not surprising given the fact that many agricultural by-products are used to feed livestock, mainly pigs and poultry. Crop cultivation is essential for pig and poultry farming. Buffaloes and cattle are mainly raised on natural grass and pastures so they are more developed in districts such as Thang Binh, Phu Ninh and Nui Thanh, where this activity is more propitious.

## *Industrial and Construction Sector*

The manufacturing sector plays an important role in the economic activities in the region. The rapid development of manufacturing in the recent years is the key factor energising the economic growth rate of the region. During the last 17 years

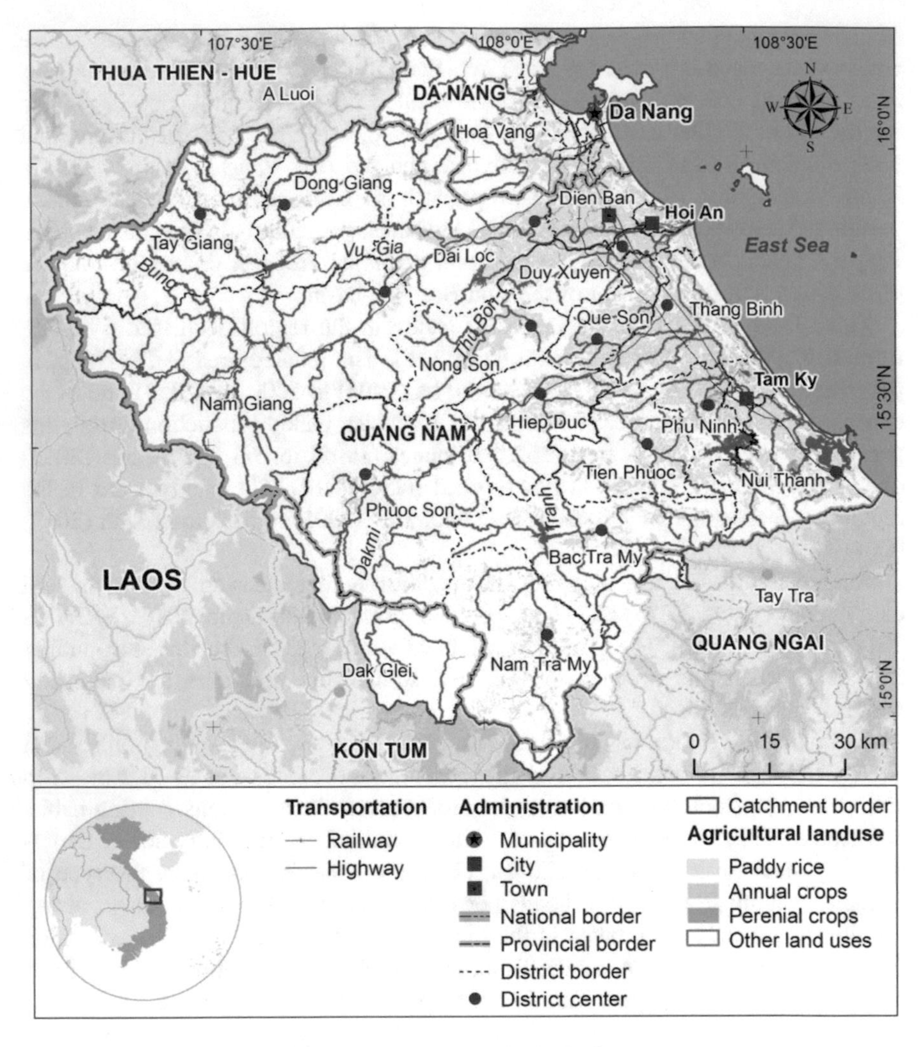


**Fig. 7** Spatial distribution of agricultural land uses in the basin

(1997–2014) since establishing the province, the industry of Quang Nam has grown by 18.7 % per year, particularly in the last 10 years. The industry of Da Nang has also grown rapidly. The growth rate of industry is much higher than the average growth rate of service and agricultural sectors and therefore the share of industry in the GDP of the region is significantly increasing. The GDP share of industry in Quang Nam increased from 19.6 % in 1997 to 42.0 % in 2014 and it has increased from 40.3 to 46.0 % in the same period in Da Nang. The proportion of labour force in industry has also significantly increased from 4.5 % in 1997 to around 23 % in 2014 (QSO 2014).

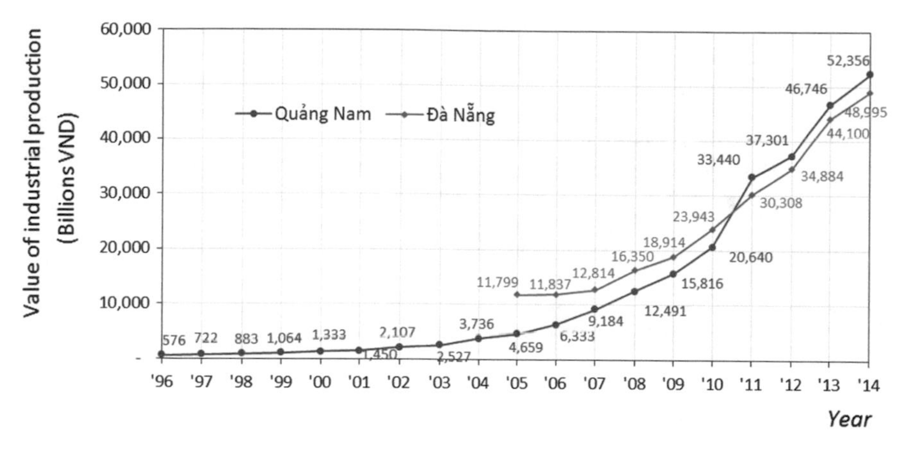


**Fig. 8** Value of industrial production of Quang Nam and Da Nang in 1996–2014 (*Data* GSO 2014)

The industrial development in Quang Nam and Da Nang from 1996 to 2014 is plotted in Fig. 8. Industrial values in Quang Nam have increased by a factor of 90 since 1996 (QSO 2005, 2014). Da Nang is the largest industrial centre in the region. The industrial production of the city is mainly concentrated in the industrial parks located in surrounding areas such as Hoa Khanh, Extended Hoa Khanh, Hoa Cam, Tho Quang and An Don. A high-tech industrial park has also been constructed in the outskirts of the city (Fig. 9).

The Chu Lai Open Economic Zone is the largest industrial centre in the Quang Nam Province. A cluster of industrial parks, developed in the southern coastal parts of the province, drives the industrial development in the region. Dien Ban and Tam Ky are the other industrial centres in the province (Fig. 9).

The major industrial products in the region are automobile products, rubber, garments, footwear, glass, processed food, software and construction materials. Handicraft goods are also developed in the region. The Tra Kieu bronze casting handicraft village and the Ma Chau silk village are the most populous traditional handicraft villages in the region.

Industry in the region is expected to continue growing at a high rate in the future thanks to the expansion of local and foreign direct investment.

## Service and Tourism Sector

The service sector accounted for 41 and 51 % GDP in Quang Nam and Da Nang in 2014, respectively (GSO 2014). In Quang Nam, the growth rate of service sector

**Fig. 9** Industrial centres and industrial parks in the basin

was always higher than the average growth rate of the GDP, and therefore the proportion of service increased continuously from 31 % in 1997 to 41 % in 2014 (QSO 2005, 2014). Other than in Quang Nam, due to the service-based economy, the proportion of service in Da Nang economy has remained high and has contributed to half of the GDP during the last ten years.

The region has high potential to develop tourism, particularly sea tourism. Hoi An and My Son are both world cultural heritage sites. Da Nang and Hoi An are two of the largest tourist centres in Vietnam and attract millions of visitors every year.

## Typical Disasters in the Research Area

The river basin is regularly influenced by natural disasters such as typhoons, tropical depressions, flooding, droughts, salinity intrusion, landslides, riverbank and coastal erosion. Flooding occurs frequently from September to December, while droughts and salinity intrusion take place during the period from February to August. Droughts do not occur in the whole basin, but are concentrated in the Thu Bon and Tam Ky sub-catchments. Low water levels in the river during the dry seasons result in saltwater intruding further inland. Riverbank erosion is a severe problem for communities along the Vu Gia and Thu Bon rivers, while coastal erosion is a large concern for the local communities along the 200 km long coastal line. Additionally, torrential rains in the mountainous areas cause flash floods. Figure 10 illustrates the different types of disasters in the region.

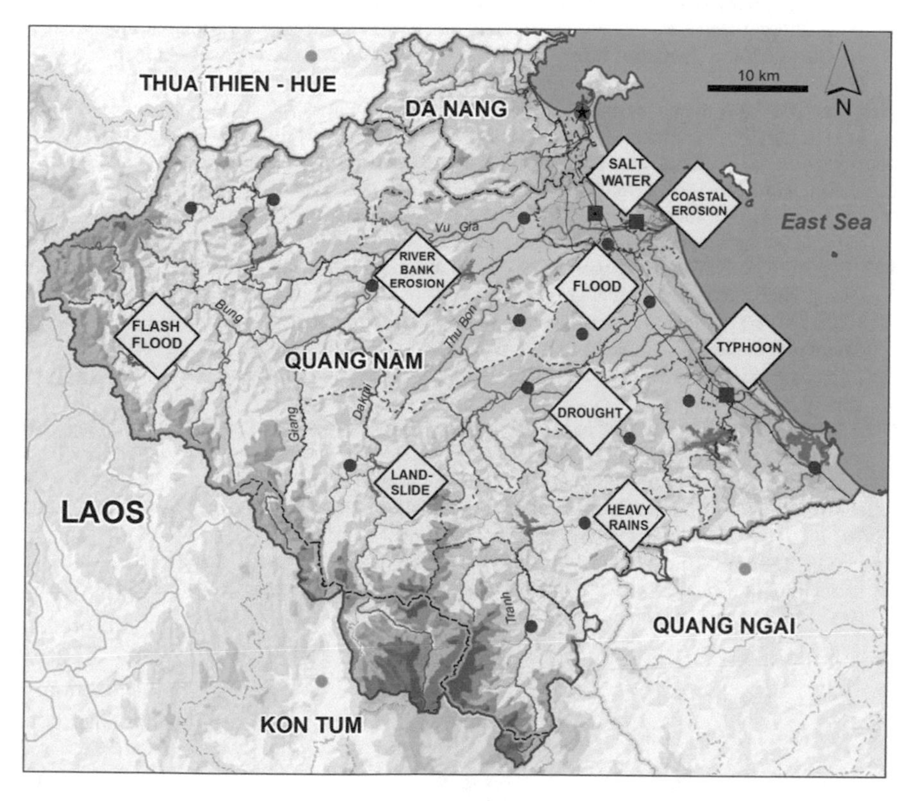


**Fig. 10** Disasters in the basin

## References

DSO (Danang Statistical Office) (2010) Da Nang: 15 years of innovation and development 1996–2010. Da Nang, Vietnam

DSO (2014) Statistical Yearbooks of Da Nang 2010–2014. Statistical Publishing House, Danang

DONRE (Department of Natural Resources and Environment Quang Nam Province) (2011) The Vu Gia Thu Bon Basin. Available: http://www.adb.org/documents/events/2010/cebu-water-basins-II/list-participants.pdf (Accessed)

GSO (General Statistics Office) (2010) The 2009 Vietnam Population and Housing census: Completed result. Statistical Publishing House, Hanoi

GSO (General Statistics Office) (2014) Statistical Yearbook of Vietnam 2014. Statistical Publishing House, Hanoi

ICEM (2008) Strategic environmental assessment of the Quang Nam Province Hydropower Plan for the Vu Gia Thu Bon River Basin. Prepared for the ADB. MONRE, MOITT & EVN, Hanoi

MONRE (Ministry of Natural Resources and Environment) (2002) Development and protection planning for water resources for the Vu Gia Thu Bon River Basin, Hanoi

MONRE (2009) Climate change, sea level rise scenarios for Vietnam, Hanoi. Available at: http://vgbc.org.vn/vi/nc/169-climate-change-sea-level-rise-scenarios-for-vietnam-2009

MONRE (2014a) National database of resources and environment. Land use map (online) in 2010 in Quang Nam Province 2010. Available at http://nredb.ciren.vn/Default.aspx?PageID=124&ID=288

MONRE (2014b) National database of resources and environment. Land use map (online) in 2010 in Danang City. Available at http://nredb.ciren.vn/Default.aspx?PageID=124&ID=259

MONRE (2014c) National database of resources and environment. Land use map (online) in 2010 in Kon Tum Province. Available at http://nredb.ciren.vn/Default.aspx?PageID=124&ID=275

MONRE (2014d) National database of resources and environment. Current land use in 2010 in Da Nang City. Available at http://nredb.ciren.vn/Default.aspx?PageID=210&ID=69

MONRE (2014e) National database of resources and environment. Current land use in 2010 in Quang Nam Province. Available at http://nredb.ciren.vn/Default.aspx?PageID=210&ID=98

MONRE (2014f) National database of resources and environment. Current land use in 2010 in Kontum Province. Available at http://nredb.ciren.vn/Default.aspx?PageID=210&ID=86

PPC Quang Nam (2014) Strategy for urban development to 2020 in Quang Nam. Working paper, Tamky, Quang Nam

QSO (Quang Nam Statistical Office) (2005) Quang Nam-30 years for development. Quang Nam Statistical Office, 110 pp

QSO (2014) Statistical Yearbooks of Quang Nam 2010–2014. Statistical Publishing House, Ha Noi

RCHM (Regional Center for Hydro-meteorology) (2013) Time series of meteo-hydrologic data in 1977–2010 in the Vu Gia Thu Bon River Basin

Trinh QV (2014) Estimating the impact of climate change induced saltwater intrusion on agriculture in estuaries- the case of Vu Gia Thu Bon. Doctoral dissertation, RURH University Bochum, Vietnam, p 175

# Vu Gia Thu Bon River Basin Information System (VGTB RBIS)—Managing Data for Assessing Land Use and Climate Change Interactions in Central Vietnam

**Franziska Zander**

**Abstract** The assessment of land use and climate change interactions requires and produces a lot of different types of data from several disciplines. In order to have a common and web-based platform to manage, share, and present such data in the context of an interdisciplinary international research project, the Vu Gia Thu Bon River Basin Information System (RBIS) was established. The modularly structured information system has a user-friendly interface with full read/write access for the management, linkage, analysis, visualization, and presentation of different types of data in the context of multidisciplinary environmental assessment and planning. Besides the management of metadata, different types of datasets can be managed (e.g., time series data, spatial geodata, documents). Standardized interfaces are provided to enable the exposure and exchange of data with other systems. Here the key features of the system are presented alongside with user related application guidance.

## Introduction

The Vu Gia Thu Bon (VGTB) RBIS was set up in 2010 based on the software framework RBIS to serve as the database for the research project "Land Use and Climate Change Interactions in central Vietnam" (LUCCi; LUCCi consortium 2016) and as an information and decision support system for local stakeholders. It is named after the two rivers Vu Gia and Thu Bon in central Vietnam, which also define the study region of the project. During the following years, the system was continuously developed based on known or raised demands. Examples for demands raised during the project are the multilingual support to reduce the language barrier

F. Zander (✉)
Institute of Geography, Friedrich-Schiller University Jena, Jena, Germany
e-mail: franziska.zander@uni-jena.de





A. Nauditt and L. Ribbe (eds.), *Land Use and Climate Change Interactions in Central Vietnam*, Water Resources Development and Management,
DOI 10.1007/978-981-10-2624-9_3

for local stakeholders, the implementation of a Catalog Web Service (CSW) for the exposure of metadata and enable the search from external platforms via a standardized interface, and the detailed description of processes to describe and linkage between datasets and the used tools (e.g., simulation software) (see section "Provided Interfaces and Services").

After an overview of general concepts, architecture and features of the system (section "Concepts, Architecture, and Features"), the description of the user and permission management (section "User and Permission Management"), the current set of managed datasets, its structure and examples are described (section "Data Management, Structure, and Examples"). For the VGTB RBIS important interfaces and provided services are part of section "Provided Interfaces and Services". The last section informs about how user acceptance and trust, and visibility was raised, and summarizes collected experiences.

## Concepts, Architecture, and Features

The underlying RBIS framework is built on open source software and standards to support reuse, extensibility, and a cost-efficient deployment and operation of the software. The internal structure is modularized to allow easy customization for each project. Thus it is applied in several research projects of different sizes worldwide. The fine-grained and detailed user and permission management ensures a full read and write access and appropriate protection of owner rights and use limitations.

### *System Architecture and Common Features*

The system is built for a Linux environment and uses the Apache web server, PHP, and PostgreSQL (PostgreSQL Global Development Group 2016) as a database management system with the spatial extension PostGIS (PostGIS Developers 2016). For the visualization of spatial data, the interactive mapping applications MapServer (Open Source Geospatial Foundation 2016) together with the JavaScript library OpenLayers (OpenLayers Contributors 2016) for displaying map data in web browsers are used. The web-based interface of RBIS uses several JavaScript libraries [jQuery (jQuery Foundation 2016), …] to create a user-friendly and easy to use graphical user interface. The interface itself is available per default in English, but to reduce language barriers it can be changed to Vietnamese or other languages (i.e., Spanish, Portuguese, or German). The common access on stored metadata is realized with a description layer (Kralisch et al. 2009) using XML files to hold all information needed for access, manipulation, visualization, and linking of datasets. This makes it easy to adapt the frontend and change or extend the underlying database schema. RBIS is able to run within a virtual environment (VirtualBox (Orcale 2016)), even without Internet access, as a copy on a local computer.

## Modules and Features

RBIS is structured in modules dealing with one type of data. A module includes the description with detailed metadata, the storage of the data itself and special features related to the data type. A special focus lies on the management, analysis, and visualization of measured and simulated time series data. Once a time series data is linked to a station (e.g., climate station) the spatial location is assigned and can be visualized in a map; as like for any other datasets with a spatial context in the form of coordinates, extent or polygons. Imported time series data can be visualized with an interactive diagram and analyzed with respect to their parameter and spatial location. During the upload, detected gaps can be filled using a rule-based gap filling toolbox. Time series data importing and exporting can be customized and automated. Spatial data in the form of raster or vector can be imported, visualized in maps, and described by metadata according to the ISO 19115 standard for geographical metadata (ISO 2003). A further module deals with the description of study sites in the form of study areas or locations as a general spatial reference. Those, as well as all other managed datasets, can be linked to each other. A special type of linking is possible within the module processing where in- and output data can be assigned. The processing module is described in more detail within section "Data Management, Structure, and Examples". Furthermore, any type file (e.g., documents, images ...) can be described by metadata or directly upload to each dataset. Other modules deal with project specific and variable sets of datasets like soil or vegetation data and their analyzed parameter.

## User and Permission Management

User accounts can be requested via the registration form. The administrator will be informed via e-mail and can create a user account based on the provided request notices. A user account will be assigned to one or more permission groups and may have an expiration date. RBIS has a two level permission management. The first level is the management of permission through the association of predefined permission groups. Those groups define view, edit, and download permissions for each data type (e.g., time series data) and are applied to all datasets within that category. The second level is optional and is datasets based. More or less restrictions can be assigned in comparison to permissions already in place due to the group permissions. The management of dataset based permissions is done by the owner. The owner will also receive permission requests if his data can only be downloaded on request (Zander and Kralisch 2016).

# Data Management, Structure, and Examples

Within the research project LUCCi a lot of different datasets from several disciplines were collected, described with metadata, and incorporated into the VGTB RBIS. Those datasets are base, interim, or result data and originate from several existing sources or were produced and measured by the LUCCi project consortium. The managed data types are as follows; the order corresponds to the menu at the left side shown in Fig. 1 which shows the layout of the Vu Gia Thu Bon RBIS and the list view of all managed time series data datasets.

## *Responsible Person and Organizations*

Person and organization datasets with contact details serve as a reference to assign a responsible party for managed data and metadata within the VGTB RBIS.

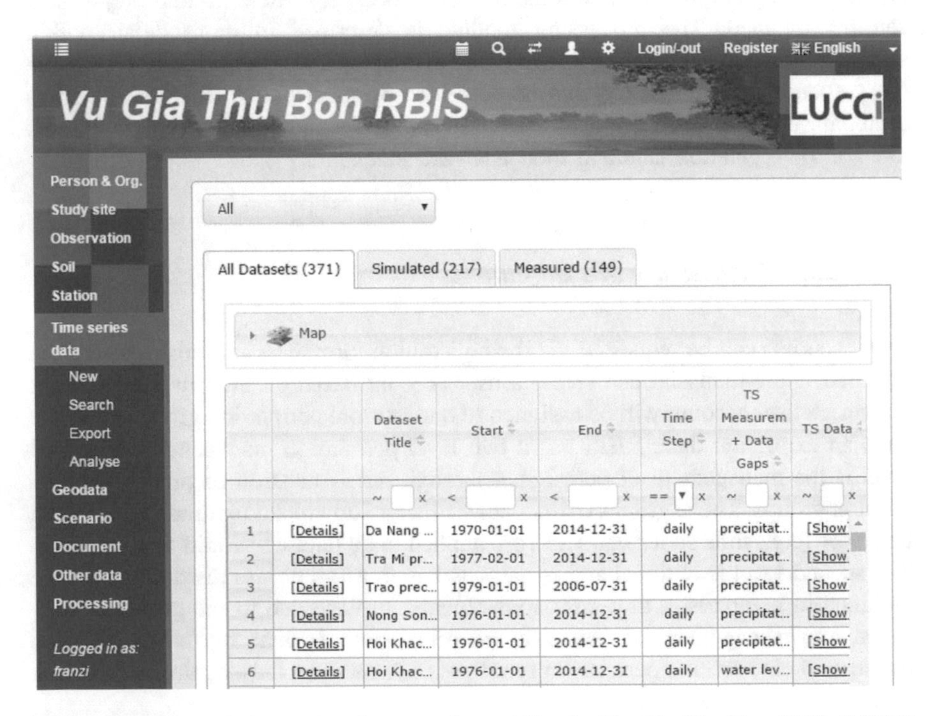


**Fig. 1** Screenshot of time series data overview within the Vu Gia Thu Bon RBIS (LUCCi consortium 2016)

## *Study Sites*

Study sites describe a location or an area where managed data can belong to. Main study areas within the project are the boundary of the whole catchment of the two river Vu Gia and Thu Bon, and the lowland part of those rivers as a study area for flood related activities. In addition to these areas, the location for greenhouse gas analysis, carbon field campaigns, soil salinization surveys, water quality sampling, and the biodiversity plots are described as study locations. The results of two field campaigns (in 2011 and 2012) for water quality sampling are linked and visualized in a map.

## *Soil Data*

During the project period, around 200 soil profiles with one or up to five analyzed soil horizons were investigated. The profile description and horizons with all analyzed parameters are stored within the soil module of RBIS. They are accessible for all registered users without restrictions.

## *Station and Time Series Data*

In the Vu Gia and Thu Bon region there are only 44 climate, precipitation or discharge stations plus two within LUCCi installed climate stations currently measuring data and accessible for the researchers within the LUCCi project. Stations are described by their name, type (climate, gauging …), spatial location, year of establishment (and end year), and responsible parties.

Time series data is described by its own metadata and should be linked to one station. Beside 150 measured time series datasets (e.g., water level, discharge, precipitation …), 217 simulated time series datasets are available in the VGTB RBIS. These data belong to ERA40, A1B, and B1 climate scenarios and were extracted for each real measurement station location in the field. In addition to the raw data, bias corrected (quantile mapping and local intensity scaling) time series data are available. Measured time series data, except for the publicly accessible data and data measured by the LUCCi project, is only accessible on request for project members and the data provider. Figure 2 shows the selection of a station via map and the linkage to one of the assigned time series data of simulated data and its visualization in an interactive diagram.

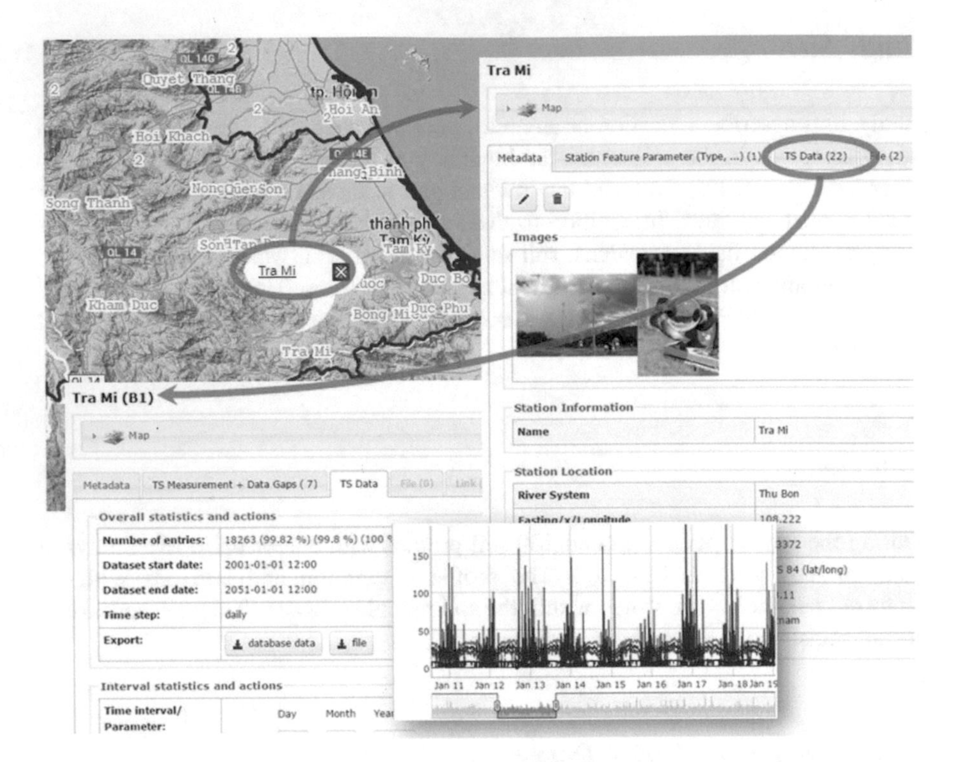


**Fig. 2** Example for the linkage between a station and assigned time series data. *Source* LUCCi consortium (2016)

## *Geodata*

Base shape and raster files, as well as interim and final results, are described according to ISO 19115 under the category geodata. Base data are for example administrative boundaries, infrastructure, or soil data. Interim and final results are a corrected DGM, HRUs for all catchments (Hydrological Response Units) or derived or projected land cover change maps based on satellite images. The derived data (e.g., land cover) from earth observations is accessible on request to keep track of who is using the data for what (Fig. 3).

## *Documents*

Presentations and posters from the final LUCCi workshop in Vietnam in 2015 in English and Vietnamese, statistical yearbooks, development plans, and publications are described under this section.

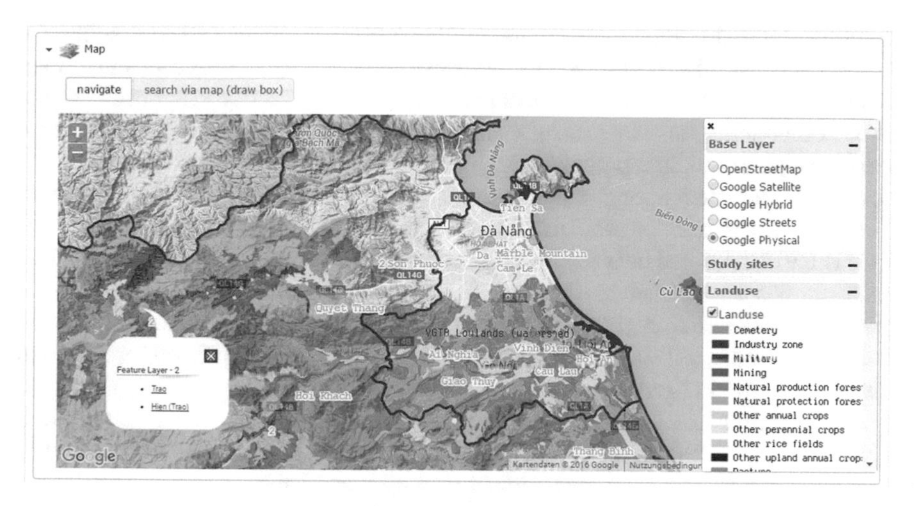


**Fig. 3** Screenshot of a map in the in the Vu Gia Thu Bon RBIS (LUCCi consortium 2016) covering the region of central Vietnam. The map shows the location of stations (*orange circles*), study site boundaries (*dark blue*) and an enabled land cover WMS layer for the region

## *Other Data*

This category summarizes all datasets not explicitly described in other modules like statistical yearbook data, socialeconomic data, within the project created risk maps or just the description of external repositories where data have been obtained during the project.

## *Calendar*

Important dates, planned field trips, workshops, or meetings are announced in the internal calendar. Beyond the normal calendar functionalities, it is possible to attach files to each event. Thus it is possible to attach the minutes of meetings or presentations of a workshop. Calendar events are only accessible and visible for core project members to keep internal project material and processes protected.

## *Processing*

One focus within the LUCCi project was the modeling of several processes or future scenarios (land use change, hydrology …) and the coupling and chaining of different simulation models. To represent those process chains and to describe the lineage process, the RBIS module "Processing" has been designed and implemented. It is described in more detail in the following section. It consists of the three components: data source, software, and processing step. A dataset (e.g., time series data) can link to a processing step to build a linkage between result data and how it was created during a processing step. A data source describes a list of datasets, which can be associated as source data. The list can include the VGTB RBIS managed data (e.g., precipitation time series data, geology ...) or with metadata described external data sources (e.g., simulated climate data). The processing step description gives information about, e.g., what was done, the rational, used data source, software version, creator, parent or previous processing step, and date of execution (Zander and Kralisch 2016).

One example for a simplified process/model chain is illustrated in Fig. 4. It shows that the ECHAM5 climate datasets (described as external data repository) of different climate scenarios serve together with the historical ERA40 reanalysis datasets (external data repository) as input data for the regional downscaling with WRF (Laux et al. 2013). The results were used as climate data input together with soil, geology, elevation datasets, and a combined projected land cover scenario for 2020 to model the impact on the water availability and sediment load in the Vu Gia and Thu Bon (VGTB) catchment with the model JAMS/J2K VGTB. Calculated amounts of water at several points inside the catchment and the outlet have been used as input for the application of MIKE FLOOD, MIKE 11, and MIKE HYDRO Basin (Fink et al. 2013).

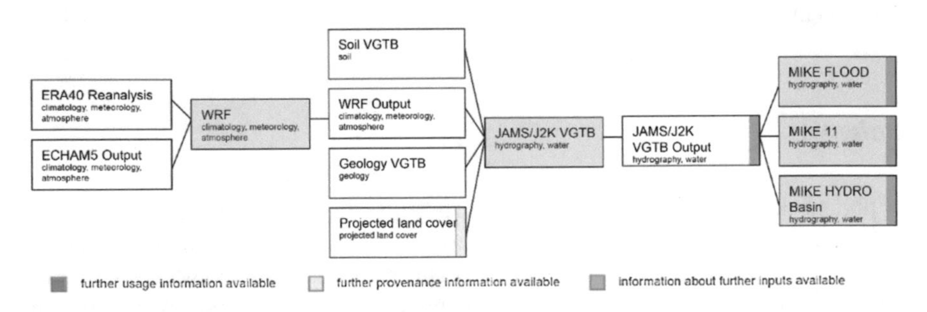


**Fig. 4** Schematic illustration of provenance information based on process descriptions within the Vu Gia Thu Bon RBIS (changed after Henzen et al. 2016)

## Provided Interfaces and Services

The Vu Gia Thu Bon RBIS provides several interfaces and services to support researchers working with the data. Important ones for the VGTB RBIS are as follows.

### *Search and Filter Functions*

A global text and spatial search is provided to get a rough overview and to search over all stored datasets. In addition, each data type, like time series data or documents, can be searched spatially via a map or text based on the attributes. Furthermore, datasets listed in the overview table can be filtered based on different filter operators for each attribute type. A temporary filter can be set globally based on study areas with an optional buffer zone. One use case would be to use the study area of the Lowlands of the Vu Gia and Thu Bon catchment to reduce the amount of displayed datasets to that specific region if this is the current area of interest (Zander and Kralisch 2016).

### *Time Series Data Import and Export*

The management of time series data plays a central role within the VGTB RBIS. The measurement network for hydro- and meteorological measurement in or nearby the project region is not very dense. Three climate stations (Da Nang, Hue, and Quang Ngai) are currently reporting to the World Weather Watch (WMO) network (WMO 2013). Based on this data the Global Summary Of the Day (GSOD) (Lott 1998) product is calculated daily. To import and keep the freely accessible GSOD data updated, there is an automated RBIS processing function. Beyond the download from the original source and import of the data, units are converted from English to SI-units (Zander et al. 2013). Besides the export functions for each single dataset, it is also possible to combine data from different datasets (e.g., of one parameter like precipitation) and download it as one file. This file is already in the format suitable for JAMS (Kralisch et al. 2009) models, which have been applied for hydrological and sediment load simulation within the context of the LUCCi project.

### *Web Map Service (WMS)*

Imported geodata datasets displayed in a map can also be published as Web Map Service (WMS). Thus the data can not only be visualized together with a base map and the detailed metadata description within the VGTB RBIS, but also in any external application with a WMS client such as the desktop GIS QGIS (QGIS community 2016). An example for land use data as published as WMS is shown in Fig. 3.

## *Catalog Web Service (CSW)*

To deliver and expose metadata information to other spatial data infrastructures, a Catalog Web Service (CSW) (Nebert et al. 2007) has been set up using pycsw (Kralidis and Tzotsos 2016) as a CSW server. This was especially requested for the LUCCi responsible scientific coordination project GLUES (GLUES 2013). Via CSW to the internet-exposed metadata information of in the VGTB RBIS stored geodata, time series data, and soil profiles are updated every night based on datasets visible for a guest account. Two examples for the access of the CSW are the GLUES Geoportal (GLUES 2016) (Fig. 5) and a CSW search (Zander 2016) of all project databases worldwide based on the underlying software framework RBIS and with a running CSW service.

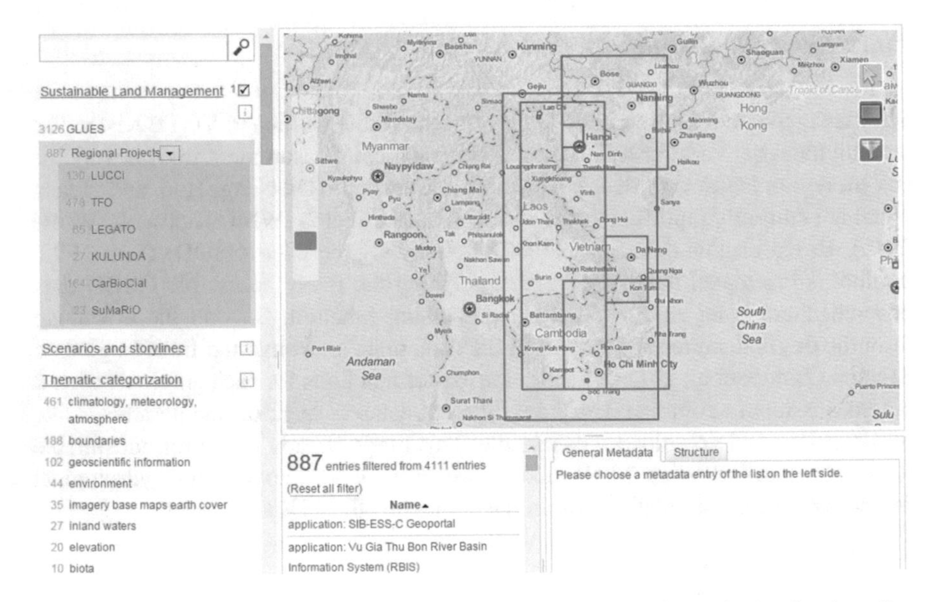


**Fig. 5** Screenshot of a GLUES Geodata infrastructure tool (GLUES 2016) showing bounding boxes of datasets available in Vietnam from different research projects of the research program "Sustainable Land Management"

## User Acceptance, Visibility and Experiences

Building trust and increasing user acceptance is one of the main keys for the success of a system like the VGTB RBIS. In order to address this within the LUCCi project, several training courses in small or bigger groups were offered and conducted for both German and Vietnamese partners and stakeholders in Cologne, Jena, Hanoi, Da Nang, and Tam Ky. The group of local stakeholders in Vietnam does not usually communicate in English, which forces not only the employment of translators during the courses, but also underlines the necessity, and later the usefulness, of the multilingual support. It has been shown during the last years that user acceptance and trust is also highly related to comprehensive support. This includes not only user account request related communications, but also support during data preparation and upload or appropriate reaction based on user feedback, e.g. related to malfunctions or suggestions for improvements or own observations (e.g., during training courses). Due to the flexible and modular underlying structure of the VGTB RBIS, it was possible to incorporate new types of data, adapt existing data description fields, or add new functions. A big step to ease the usability of RBIS was the change of the web-based user interface from a plain HTML to a dynamic WebApp with HTML and JavaScript. Since the setup of the VGTB RBIS, there are in average 1.2 logins per day and currently 160 registered user accounts. Most of them are not part of the LUCCi consortium. In addition to this, there are around 50 rejected account requests. This is related to the former strategy to provide only user accounts for project members and not to provide accounts to requests that did not clearly indicate the purpose of use. Since activation of the second-level permission management based on dataset level by the end of the project mid of 2015, user account granting is less restricted (and less time-consuming). Another reason for the still high interest and visibility is the absence of comparable accessible information systems and data collections in the region and in combination with the easy access for search engines like Google to index metadata due to the automated guest login. The offered transfer of the VGTB RBIS to be hosted and maintained in Vietnam is postponed due to a lack of available hosting responsibilities and maintenance resources. Nevertheless, attempts are being made to transfer and implement the developed system for other catchments in Vietnam. Due to the flexible and modular-based underlying structure, it will be easy to incorporate additional types of data, adapt existing data descriptions, or add new functions.

**Acknowledgments** The authors acknowledge the support of the German Ministry of Education and Research which has funded the RBIS development within various research programs (grant numbers 03IP514, 01LL0912, 01LB0801, 01LL0908D, 01LG1201E).

## References

Fink M, Fischer C, Führer N, Firoz A, Viet T, Laux P, Flügel W-A (2013) Distributed hydrological modeling of a monsoon dominated river system in central Vietnam

GLUES (2013) GLUES-GDI—Geodata infrastructure

GLUES (2016) GLUES Geoportal. http://geoportal-glues.ufz.de/index.php. Accessed 20 Mar 2016

Henzen C, Mäs S, Zander F, Schroeder M, Bernard L (2016) Representing research collaborations and linking scientific project results in spatial data infrastructures by provenance information (under review). Helsinki, Finland

ISO (2003) International Standard ISO 19115 Geographic information—Metadata

jQuery Foundation (2016) jQuery—write less, do more. https://jquery.com/. Accessed 12 Mar 2016

Kralidis T, Tzotsos A (2016) pycsw—Metadata publishing just got easier. http://pycsw.org/. Accessed 12 Mar 2016

Kralisch S, Zander F, Krause P (2009) Coupling the RBIS environmental information system and the JAMS modelling framework. In: Anderssen R, Braddock R, Newham L (eds) Proceedings of the 18th World IMACS/and MODSIM09 International Congress on Modelling and Simulation. Cairns, Australia, pp 902–908

Laux P, Lorenz C, Thuc T, Ribbe L, Kunstmann H et al (2013) Setting up regional climate simulations for Southeast Asia. In: High performance computing in science and engineering '12. Springer, pp 391–406

Lott N (1998) Global surface summary of day

LUCCi consortium (2016) Vu Gia Thu Bon RBIS. http://leutra.geogr.uni-jena.de/vgtbRBIS/. Accessed 10 Mar 2016

Nebert D, Whiteside A, Vretanos P (2007) OpenGIS catalogue services specification 2.0.2

Open Source Geospatial Foundation (2016) Welcome to MapServer—MapServer 7.0.1 documentation. http://mapserver.org/. Accessed 12 Mar 2016

OpenLayers Contributors (2016) OpenLayers 2. http://openlayers.org/two/. Accessed 12 Mar 2016

Orcale (2016) Oracle VM VirtualBox. https://www.virtualbox.org/. Accessed 12 Mar 2016

PostGIS Developers (2016) PostGIS—Spatial and geographic objects for PostgreSQL. http://postgis.net/. Accessed 10 Mar 2016

PostgreSQL Global Development Group (2016) PostgreSQL: the world's most advanced open source database. http://www.postgresql.org/. Accessed 10 Mar 2016

QGIS community (2016) QGIS als OGC Datenclient. http://docs.qgis.org/2.8/de/docs/user_manual/working_with_ogc/ogc_client_support.html. Accessed 6 Apr 2016

WMO (2013) World Weather Watch WMO. http://www.wmo.int/pages/prog/www/index_en.html . Accessed 22 Mar 2013

Zander F (2016) RBIS CSW global search. http://leutra.geogr.uni-jena.de/RBISsearch/metadata/global_search.php. Accessed 20 Mar 2016

Zander F, Kralisch S (2016) River Basin Information System (RBIS)—open environmental data management for research and decision making. ISPRS Int J Geo-Inf 5:123

Zander F, Kralisch S, Flügel W (2013) Data and information management for integrated research–requirements, experiences and solutions. In: Proceedings of the 20th International Congress on Modelling and Simulation. Adelaide, Australia, pp 1–6

# Forest Change and REDD+ Strategies

**Valerio Avitabile, Michael Schultz, Giulia Salvini, Arun Kumar Pratihast, Astrid Bos, Nadine Herold, Pham Manh Cuong, Vu Quang Hien and Martin Herold**

**Abstract** In recent years the United Nations initiative on Reducing Emissions from Deforestation and forest Degradation (REDD+) program gained increasing attention in the policy arena, representing a valuable incentive for developing countries to take actions to reduce greenhouse gas emissions and at the same time promote sustainable forest management and improve local livelihoods. To design an effective REDD+ implementation plan at the local level it is crucial to make an in-depth analysis of the international and national requirements, analyse the forest change processes and related drivers at sub-national scale, and assess the local management options and constrains to ultimately select the appropriate policy mix and land management interventions. The present chapter first describes the state and historical changes of forests in Vietnam, identifies the direct and underlying drivers of deforestation and forest degradation at national scale, discusses the role of forests for climate change mitigation and indicates the key activities for reducing carbon emissions in Vietnam. Second, the main biophysical parameters and processes are assessed at sub-national scale for the Vu Gia Thu Bon river basin. The land cover and carbon stocks are mapped and quantified for the year 2010 and the land cover change and related carbon emissions are estimated for the period 2001–2010, allowing to model the

V. Avitabile (✉) · M. Schultz · G. Salvini · A.K. Pratihast · A. Bos · M. Herold
Laboratory of Geo-Information Science and Remote Sensing,
Wageningen University & Research, Droevendaalsesteeg 3,
Wageningen 6708 PB, The Netherlands
e-mail: vale.avi@gmail.com

N. Herold
Soil, Water & Land Use, Wageningen University & Research,
PO Box 47, Wageningen 6700 AA, The Netherlands

P.M. Cuong
Department of Agriculture and Rural Development of Tuyen Quang Province,
108 Nguyen Van Cu, Minh Xuan, Tuyen Quang, Vietnam

V.Q. Hien
Training and International Cooperation Division, Forest Inventory
and Planning Institute, Thanh Tri district, Hanoi, Vietnam



A. Nauditt and L. Ribbe (eds.), *Land Use and Climate Change Interactions in Central Vietnam*, Water Resources Development and Management,
DOI 10.1007/978-981-10-2624-9_4

land cover change and predict deforestation risks until the year 2020. Among the areas at higher risk of deforestation, the Tra Bui commune located in Quang Nam province is selected to design a sub-national REDD+ implementation plan in the third part of the chapter. The plan is based upon an in-depth analysis of the local context and land cover change dynamics, the local drivers of deforestation and a conducted stakeholder involvement process in the commune. Based upon this analysis, the last section provides recommendations about the local land management strategies that could be introduced in the commune and discusses the policy interventions are likely to enable their implementation.

# Introduction

## *Forests and Climate Change in Vietnam*

### Historical Forest Change in Vietnam

Vietnam's forest cover declined from 41 to 27 % between 1943 and 1990 (FAO 2010). Since then, Vietnam has made considerable efforts to increase its overall forest cover, which reached 13.5 million ha or 39.7 % in 2011 (MARD 2012). The increase has been mainly due to new plantations, which account for 2.9 million ha. However, albeit some net increase in forests is observed, serious deforestation and extensive degradation occurs in some regions, including parts of the Central Highlands, the Central Coast and the Southeast region (Meyfroidt and Lambin 2008). Furthermore, it is generally acknowledged that natural forests are increasingly becoming fragmented and degraded. Over two-thirds of Vietnam's natural forests are considered poor or regenerating, while rich and closed-canopy forests constitute only 4.6 % of the total. Between 1999 and 2005 the area of natural forest classified as rich decreased by 10.2 % and the amount of medium quality forest declined by 13.4 %. Lowland forests supporting their full natural biodiversity have been almost entirely lost, while Vietnam's mangrove forests have been significantly degraded. These statistics clearly indicate that the forests of Vietnam are under serious threat and appropriate policies and actions for their protection are urgently required.

### Forest and Climate Change Policies in Vietnam

Vietnam is considered one of the most vulnerable countries to the adverse effects of climate change, facing potential extensive economic damage and loss of life (UN 2012; MONRE 2009; DANIDA 2008). Therefore, the country has much to gain by joining the international challenge to mitigate greenhouse gas emissions, even though Vietnamese emissions per capita remain very low. Forests play a relevant role for the mitigation and adaptation to climate change in Vietnam. The National Strategy on Climate Change (2011) and National Strategy on Green Growth (2012) aim for the "protection, improvement in quality and sustainable use of existing forests, expansion of forests into non-forested areas, increasing carbon removals, biodiversity conservation and maintaining the environmental protection functions of the forest" as key national priorities.

The current national strategy for the forest sector in Vietnam is the National Forest Development Strategy for the period from 2006 to 2020. It builds on previous strategies and programs, setting out ambitious targets for policy reform, plantations, financial support for forest protection and plantations and a greater role and responsibility for the local communities. It seeks to modernize forestry, so that forestry can play its part in the industrialization and modernization of rural agriculture, in hunger eradication, in poverty reduction for people in mountainous areas, and in environmental protection. On this basis, the National Action Plan on Forest Protection and Development for the period 2011–2020 is aimed to: (a) protect existing forest areas; (b) increase forest coverage to 42–43 % by 2015 and to 44–45 % by 2020; improve yield, quality and value of forests; re-structure the forestry sector to generate more value-added products; (c) generate employment and income opportunities, improve livelihoods of local people and contribute to poverty alleviation. The Plan shows the continuous and consistent government policies on forest protection and development, with substantial investment from state budget and support from international development partners.

## Drivers of Deforestation and Degradation

The factors driving deforestation and degradation in Vietnam have changed in the recent decades. While during the period 1943–1993 deforestation and degradation were mostly a result of war and agricultural expansion by people migrating into forested areas, the current main direct causes of deforestation at national level are: (i) conversion to agriculture, particularly to annual crops and industrial perennial crops; (ii) unsustainable logging (notably illegal logging); (iii) infrastructure development; (iv) forest fires.

Other direct drivers, such as invasive species, mining, biofuels and climate change, exists but are not significant at present. Although the impact of these factors may grow in the future, currently their impact is deemed less significant. Besides the key drivers of deforestation and degradation, there are also more general factors within the forestry sector and other sectors to be addressed. There are discrepancies among policies and programs within and among sectors including forestry, agriculture, natural resources and environment, transportation and construction. In addition, the current logging ban in natural forests of some provinces and the setting of harvesting quotas at very low levels also contribute to illegal extraction of timber.

## Conversion of Forest to Agricultural Land

Vietnam continues to be one of the world leaders in the export of agricultural commodities, including coffee, cashew, pepper, shrimps, rice and (increasingly) rubber. Over the past 10 years the expansion of industrial crops has grown faster than planned, increasing from 1.634 million ha in 2005 to 1.886 million ha in 2008; the area of rubber plantations in 2013 was over 900,000 ha and was 100,000 ha greater than 2020 target. There has also been a large increase in the area used for aquaculture, primarily shrimp farms in the deltas. Between 1991 and 2001, the total area of coastal and marine aquaculture in Vietnam almost doubled. The approved program on restructuring the agricultural sector intends to stabilize the total area for coffee at 500,000 ha, rubber areas are projected to expand more than 120,000 ha to reach an area of 800,000 ha, and cashew crops are expected to expand 30,000 ha to reach the target of 430,000 ha.

Most of the recent expansion in the perennial industrial crops has concentrated in the Central Highlands and the Southeast region, which have comparative advantages and present the richest and largest natural forests. Over the past 10 years these provinces have experienced some of the highest levels of deforestation. Similarly, in coastal areas government policies and market signals have directly or indirectly led to the large-scale conversion of rice lands and coastal mangrove forest areas to shrimp farms. With relatively high forest cover remaining, good soils and good market signal on exporting perennial crops, such areas will come under further pressure from agricultural expansion and are highly relevant for forest protection initiatives such as the UN Reduced Emissions from Deforestation and forest Degradation (REDD+) initiative.

## Unsustainable Logging

Forest degradation is primarily caused by unsustainable logging, which is often a result of poor management practices and/or illegal activities as well as timber harvesting by rural households for local consumption. The scale of illegal practices is difficult to estimate but it is generally regarded as an important driver of the forest loss in Vietnam, which may lead to a severe impoverishment of the country's forest estate. Some illegal practices are performed by local households driven by poverty and desperation, but more often they are driven and controlled by criminal gangs and networks.

There is a high demand for timber for inexpensive furniture made from tropical hardwood. Vietnam has become a major hub for the export of furniture, making wood products Vietnam's fifth largest export earner. Revenue from wood processing industry in 2013 was over US$5.3 billion. The issue of the illegal trade in timber as well as the illegal extraction in Vietnam has serious implications for the future of the industry as well as the potential benefits from REDD+. With stricter requirements to show proof of legal provenance, there is a growing need for Vietnam to stamp out the use of timber from illegal sources and comply to the US

Lacey Act and the EU FLEGT. Therefore, possible leakage will need to be taken into account under any REDD+ scheme.

In addition, there are some more general factors within the forestry sector that are leading to unsustainable wood extraction and are related to some of the current forest policies and programs. Examples include the forestland classification process, which opens up possibilities for the unnecessary removal of natural forested areas, the current logging ban and the setting of the harvesting quota system at such low levels that it encourages the illegal extraction of timber.

## Infrastructure Development

Infrastructural development is required for Vietnam to be able to keep developing. Of all the potential infrastructural developments, road building and dam construction have the strongest impact in terms of forest loss. Vietnam's roads have more than doubled in length since 1990. While the forest area cleared to make way for the construction may not be significant, the greater accessibility of such areas to encroachment and unsustainable exploitation has a highly detrimental impact. In the pursuit of rapid economic growth and efforts to curb poverty, roads and dams need to be built but to ensure an optimal use of resources it is important that environmental and social costs are accounted for and that efficient and sustainable forms of economic growth are promoted in order to maximize social welfare.

Hydropower plays an important role in electricity generation in Vietnam. In 2010 hydropower provided 9412 MW out of a total 26,209 MW, and there are plans to increase the hydropower capacity to 10,766 MW by 2025, exploiting most of the remaining potential of the country. The regions of North West and Central Coast are the areas with the highest current production, as well as offering the greatest potential for future development. The underlying factor driving hydropower development is the burgeoning demand for electricity, which is currently projected to grow at 11 % per year. In order to meet this rapidly growing demand, the power industry has struggled to expand and improve the power system. The construction of dams along the Dong Nai has already destroyed more than 15,000 ha of natural forest. The ADB Environmental Assessment of the Hydropower Master Plan estimated that the 21 planned large-scale dams (with a capacity over 4610 MW) will lead to a land loss of around 21,133 ha, including 4227 ha of natural forests and 1367 ha of plantations. The total resource value of the forest lost (including environmental service functions) was estimated to be 72.4 million US Dollars. This study also estimated the indirect impacts on the forested areas from in-migration and resettlement of people with an expected 61,571 people being displaced from the 21 schemes. Besides these large schemes, there are several medium and small hydropower schemes currently planned, which have lower requirements and scrutiny in terms of Environmental Impact Assessments. Small hydropower stations and pumped storage are estimated to produce 3860 MW up till 2025 and are expected to impact smaller areas but in many locations.

### Forest Fires

About 6 million ha of Vietnam's forests are considered to be vulnerable to fire, mostly located in the Northwest, the Central Highlands (Kon Tum, Lam Dong and Gia Lai), the Southeast and the Mekong Delta. According to government statistics, between 1992 and 2002 an average of 6000 ha per year of forest was lost due to fires. About 3100 ha per year were damaged by fire between 2004 and 2008, which reduced to 1500 ha/year in 2009 and then increased to about 3000 ha/year in 2010 due to the dry weather associated with El Nino phenomenon. However, the forest lost was substantially reduced in 2011–2013 (average of 1500 ha per year) due to improvement in forest fires protection and cooperation among forest owners, forest protection forces, the police and army and local stakeholders, including local communities.

## *Opportunities for REDD+ in Vietnam*

### The Role of Forests for Climate Change Mitigation

Forests play a major role in mitigation of climate change via carbon storage as well as providing important ecosystem services such as water storage, soil fertility regulation and biodiversity preservation. In recent years the United Nations (UN) collaborative initiative on Reducing Emissions from Deforestation and forest Degradation (REDD+) program gained increasing attention in the policy arena, representing a valuable incentive for developing countries to take actions to reduce GreenHouse Gas (GHG) emissions from forest land and at the same time promote sustainable forest management and improve local livelihoods.

In Vietnam, national statistics indicate that about 70 % of the total population is living in the rural areas and their livelihoods are heavily dependent upon the agricultural cultivation. Although the national forest coverage increased, the quality of forests is still quite low while the demand of timber as well as the conversion of forests into agriculturally cultivated land and other uses is increasing over time. Hence, the threat to the existing forests and challenges for their protection are remaining. Vietnam has adopted a nationwide and landscape-based approach for the REDD+ implementation. Therefore, there are clear opportunities for REDD+ in the agricultural sector. REDD+ is expected to contribute significantly to obtain the National Program objective of reducing 20 % of emissions from the agriculture and rural development sector in the period 2011–2020 (REDD Vietnam 2012). Key activities in the forest sector to achieve this objective are identified as follows:

- Strengthen forest plantation, restore forestry, reforest and enrich forest in planned areas according to the forestry development strategy for the period 2010–2020. This action is being implemented on 2.6 million ha plantation forest across the country and has the potential to sequester 702 million ton $CO_2$ equivalent ($CO_2$e);

- Protect, develop and sustainable use of forest land to increase carbon sequestration and eliminate GHG emission from forestry. This action is implemented on 13.8 million ha on forest areas across the country and has the potential to sequester 669 million ton of $CO_2e$;
- Other activities including strengthen communication campaign and capacity building on awareness to protect and sustainable forest utilization, forest fire prevention, strengthen international collaboration to promote carbon credit market in the forestry sector.

Compared to the forest sector, the agriculture sector forecasts lower emission reductions. Crop production is aimed to reduce 5.72 million ton $CO_2e$ (equivalent to 10 % of the forecast emissions from the crop production sector up to 2020) and the livestock production is expected to reduce 6.30 million ton of $CO_2e$ (equivalent to 26 % of total forecasted GHG emission in the livestock sector up to 2020).

## Strategies for Local REDD+ Implementation

Drivers of deforestation and forest degradation occur from global to local scale, and thus strategies to address drivers need to be considered at all scales. The assessment of the most appropriate scale for intervention must be considered by policy and decision-makers. To realize an effective REDD+ policy design at the local level it is crucial to adequately take into account the complexity and interconnected nature of the diverse actors and the underlying roots of deforestation processes. This would allow formulating effective policy interventions that satisfy the needs of all stakeholder groups and enhance/maximize carbon stocks.

The first step in designing and realizing forest conservation polices is to understand the drivers of deforestation. Angelsen and Kaimowitz (2001) describe deforestation as determined by drivers at different levels. At one level are the agents (individuals, households or companies), who are the sources of deforestation. The main agents of deforestation are subsistence farmers, cash crop smallholders and large companies. At the second level there are decision parameters such as prices, access to markets, agricultural technologies and agroecological conditions that influence the choices made by the agents. These decision parameters are the direct causes of deforestation. At a third level, these decision parameters are affected by national and international policies, which are the underlying causes of deforestation.

The description of this system underlines the complexity of the land cover change dynamics, which are the result of the interrelations between these three different levels and their components. For this reason it is crucial to identify the policy mix that tackles the causes of deforestation at the local level and encourages local actors to change their land use, by influencing/regulating their land use decision. This refers to Policies And Measures (PAMs) to address the underlying drivers of forest carbon change, such as incentives, disincentives and enabling measures, tenure reforms, land use planning, better governance and command and control measures. In order for REDD+ to be effective at the local level, these policies should be site-specific, taking in consideration the existing policies, the socio-economic context and the potential consequences on different forest users. They should promote equitable benefit sharing, consider possible side effects on land exploitation and ensure long-term local engagement. Hence for REDD+ to be successful it will need to reach the actors responsible for addressing the drivers of deforestation and for shifting land use. These actors span all scales, from international commodity buyers to forest-dependent communities. Enabling factors such as effective information systems to guide decisions, institutional capacity, transparency and accountability, political will and consultation with stakeholders underpin any strategy to affect drivers.

Defining pathways to enable the effective implementation of REDD+ policies and measures will be critical, taking into account several factors including weak forest sector governance and institutions, conflicting policies beyond the forest sector and illegal activity as critical underlying drivers of deforestation and forest degradation. Hence the development of REDD+ strategies that focus solely on affecting direct drivers in order to demonstrate quantifiable emissions reductions may place less emphasis on addressing the critical underlying factors that will determine whether direct driver interventions can be successful in achieving the intended emissions reductions. Most often, a mix of incentive investments, disincentives and enabling measures, under a comprehensive REDD+ strategy, will provide greatest leverage to affect drivers.

## *The LUCCI Framework for REDD+*

Any change in the terrestrial carbon stocks as a result of direct or indirect human activities has an impact on the climate, and tropical deforestation and forest degradation cause a significant contribution to the increase of GHG in the atmosphere. For this reason, reducing the emissions from forest land to the atmosphere and maintaining or increasing the terrestrial carbon pool is an important climate change mitigation activity, and the REDD+ program currently represents the key international mechanism to reduce greenhouse gas emissions from forest in developing countries.

The implementation of the REDD+ program requires to first quantify and map the historical carbon dynamics due to land use change. According to the Intergovernmental Panel on Climate Change (IPCC) Guidelines (IPCC 2006), GHG emissions and removals occurring on a certain piece of land can be calculated on the basis of two inputs: Activity Data and Emission Factors. Activity Data consist on the areal extent of an activity that causes emissions or removals, usually referred to as area change data. Emission Factors consist on the amount of emissions or removals per unit area related to a certain activity, usually referred to as changes in stocks between the two land cover types. Such dynamics (including carbon emissions and removals) may then be used to estimate the reference emission levels against which the REDD+ emission reductions can be accounted for. The amounts of carbon stored in the terrestrial pools (i.e., vegetation biomass and soil carbon) and their changes in a historical reference period are therefore the key measures for quantifying the role of a forest in the REDD+ mechanism.

In order to support the design, development and implementation at sub-national scale of the Vietnam national REDD+ program, within the LUCCi project a dedicated analysis was carried out to estimate the emissions and removals of carbon stored in vegetation biomass and soils due to changes in land cover occurred between 2001 and 2010 in the Vu Gia Thu Bon (VGTB) river basin (see Land change and carbon emissions in the VGTB). Activity data were obtained by mapping land cover and land cover changes, while emission factors were derived from forest inventory data. The analysis allowed to assess what are the main change processes and carbon pools driving the net carbon emissions from the land sector in the VGTB. These information and related datasets were the basis to design a case-study for the REDD+ implementation strategy at local level (see Design of a local REDD+ implementation strategy).

## Land Change and Carbon Emissions in the VGTB

### *Land Cover in the Year 2010*

In order to estimate the carbon emissions from land cover changes, the first step was to characterize and map the land cover of the Vu Gia Thu Bon (VGTB) for the year 2010. The land cover map was created using satellite data (Landsat images), ancillary datasets and field data, and included the six Intergovernmental Panel on Climate Change (IPCC) land use classes: Forest, Grassland, Cropland, Other Lands, Settlements and Wetlands/Water. Forest identifies areas with tree density higher than 10 %, Grassland includes areas dominated by grass, shrubs and woody regrowth with tree density lower than the 10 % forest cover threshold and Wetlands/Water identifies any land that is covered or saturated by water for all or part of the year. In the VGTB, Wetlands mostly corresponds to water bodies and reservoirs for the production of hydroelectricity and is referred to as Water.

The land cover map presents a spatial resolution of 30 m and was obtained from the analysis of 12 Landsat scenes using up-to-date image processing techniques. First, the availability and quality of existing satellite images and ancillary spatial datasets was assessed in order to select the most appropriate data with regard to the specific characteristics of the VGTB, such as the phenological changes of vegetation due to seasonality. Second, the selected datasets were pre-processed and harmonized using topographic correction and radiometric normalization techniques. A correction of topographic illumination effect was carried out to minimize classification errors caused by shadow effects due to the terrain, and then the images were adjusted in order to meet the same radiometric scale. A geometric correction was not necessary since the available Landsat data were already pre-processed to Level 1T. Since the Landsat Enhanced Thematic Mapper (ETM+) data used for the analysis were affected by the Scan Line Corrector (SLC) failure, extensive gap-filling processing techniques, based on image composition techniques and geo-statistics employing multiple scenes of the same season, were carried out to minimize the data gaps. Third, the clouds present in the image mosaic were identified and masked on the basis of appropriate image indexes. Four, the satellite data were classified according to the IPCC classes using the non-parametric Random Forest algorithm (Breiman 2001). An extensive database consisting of information from different sources and sensors was built and used as input for the classification. The algorithm was trained with a set of 168 training polygons acquired during a field trip in March 2012. Further training was provided through visual analysis of high-resolution (2.5 m) remote sensing imagery (SPOT 5). The classification took into consideration the effects of seasonal changes, which are especially marked for croplands characterized by intra-annual crop rotations, and included additional datasets as the FIPI 2010 Forest map and GIS data (river and road network). The detailed processing chain leading to the 2010 VGTB Land Cover Map is described in Schultz and Avitabile (2012a).

The 2010 VGTB Land Cover Map is presented in Fig. 1. The VGTB river basin comprises 12,577 $km^2$ and in the year 2010 almost half of the land area was covered by Forest (5817 $km^2$, 47 % of land area), followed by Cropland (3160 $km^2$, 26 % of land area) and Grassland (2454 $km^2$, 20 % of land area), while smaller areas were occupied by Settlements (316 $km^2$, 3 % of land area), Water (330 $km^2$, 3 % of land area) and Other Lands (266 $km^2$, 2 % of land area) (Fig. 1). 43 $km^2$ (0.3 % of land area) could not be classified due to missing data (Avitabile et al. 2016).

The map was validated according to recommendations from Olofsson et al. (2014), thus including the sub-categories of the Sampling Design (which determines the setup of the distribution of the reference points across the data set), the Response Design (which determines the extraction of the reference information from a reference point) and the Analysis (which determines the algebra used for the calculation of the class and the overall accuracies). High resolution 5 satellite data (2.5 m resolution) acquired in 2011, 2010 and 2008 provided by the Planet-Action initiative were used as reference data for the accuracy assessment. The map accuracy was assessed for 300 validation points stratified by class proportions and landscape heterogeneity through visual interpretation of the SPOT 5 images. The sampling units were defined by a spatial extent of 30 × 30 m and each reference point was visually interpreted for the major Land Cover class. The overall accuracy of the 2010 Land Cover map resulted to be equal to 82 % (Avitabile et al. 2016).

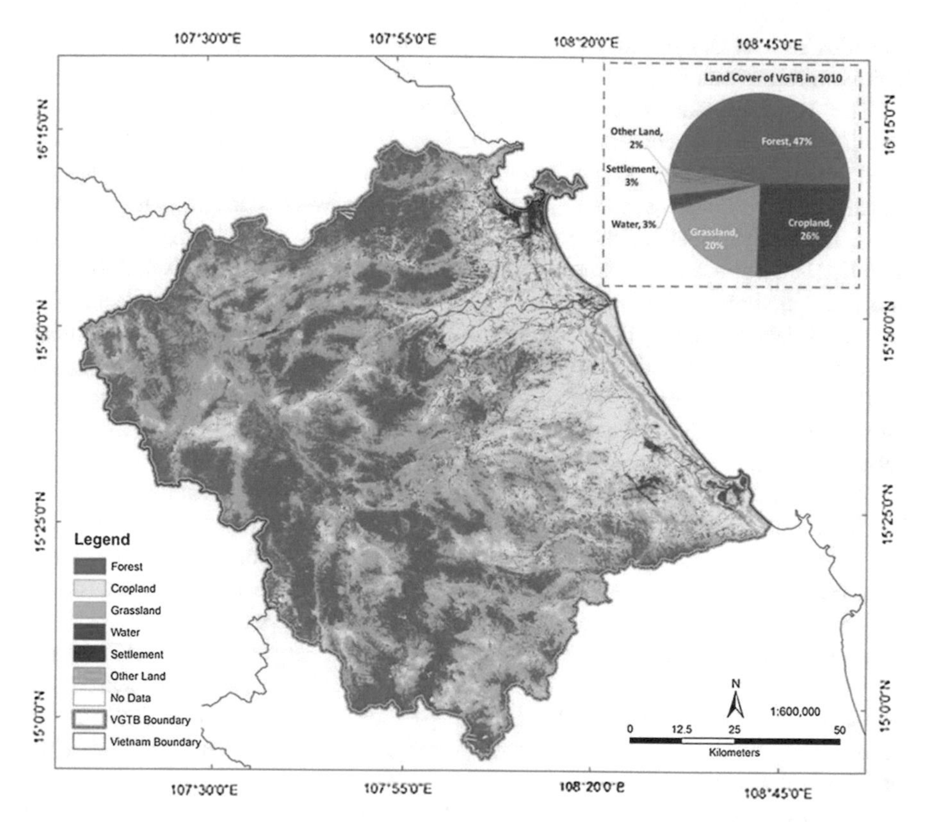


**Fig. 1** Land cover map of the VGTB for the year 2010. *Inset* Distribution of the land cover classes in the VGTB

## *Land Cover Change Between 2001 and 2010*

The land cover change processes occurring in the VGTB river basin during the period 2001–2010 were mapped using Landsat data. Since the VGTB is a highly dynamic area where land change process may occur rapidly, the change detection was performed for the two subperiods 2001–2005 and 2005–2010. The base year was set to 2001 instead of 2000 because the available satellite images were highly affected by cloud coverage and haze contamination in the year 2000. The satellite images were classified on the basis of expert knowledge acquired during the field campaigns and through discussions with local and national FIPI experts. The change categories were defined as combinations of the six IPCC Land Cover classes, which can be reliably mapped using remote sensing data. Changes occurred between grasslands, shrublands and regrowth (areas with tree density lower than the forest cover threshold) were not mapped since these three vegetation types are included in the same IPCC class (Grassland). Similarly, the IPCC class Forest includes natural forest and plantation, and therefore changes from forest to plantation were not mapped.

The Landsat data were first pre-processed for cloud and shadow removal, gap-filling and change/no-change thresholding. Intra-annual changes, i.e. seasonal land cover changes that do not lead to a change in stable land cover class and are mostly due to regular seasonal agricultural practices, were identified and removed from the analysis. Then, the images were transformed and classified using the Iterative reweighted Multivariate Alteration Detection approach (Nielsen 2007) and the Random Forest (Breiman 2001) classification algorithm. The classification of the input datasets for 2001–05 and 2005–10 were trained using expert knowledge acquired through an understanding of the land cover change processes during the 2012 field trip and with the support of local and national FIPI experts. The training data were obtained through visual interpretation of the 2001, 2005 and 2010 Landsat mosaics and from the 2010 VGTB Land Cover map.

The results of the change detection indicated that the change rate was equal to 1.6 % during the period 2001–2005 and 3.1 % of the land area during the period 2005–2010. The gross deforestation rate resulted to be about 2.2 % of land area over the two 5-year periods, hence the average annual deforestation rate was about 0.22 %. Gross deforestation was partly counterbalanced by forest regrowth on croplands, which occurred on about 0.5 % of land area between 2005 and 2010 and only 0.1 % between 2001 and 2005. Specifically, the area under deforestation was equal to 136 and 135 $km^2$ during the periods 2005–2010 and 2001–2005, respectively, while the area under reforestation was equal to 59 $km^2$ for 2005–2010 and 13 $km^2$ for 2001–2005.

The largest single land cover change category during both periods consisted on the transition of Forest to Cropland, which accounted for 95 and 82 $km^2$ (equal to 49 and 21 % of the change area) for 2001–2005 and 2005–2010, respectively. Consistent changes were also detected for the transition of Cropland to Forest, which occurred on 46 $km^2$ between 2005 and 2010, while between 2001 and 2005 the reforestation rate on Cropland was lower (8 $km^2$). Another prominent change mapped was the transition of Grassland to Cropland, with more than 80 $km^2$ of Cropland established on Grasslands from 2005 to 2010. Urbanization rate was very high and large areas were converted to Settlements, and apart for some artificial lakes almost no decrease of settlement was mapped.

Considering only the net changes occurred between 2001 and 2010, forest areas decreased from 6016 to 5817 $km^2$ while croplands increased from 2980 to 3160 $km^2$ (Fig. 2). Grasslands presented a net small decrease (from 2491 to 2454 $km^2$) while settlement areas and water bodies showed a net increase (from 269 to 316 $km^2$ and from 321 to 330 $km^2$, respectively), and other lands had no net change.

The land cover change map was validated using 340 reference data stratified by change class, after aggregating the classes Settlement, Other Land and Water in the class “No vegetation”. The reference data were extracted from Landsat images and higher resolution data (Spot5, Rapideye, GoogleEarth) when available, and interpreted by visual analysis to identify the most common change class. The validation

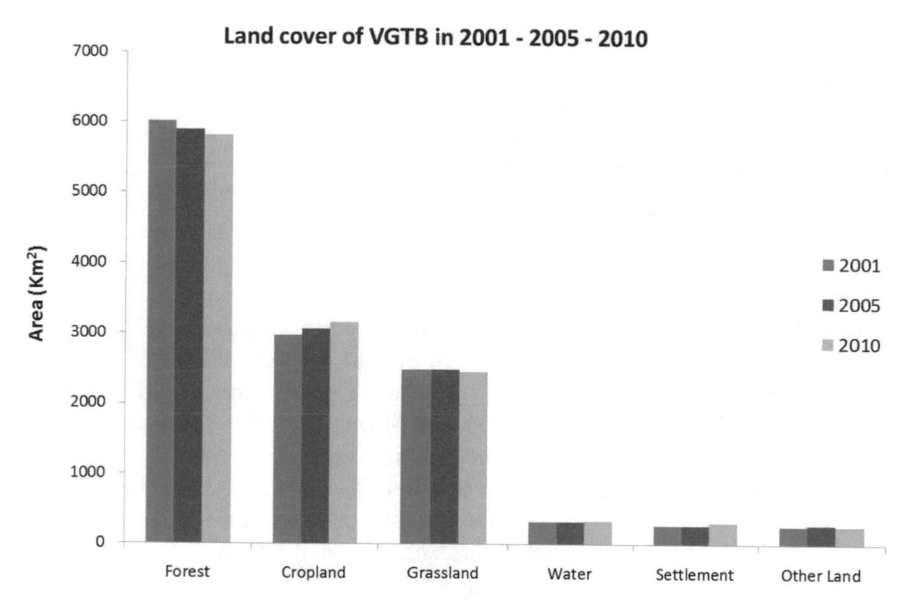


**Fig. 2** Change in area of the land cover classes of VGTB during the period 2001–2005–2010

of the land cover change map indicated that the overall accuracy was 81 % and the producers and users accuracies were higher than 70 %. The complete description of the data processing and the respective results are provided in Schultz and Avitabile (2012b) and in Avitabile et al. (2016).

## *Carbon Stocks in the Year 2010*

### Carbon Stocks in Biomass

The amount of carbon stored in forest ecosystems is a key parameter for quantifying their climate change mitigation capability and is an essential information to design and implement REDD+ activities. The assessment of forest-related carbon emissions and removals by sources and sinks due to direct or indirect human activities (also called "emissions factors") are obtained on the basis of estimates of the carbon stock density in each land cover classes. Hence, knowledge of spatial distribution of carbon stocks is critical to obtain better estimates of emissions at specific locations. In addition to REDD+ goals, carbon stock is a critical information for land management because it is a proxy for several ecosystems services, affects the local and regional climate and provides a variety of products such as timber, fuelwood and biofuel.

In most tropical ecosystems, the vast majority of the carbon is stored in aboveground tree biomass. The amount of carbon stored in vegetation is equivalent to about 50 % of their biomass density (Penman et al. 2003), hence carbon assessment consists on quantifying biomass, from which carbon stock of vegetation is directly derived. Biomass is defined as the total amount of aboveground and belowground living organic matter in woody vegetation expressed in units of dry weight. Biomass density represents the mass per unit area and is usually expressed as tons (Mg) per hectare, while total biomass of a certain region is obtained by multiplying its mean biomass density with the corresponding area.

Due to the large spatial extent and the considerable ecological variability of the VGTB river basin, a relatively large amount of field measurements were needed in order to have samples representative of the different ecosystems. In addition, in the context of REDD+ some field samples should also be located at locations expected to be deforested to eventually obtain more accurate estimates of the potential carbon emission reductions. For these reasons, two field biomass datasets including more than 3000 field plots were used. The first dataset consisted of the data collected by the National Forest Inventory (NFI) of Vietnam in the Quang Nam province. The NFI data are collected throughout the country at fixed, regular locations according to a systematic sampling strategy. The NFI dataset included 2994 field plots (plot area of 0.05 ha) measured between 2007 and 2009. The second dataset was acquired within the LUCCi project during dedicated field campaigns in 2011 and 2012 at specific locations, selected because of their representativeness of different forest types or change processes and for their vicinity to hot-spots of deforestation. This dataset consisted of 89 field plots (plot area of 0.126 ha) that allowed to obtain accurate estimates of carbon stock changes due to deforestation.

For all plots, the Diameter at Breast Height (DBH) and species were identified for the trees with DBH >5 cm. The wood density of the tree species was identified using the mean value provided in the Global Wood Density Database (Zanne et al. 2009). The aboveground biomass of the field plots was estimated from the tree parameters (DBH and wood density) using the generalized allometric equation for moist tropical forest (Chave et al. 2005). The belowground compartment was derived from the aboveground biomass using the average IPCC root-to-shoot ratios in the tropical moist ecoregion: 0.205 for aboveground biomass <125 Mg/ha and 0.235 for aboveground biomass >125 Mg/ha (Mokany et al. 2006). The tree biomass (aboveground and belowground) was converted to carbon units using the conversion factor of 0.5 and used to calculate the carbon density of each plot, considering the plot nested approach and the terrain slope (Pearson et al. 2005).

The field data were then stratified by spatial maps aimed to represent land cover classes with similar carbon stock density values. These strata were obtained by combining the 2010 VGTB Land Cover map and the 2010 Forest Map produced by the Forest Inventory and Planning Institute (FIPI) of Vietnam. The fusion of the two maps allowed to obtain a Stratification Map with improved spatial resolution (30 m) and higher thematic detail, merging the IPCC and the FIPI forest legends in a new combined legend including 10 classes: Rich Forest, Medium Forest, Poor Forest, Forest Regrowth, Plantation, Grassland, Cropland, Settlements, Other Land,

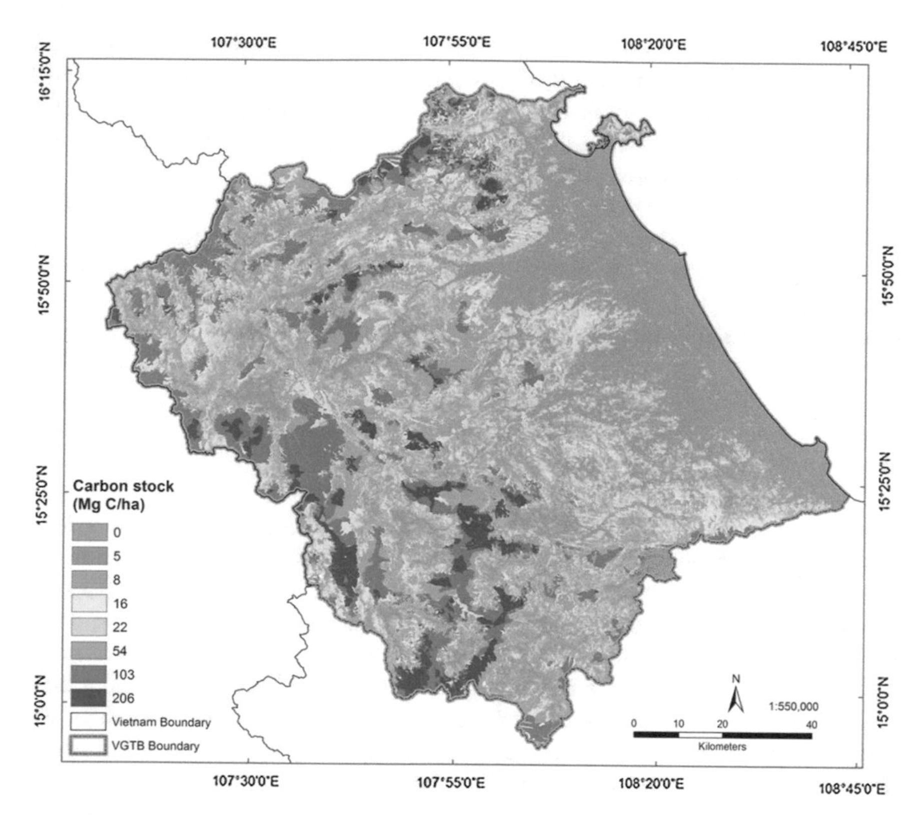


**Fig. 3** Map of carbon stock density in aboveground and belowground biomass in the year 2010 for the VGTB

Water. Then, the average carbon stock per strata was computed and applied to the Stratification Map to produce a map of spatial distribution of carbon stock density and quantify the total carbon stock of the study area (Fig. 3). Since no plots were located in the strata Cropland, Water, Settlement and Other Land, the IPCC Tier 1 Default value (IPCC 2006) were applied for Cropland (Table 5.9 in IPCC 2006) and the carbon density was considered equal to zero for Settlement, Other Land and Water.

In total, the vegetation of the VGTB river basin stores about 43 million tons of carbon in its woody biomass, with 35 million tons in aboveground biomass and 8 million tons in belowground biomass. The carbon stock density varies considerably among the vegetation strata: Rich Forests store about 206 Mg C/ha, Medium Forest 103 Mg C/ha, Poor Forest 54 Mg C/ha, Regrowth Forest 22 Mg C/ha, Plantation 16 Mg C/ha, Grassland 8 Mg C/ha and Cropland 5 Mg C/ha (Avitabile et al. 2016) (Fig. 3). Almost 80 % of the total carbon stock is stored in Rich, Medium and Poor forests on about 28 % of land area. Rich forests alone store 28 % of carbon on 5 % of land area and therefore identify the areas where larger emissions could potentially be caused by deforestation activities. Regrowth areas store

10 % of carbon on 15 % of land area, while Plantation contains only 3 % of carbon stock on 4 % of land area. Cropland and Grassland cover almost half of the VGTB (46 % of land area) and store only 9 % of total carbon, while the remaining 7 % of land (Settlements, Other Land and Water bodies) is considered having no carbon stock in biomass (Avitabile et al. 2016).

The results show that the carbon stock varies considerably among the vegetation strata and that the differentiation of forest in five forest types represents appropriately different carbon stock density values. However, the variability within each strata was relatively large. The complete description of the field data acquisition, processing and results can be found in Avitabile et al. (2016).

## Carbon Stocks in Soils Under Forest Change

Soil organic carbon (SOC) comprises about two-third of the terrestrial carbon storage and can act as a sink or source of atmospheric carbon dioxide. In order to support the implementation of the overall REDD+ goal, it is necessary to understand the extent of soil carbon stock change due to land cover change processes and quantify the related amount of carbon emissions from the soil. However, field data or literature values on SOC in the VGTB were not available and due to time and cost constrains it was not feasible to acquire field observations for all land change categories. Hence, the magnitude of carbon emissions from soil and litter was assessed through a dedicated soil sampling campaign only for the main change process observed in the VGTB, which is the conversion of forest to cropland. In addition, some samples were also acquired to assess if relevant carbon stock changes occur with the conversion of forest to plantation (a change category that was not included in the land cover change map) and forest to bare soil (i.e., after deforestation but before a new land use is implemented).

A stratified random sampling design based on land cover, land cover change, forest area and soil type was used to optimally locate the field measurements. The plots were acquired on Acrisol, which resulted to be the dominating soil group occurring on 74.7 % of the study area. In total, 27 plots located in three sites in the VGTB were sampled during a field campaign in 2012 in the following sampling strata: forest, plantation, cropland and recent deforestation. The sampling plots had an area of 20 × 20 m, within which five individual soil cores were taken from the first 30 cm of the soil. The mineral soil samples were dried at 40 °C and litter samples at 70 °C after removal of coarse roots and stones with a diameter of >2 mm. The total and inorganic carbon concentrations were determined by dry combustion, and organic carbon concentrations of the mineral soil were obtained as difference between total and inorganic carbon. Bulk density of the mineral soil was calculated with the mass of the oven dry soil (105 °C) and the core volume. The organic carbon was then obtained on the basis of the bulk density, organic carbon concentration, concentration of fine earth material (soil <2 mm) to total soil mass, and layer thickness. The mean organic carbon in the soil and litter per plot was used to calculate the average carbon stock and related standard error for each sampling strata.

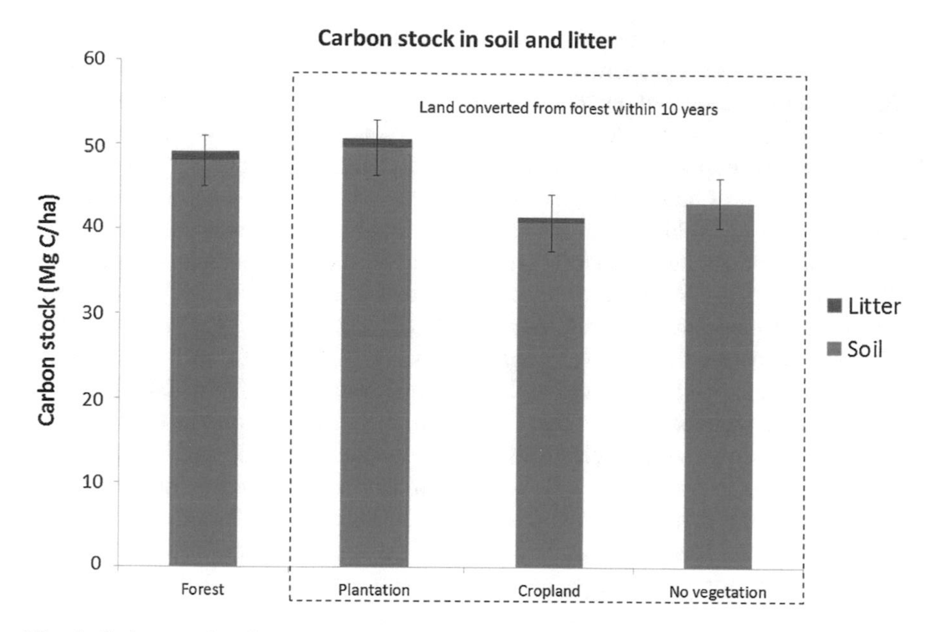


**Fig. 4** Carbon stock values in soil (*blue*) and litter (*red*) in forest areas and in areas recently converted from forest to plantation, cropland and bare soil (no vegetation). The *dashed line* indicates the carbon values that do not refer to stable land cover classes but to areas recently deforested (<10 years). The *error bars* represent the standard errors for both litter and soil values

The SOC stocks of Forest (including plantations) was equal 48.7 ± 3.8 Mg C/ha (mean value ± standard error). The SOC stocks of Forest (excluding plantations) was 48.1 ± 4.9 Mg C/ha and the SOC of Plantations established during the period 2001–2010 was 49.6 ± 5.4 Mg C/ha, indicating no significant difference between these two vegetation types (Fig. 4). Hence, the emissions due to the conversion of Forest to Plantation can be considered negligible during the first 10 years from the change process. Instead, the SOC stock was 43.2 ± 4.8 Mg C/ha after deforestation (forest converted to bare soil) and decreased further to 40.8 ± 5.5 Mg C/ha in forest converted to croplands. Similarly, the organic carbon in the litter layer was highest (and not significantly different) in the Forest (1.10 ± 0.14 Mg C/ha) and Plantation (1.09 ± 0.06 Mg C/ha), while it was lower in the Cropland (0.65 ± 0.09 Mg C/ha) and usually absent in recently deforested areas (as a consequence of the burning practices). On this basis, the conversion of Forest to Cropland resulted in a carbon loss of 7.9 ± 4.0 Mg C/ha in the mineral soil and 0.5 ± 0.1 Mg C/ha in the litter layer, while the emissions immediately after deforestation were lower in the mineral soil (5.4 ± 3.6 Mg C/ha) and higher in the litter layer (1.10 ± 0.08 Mg C/ha, assuming that the litter layer is burned) (Avitabile et al. 2016).

The results indicates that, within a time frame of 10 years, the transition of forest to cropland causes higher carbon emissions from the soil compared to the conversion of forest to plantation. Moreover, the transition of forest to cropland represents the highest proportion of the overall land cover changes occurring in the VGTB, and it is expected that the long-term (i.e., more than a decade) transition of forest to cropland would decrease further the SOC content. Low SOC stocks in croplands have important implications for crop production, since SOC is essential for the retention of nutrients in the highly weathered soils of the humid tropics and in unfertilized agricultural land it is a major source of nutrients taken up by crop plants. However, different management practices including organic amendments, tillage or the selection of crop species can result in different SOC stocks, and improved cropland management may partly offset the loss of SOC due to deforestation (Lugo and Brown 1993). Instead, the conversion of forest to plantation resulted in a negligible SOC loss with the monitoring period (2001–2010). This is in accordance with the study of Sang et al. (2012), which reported that SOC stocks of regrowth forests, Acacia plantations and grasslands did not significantly differ from each other in North and South Vietnam under the dominant soil group Acrisol. However, as for croplands, it is likely that in longer time frames the SOC of plantations may decrease further. Moreover, if the total value of forest beyond soil carbon sequestration and timber production is considered, natural forests provide important ecosystem services and biodiversity compared to plantations. The complete description of the field data processing and the results on SOC in VGTB can be found in Herold (2012) and Avitabile et al. (2016).

## *Carbon Emissions Between 2001 and 2010*

On the basis of the land change analysis and carbon stock assessment it was possible to estimate the amount and spatial distribution of the carbon emissions and removals due to changes in land cover occurred between 2001 and 2010 in the VGTB river basin, Vietnam. The carbon pools included in the estimates for all land changes were aboveground and belowground biomass of living vegetation, while carbon emissions from soils were estimated only for the main land cover change category (i.e., conversion of Forest to Cropland). Carbon stock changes in the dead wood and litter, which contribution to total emissions are usually negligible, were omitted because not sufficient reliable data were available. The carbon emissions and removals were converted to $CO_2$ equivalent units ($CO_2$) using the conversion factor of 44/12. Carbon dynamics that did not result in land cover changes but occurred within a stable land cover class (as forest degradation or forest growth) were not considered in this study.

The estimation of the carbon emissions and removals from biomass were obtained by combining the Land Cover Change map for the period 2001–2010 with the emission factors derived from the Carbon Stock map for the year 2010. The emission factors represent the carbon emissions and removals per unit area of each change category due to direct or indirect human activities. The emission factors were obtained using the stock-change approach by subtracting the mean carbon stock of the class after the change to that of the class before the change (Penman et al. 2003). Positive emission factors indicate release of carbon from land to the atmosphere (emissions) while negative emission factors indicate absorption of carbon from the atmosphere to the land (removals). The carbon emissions and removals for each land cover change category were estimated separately for the two sub-periods (2001–2005 and 2005–2010), and then summed to obtain the carbon dynamics for the monitoring period 2001–2010.

Since the carbon stocks in the forest areas of the VGTB river basin vary largely by forest type, ranging from 206 Mg C/ha for Rich Forest to 16 Mg C/ha for Plantation forests, the carbon dynamics and related emission factors also depend largely on the type of forest affected by the change activity. This information was not available from the Land Cover change map (which identifies only one generic forest class) and was obtained from the 2010 FIPI Forest Map (which identifies the five forest types) using the proximity criterion, attributing each patch of forest converted to another land cover to the spatially nearest type of forest. In this way it was possible to separate emissions from carbon-rich and carbon-poor forest changes, improving considerably the accuracy of the estimates compared to those based only on an average-carbon forests. Instead, changes from non-forest to forest always assumed that the conversion was to Regrowth forest, since other forest classes can develop only in time frames longer than 10 years (the monitoring period of this study). The land cover classes Settlement, Other Land and Wetland presented similar carbon stocks and emission factors, and were grouped into the class “No Vegetation”.

The conversion of forest to non-forest land caused the emissions of 2.06 million Tons of $CO_2$ from biomass and further 0.15 million Tons of $CO_2$ were emitted in land changes occurring among non-forest classes, while the conversion of non-forest to forest land removed from the atmosphere about 0.45 million Tons of $CO_2$. Hence, all land cover changes occurred in the VGTB river basin between 2001 and 2010 caused the net emissions of about 1.76 million Tons of $CO_2$ equivalent from above and belowground biomass. The conversion of Forest to Cropland was the single most important source of $CO_2$, accounting for 57 % of forest-related emissions, followed by the conversions of Forest to Grassland and Forest to No Vegetation (i.e., Settlement, Other land and Water), which were responsible for 27 and 16 % of forest-related emissions, respectively (Fig. 5). Most of the carbon removals were due to the conversion of Cropland to Forest, which absorbed 16 % of forest-related emissions, while carbon releases due to conversion among Non-Forest classes were mostly due to the conversion of Grassland to Cropland (Avitabile et al. 2016).

Carbon emissions from biomass were equally distributed during the two sub-periods, with 51 % net emissions occurring between 2001 and 2005 and 49 % between 2005 and 2010. Gross emissions in the 2005–2010 period were higher than those in the 2001–2005 period but they were compensated by higher removals, resulting in similar net carbon change over the two 5-years periods. During the

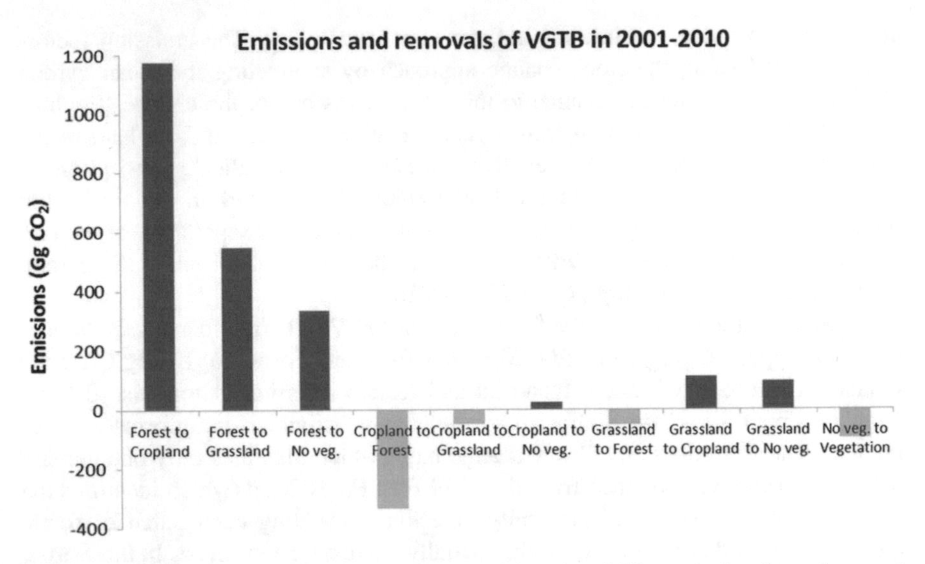


**Fig. 5** Carbon emissions (*red*) and removals (*green*) in units of $CO_2$ equivalent from above and belowground biomass for aggregated change classes in the VGTB during the period 2001–2010. The class "Forest" includes rich, medium, poor, regrowth and plantation forests; the class "Vegetation" includes forest, cropland and grassland; the class "No Vegetation" (No veg.) includes settlement, other land and water

2001–2010 period, 83.6 % of the net emissions were from aboveground biomass and the remaining 16.4 % was due to belowground biomass, approximately reflecting the root-to-shoot ratios used to compute the belowground component from the aboveground biomass. Emissions from soil were estimated for the conversion of Forest to Cropland (the main change class in terms of area and emissions), and were equal to 0.51 million Tons of $CO_2$, which correspond to 43 % of emissions from biomass. When considering the three carbon pools (aboveground biomass, belowground biomass and soil carbon), soil carbon accounted for 30 % of net emissions (Avitabile et al. 2016).

The results show that substantial carbon emissions were produced in the last decade in the VGTB river basin, mostly as a consequence of the expansion of croplands on forest land. However, most of deforestation events mapped in VGTB occured on regrowth forests or plantations that present very low C stock (16 and 22 Mg C/ha, respectively) and consequently the emissions per unit area due to their conversion to non-forest classes were also much lower compared to emissions associated to deforestation of Medium or Rich Forest (storing 103–206 Mg C/ha, respectively). This finding indicates that the stratification by forest type is a critical factor with a large impact on the estimations. In fact, the net emissions from biomass would be about 4 times higher if only one forest class with average carbon stock (81 Mg C/ha, calculated as simple average of all field plots in forest areas) was considered without further stratification by forest type. The other critical factor identified in this study is related to the use of local field data. Using reference literature values for carbon stock in aboveground biomass such as those provided by the Amazon Fund (100 Mg C/ha) (BNDES 2009) or the IPCC Tier 1 values (140 Mg C/ha in the tropical rain forest ecozone in continental Asia) (IPCC 2006) would increase the net emissions in the VGTB of about 5–7 times, respectively. Higher emissions were also estimated from two existing tropical deforestation emission databases. According to Zarin et al. (2016) and Harris et al. (2012), the gross deforestation emissions during 2001–2010 in the VGTB were about 6.4 and 5.3 times higher than our estimates, respectively. In both cases, the differences were mostly due to the fact that the emission databases estimated higher forest carbon stocks and were not calibrated for the specific conditions in VGTB, where deforestation events occurred mostly in forest areas with very low carbon stocks (Avitabile et al. 2016). The complete description of the methods and results of the carbon emissions in VGTB during the period 2001–2010 can be found in Avitabile et al. (2016).

## *Land Cover Change Scenarios for 2020*

Essential for the aim of forest management is not only to have an in-depth understanding of the past and current drivers of deforestation and forest degradation but also to project the impact of current land use dynamics on the forest cover in the future. Land use and land cover processes are interconnected in a complex way. Therefore, predicting future land changes demands for a clear understanding of these processes. In a sense, modelling the future is often a backward route, in which one first has to understand the past land cover patterns and land use characteristics in order to define rule sets as input for the model. Of particular importance is the assessment of the rate of forest-cover change and the identification of the areas under particular deforestation risk, which would result from the continuation and expansion of current land uses or Business As Usual (BAU) scenario. To this aim, the role of land use change modelling is central, allowing the assessment of both the shape and pattern of the deforestation observed (location, size, fragmentation), as well as their relationship with spatial factors influencing forest change. Land cover change modelling was performed with two approaches. Firstly, the future land changes were modelled using only bio-physical parameters with a well-known land change model. Secondly, socio-economic data (i.e., population density) were also included using an agent-based modelling approach, which allowed to model scenarios with REDD+ interventions.

### Land Cover Change Modelling

A land change model was used to analyse and project the location of land cover change in the VGTB river basin in the year 2020, using land cover maps of the years 2001 and 2010 (Schultz and Avitabile 2012a, b). The Land Change Modeler

(LCM), a tool integrated in the software Idrisi Selva, was used for the land cover change assessment, the projection of land cover change dynamics and the production of a land cover map of 2020. Statistical relationships were calibrated within the modelling environment between landscape determinants of land-use changes and recently observed spatial patterns of forest-cover change. The explanatory variables tested were distance to roads and distance to specific land cover classes: Forest, Grassland, Settlements and Cropland. The output of the model is the land cover map of the VGTB river basin in the year 2020 at 30 m spatial resolution, as a result of the projection of the past land cover change dynamics in the BAU scenario (Fig. 6). By comparing the 2010 and 2020 land cover maps it was possible to identify the areas that are most likely to be affected by land cover change in the basin in the next decade if the historical trends continue in the future.

The results indicate that the land cover expected to decline in the major extent is Forest, with a projected net decrease of 201 $km^2$, followed by Grassland, which is expected to decline of 15 $m^2$. Land cover classes that are expected to increase in area are Cropland and Settlement, with an area of 194 and 21 $km^2$ respectively. It is important to notice that the modelling and related results do not include

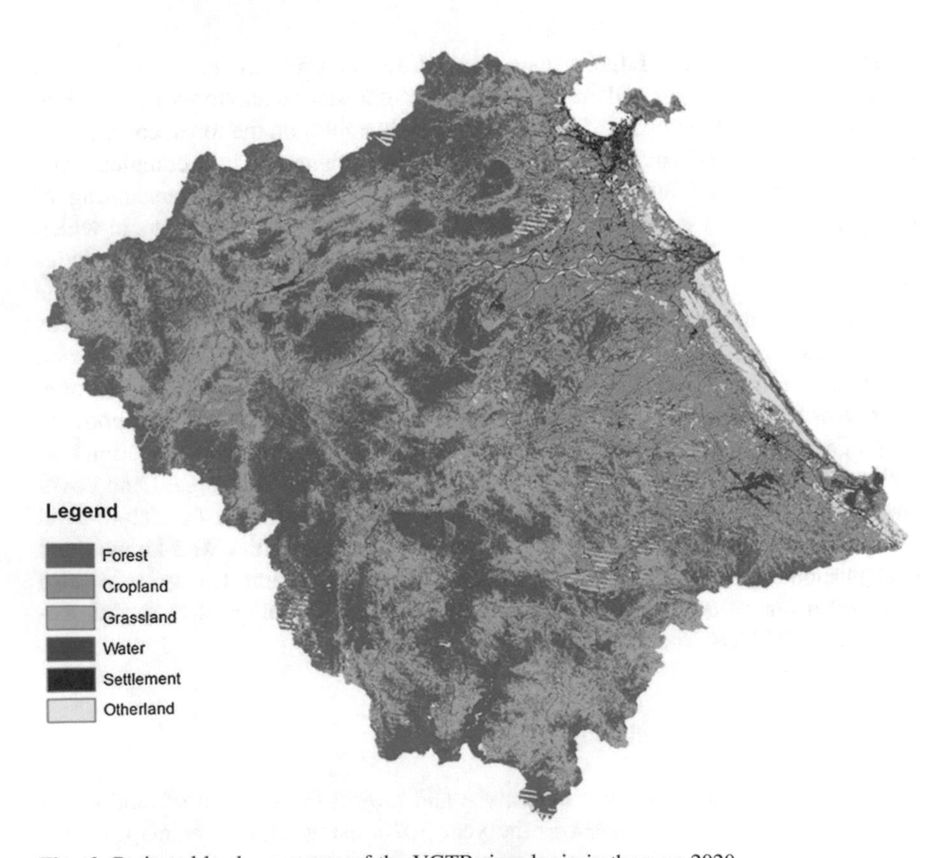


**Fig. 6** Projected land cover map of the VGTB river basin in the year 2020

socio-economic factors or governance parameters, and are based on the assumption that the BAU scenario in the current decade will mostly follow the historical trend observed in the last decade. The results indicated that distance to Settlement is the most important determinant of land cover change in the VGTB river basin. The complete description of the data processing and the modelling results can be found in Salvini and Avitabile (2013).

## Deforestation Modelling

A dedicated research was carried out to understand the drivers and processes involved in deforestation and forest degradation and to model these processes in order to provide insights in future deforestation risk areas. Firstly, the land cover maps from 2001, 2005 and 2010 were used in combination with spatial data (elevation, slope and distance to cropland, grassland, small settlements, large settlements, all roads and paved roads) to identify spatial factors potentially correlated to deforestation. All tested factors showed a significant correlation with deforestation, with the most important factors being distance to cropland and distance to small settlements. This indicated that most of the cleared forest areas are likely to be converted to cropland and that areas with a higher population density present greater deforestation risk compared to remote areas that are more difficult to access. Then, these results were used as input to develop an agent-based model, called Simulation of Deforestation Risk Areas (SoDRA), based on a NetLogo programming environment. Agent-based models provide a tool for revealing large-scale patterns that are induced by micro-level actions. The model simulates future deforestation risk areas under a business-as-usual scenario and the effects of REDD + measures on the projected deforestation for 2010–2020. The agents represent rural households, who are considered to be the key decision-making entities regarding land use and land use change. The model was calibrated and parametrized using the correlation and other statistical results from the previous phase. Inherent to modelling is the simplification of factors and complex processes experienced in reality. A major assumption in the model is that future deforestation develops in similar ways as past deforestation, hence that past drivers and processes are the same drivers and processes involved in future deforestation. The SoDRA model did not include socio-economic data on agent behaviour that are needed to distinguish different agent types, and it can be locally improved in dedicated study area and provide locally relevant modelled scenarios.

The model achieved a good quantitative representation of deforestation when comparing the model results of the business-as-usual (BAU) scenario of 2001–2010 to the corresponding measured deforestation using the land cover data, while scattered deforestation patches occurring in remote areas were partially underrepresented in the model (Fig. 7). For the 2010–2020 era, the SoDRA model predicted for the BAU scenario a scattered pattern of deforestation in the VGTB area, with the highest concentrations in the north-west and centre of the region. The results of the combined REDD+ scenario showed that deforestation rates can be reduced by over

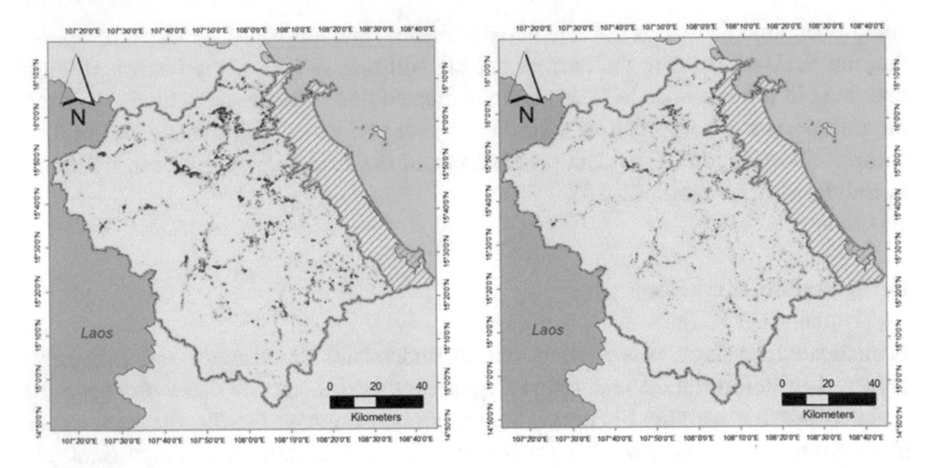


**Fig. 7** Modelled deforestation areas (in *red*) in VGTB for 2010–2020 according to the business as usual scenario (*left*) and the reduced deforestation (REDD+) scenario (*right*)

45 % when compared to the BAU scenario. This can be achieved under the following conditions: full protection of national parks and high carbon stock areas, restricted deforestation in middle carbon stock areas, and deforestation quota within 4 ha per household per 10 years. When considered separately, measures setting a quota of maximum deforestation per household had the largest impact compared to the BAU scenario while measures enforcing prohibition or reduction of deforestation in protected or high carbon stock areas resulted to have limited influence on reducing deforestation.

Although in some cases these conditions may be impractical to pursue, the REDD+ scenarios do provide insights in their relative effectiveness, and may facilitate decision-making processes regarding the selection and implementation of particular REDD+ policies. Law enforcement in remote forest areas may be difficult, making full prohibition of deforestation or successfully implementing tree quota challenging. However, focussing REDD+ measures on the most important forest areas, i.e. the areas containing the largest carbon stock, may help to set priorities, along with other considerations. Moreover, REDD+ measures to reduce deforestation should not be focussing on restrictive measures exclusively but may also focus on (financial) incentives, capacity building and technology transfer for stimulating (alternative) sustainable livelihood activities and strategies.

## Design of a Local REDD+ Implementation Strategy

The REDD+ program represents a valuable incentive for developing countries to take actions to reduce GHG emissions and at the same time promote sustainable forest management and improve local livelihoods. The government of Vietnam is designing policies to reduce deforestation and preserve forest stands through the national REDD+ strategy, which encourages specific local implementation actions and include sub-national implementation strategy plans and policies. These policies are unlikely to succeed if the main drivers of deforestation, such as the expanding need of land for agriculture, persist. Hence it is crucial to find landscape management strategies that secure food production (adapting to climate change impacts) and reduce emissions from deforestation (mitigating climate change) in a synergic way. As in several other countries, this integrated management is challenging for the government of Vietnam as they apply a sectorial approach: the Forestry department, part of the Ministry of Agriculture and Rural Development (MARD), manages the forest lands while the District office of Agriculture and Rural Development (DARD) is in charge of agricultural land management, with limited coordination between the two institutions (Salvini et al. 2016).

In this context, a dedicated study aimed at assess possible strategies to design a successful sub-national REDD+ implementation plan, which needs to be compliant with UN requirements and the Vietnamese National REDD+ plan. To design an effective REDD+ implementation plan at the local level it is crucial make in-depth analysis of the local management options and constrains to select the policy mix and land management interventions that best fit to the local setting.

The study consisted of four steps. Firstly, among the areas suitable for REDD+ activities, a local study area was selected. Secondly, according to the UN requirements, the local drivers of deforestation and forest degradation, and the safeguards for REDD+ activities were assessed. Thirdly, forest monitoring strategies in the form of community based monitoring were explored and their capabilities were assessed through local case-studies. Fourthly, the existing and planned adaptation and mitigation policies were analysed by engaging local stakeholders in participatory scenarios development on the landscape dynamics. As a result of this analysis, specific land management strategies for REDD+ activities were identified and proposed for implementation at commune level in the VGTB. Lastly, the results of the participatory scenarios were communicated to the policy makers, allowing them to reformulate suitable local land management strategies and policy interventions that consider the local drivers of deforestation and are likely to reducing GHG emissions while improving local livelihoods.

## *The Study Area*

### Identification of Potential Areas for REDD+ Activities

In Central Vietnam migration from the populated lowland areas to the Central Highlands and other remote and forest-rich areas has significantly contributed to deforestation during the last decades. In the uplands of the VGTB agriculture is characterized by sloping environments, with a long tradition of shifting cultivation practices. The current increasing population is inducing shorter fallow cycles leading to a decline in soil fertility, reduced productivity and hence extensive agriculture. This leads farmers to seek new productive land and further encroach into the forest, leading to widespread deforestation. In addition, in Quang Nam there is currently growing pressure on forestland for hydropower plant construction, infrastructure development and cash crops, especially coffee, pepper, rubber and cashews.

Within the LUCCi project such land processes were modelled and land use change scenarios for the period 2010–2020 were built on the basis of historical trends and bio-physical parameters, in order to provide an overall view of the expected main land change processes according to the business as usual scenario (see section on Land cover change modelling). Furthermore, a more detailed analysis was conducted for forest-related changes, where future deforestation processes were modelled using basic socio-economic data and an agent-based modelling approach, considering also alternative REDD+ scenarios aimed at reducing emissions from deforestation (see section on Deforestation modelling). The results of these modelling activities identified potential areas for local REDD+ implementation. Among these areas, the Tra Bui commune was selected as case-study for the design of a local REDD+ implementation plan. This commune was selected because it is characterized by drivers of deforestation and degradation that have been reported in the REDD+ Readiness Preparation Plan (RPP) of Vietnam: Conversion of forest lands for agriculturally cultivated land, unsustainable logging, infrastructure development and forest fires, and because it is characterized by high biomass loss. Additionally, it was selected because it is not an isolated case: due to the current plan of expanding the hydropower plants, other communities will be resettled in the highlands, claiming for REDD+ policy interventions aimed at reducing the pressure on forests and promoting local livelihoods.

## The Study Area: Tra Bui Commune

The study area is the Tra Bui Commune, located in the Quang Nam Province, where a community of about 500 households has been resettled in 2008, due to the construction of the "Song Tranh 2" hydroelectric dam (Fig. 8). The consequence of this resettlement was deforestation of primary forests located in the surrounding areas, as the resettled farmers needed land for crop production and logging activities (Tranh 2011). Crops (mainly rice, corn, cassava and banana) are cultivated with slash-and-burn practices, which include very few techniques for improved soil retention and agriculture yield. Due to climate change, the agricultural yield in the study area, located in the Central Highlands agro-ecological zone (IFPRI 2010), is projected to decrease up to 5.9 % in 2030. This decrease in agricultural productivity is likely to enhance further deforestation, as farmers will seek additional agricultural land.

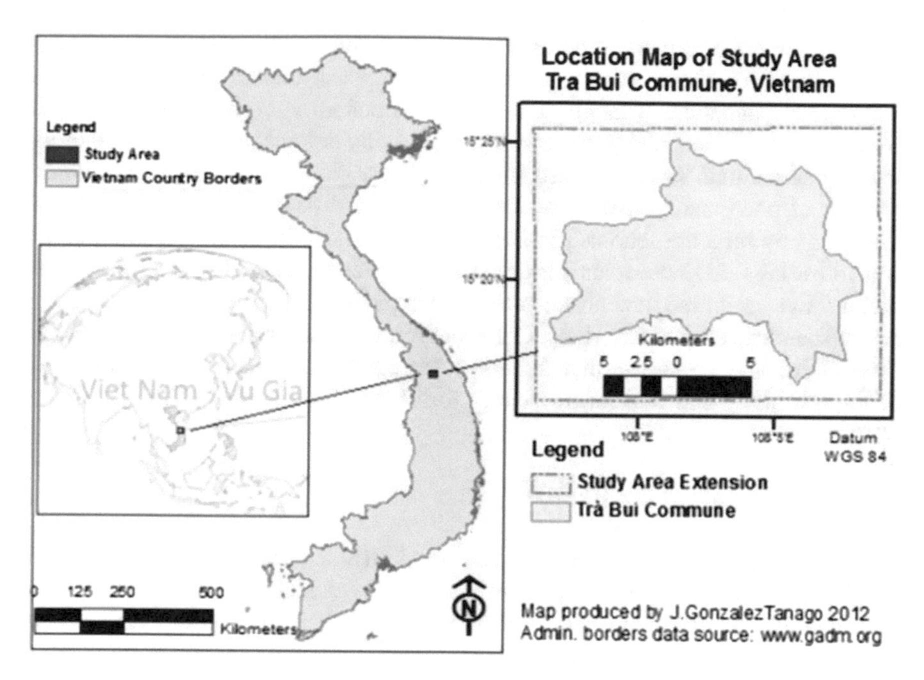


**Fig. 8** Location map of the study area: Tra Bui Commune, Central Vietnam (Reproduced from Salvini et al. 2016)

## *Assessment of the Local REDD+ Context*

### Local Drivers of Deforestation and Forest Degradation

In order to define proper policies and national REDD+ strategies and implementation plans, it is necessary not only to quantify forest change but also to understand the forces driving such change. Drivers of deforestation and forest degradation in the Tra Bui Commune are: (i) Infrastructure development, (ii) Conversion of forest to agriculturally cultivated land, (iii) Unsustainable logging, and (iv) Forest fires. These are also the main drivers that occur in the whole Vietnam, as described in the REDD+ Readiness Preparation Plan (see Drivers of deforestation and degradation). A specific assessment of each driver was done for the case of Tra Bui.

The main underlying (indirect) driver of deforestation in Tra Bui is related to infrastructure development, and specifically to the resettlement of about 500 households for the construction of “Song Tranh 2” hydroelectric dam. Hydropower plays an important role in electricity generation in Vietnam and, in order to meet the rapidly growing demand, the power industry has substantially expanded dam constructions, leading to large-scale deforestation. The impact of dam construction on deforestation is not limited to the power plant itself, but it includes indirect impacts on the forested areas from in-migration and resettlement of people.

The main direct drivers of deforestation in Tra Bui are the illegal conversion of natural forest to shifting cultivation for subsistence and cash crop production, and for timber plantations (mainly acacia and cinnamon). Agriculture techniques adopted are inefficient and provide low yield, driving deforestation to acquire fertile land for agriculture. Tra Bui is also affected by unsustainable logging that occurs as a result of poor management practices and/or illegal activities as well as timber harvesting by rural households for direct consumption. With regards to fires, in Tra Bui forest fires originate mainly from slash and burn practiced by the local community. Even with the new program to build capacity on fire management, there is no sufficient capacity to fight fires. Collaboration with the local communities can be an effective way to ensure that there is local control and a sufficient number of people patrolling and prevent or identify forest fires.

### Assessing Safeguards for REDD+

The primary goal of REDD+ is the reduction of GHG emissions, but this cannot be considered as a standalone objective: REDD+ is expected to foster the implementation of a sustainable production system able to benefit country development. Amongst others, benefits may include poverty alleviation, protection of indigenous rights, improved community livelihoods, technology transfer, sustainable use of forest resources and conservation. Additionally, the UNFCCC highlighted that, while implementing REDD+, co-benefits should be promoted and that the needs of local and indigenous communities should be addressed when action is taken to implement REDD+. In the process of selecting REDD+ policies that are suitable to the local conditions of Tra Bui, the importance of considering these safeguards was carefully followed. To this aim, local stakeholders were engaged directly via interviews and Participatory Rural Appraisal activities that helped understanding the decision making process of local farmers in land use and the types of incentives that would entice them to change land use practices and protect the forest. This helped identifying land use practices (see Landscape management strategies) that, besides improving carbon storage, can contribute to improve livelihood and alleviate poverty.

## *Community Based Forest Monitoring*

Countries participating in a REDD+ mechanism are required to set up a reliable, transparent and credible system of Measuring, Reporting and Verifying (MRV) changes in forest areas and forest carbon stocks. According to the REDD+ monitoring and implementation requirements, it is important to involve local community groups and societies to carry out forest monitoring, in particular if there is any prospect of payment and credits for environmental services. A variety of practical experiences from developing countries, such as Nepal, Tanzania,

Cameroon, India and Mexico, have demonstrated that local communities can play an essential role in forest monitoring and management program (Danielsen et al. 2011). Moreover, if communities are involved in measuring the aboveground biomass carbon pool (which can be used in calculating the carbon stock changes) in the forests they manage, they may establish 'ownership' of any carbon savings, strengthen their stake in the REDD+ reward system and greatly increase transparency in the intra-national governance of REDD+ finances (Pratihast et al. 2012).

While the MRV system is designed and implemented at national level, in the case of Tra Bui we tested and developed the capability of Community Based monitoring (CBM) at local scale as additional support to the national MRV infrastructure. CBM can play a useful role when it comes to locally driven change activities and causes of small scale forest degradation due by, for example, subsistence fuel wood collection, charcoal extraction and grazing in the forest, which was the case for Tra Bui commune. The impacts of these activities are rarely captured accurately in national databases or from remote sensing. In these cases, data acquired by communities is often essential, and can include reporting on incidence of change events, as well as ground measurements on carbon stock changes for tracking and reporting on local REDD+ implementation activities.

Two field visits were conducted in 2011 and 2012 to train a group of local people (mostly ethnic minority), local experts (local forest rangers) and national experts (regional/national forest rangers) to measure forest aboveground biomass and report forest changes. The training focused on the use of smartphones and Global Positioning System (GPS) to collect simple measurements such as tree density, diameter at breast height (DBH) and forest changes and human activities affecting forest carbon. This system is cost-effective and sustainable because of the limited cost of these devices and the direct involvement of local communities.

The capacity building exercise proved that local communities can make accurate forest inventory measurements (86 % accuracy) at low cost (1.2 USD/ha, compared to 6.4 USD/ha for national experts) if properly trained (Fig. 9). Local communities can not only measure and report the basic tree variables such as DBH, tree species and tree count but, most importantly, can repeat the measurements on a regular

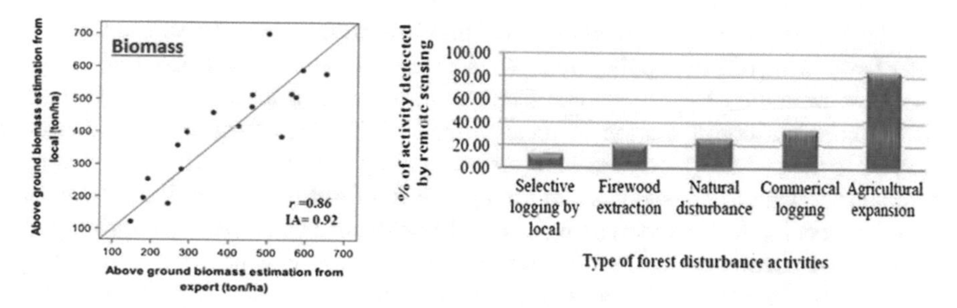


**Fig. 9** Comparison of biomass estimates from local communities and forestry experts (*left*) and assessment of deforestation drivers according to local communities (*right*) in Tra Bui (Adapted from Pratihast et al. 2012)

basis. The collected data has proven to be of a level of precision comparable to that produced by professional forest inventory staff (Pratihast et al. 2012).

Moreover, REDD+ is not only about estimation of biomass but also about tracking forest disturbance, which provides an estimate of the rates of deforestation and forest degradation. The number of forest disturbance events, their size and the timing of events were recorded by community members and were compared with remote sensing observations. The comparison of forest disturbance areas due to agriculture expansion captured by the local community and by remote sensing showed that there was high agreement for small and medium events but that local people underestimated large-area deforestation events compared to RS based estimates. Conversely, only 18 % of community-reported selective logging events could be visually identified using remote sensing, whereas around 88 % of subsistence agricultural expansion was recognized in the satellite data, showing that local people are better able to identify the small-scale forest disturbances (Fig. 9). In addition, local interviews provided information on the drivers of deforestation (cropland expansion, timber for sale and local construction, firewood), which quantitative assessment is crucial to design policies to reduce deforestation (Pratihast et al. 2012).

The analysis proved that in Tra Bui the MRV of forest areas can be supported by local communities, and the CBM data can be directly linked to national MRV in the prospect of data demand, supply, management, reporting and quality assurance. The complete description of the data processing and the results can be found in Pratihast et al. (2012).

## *Landscape Management Strategies*

### Assessment of the Local Context

Policy scenarios at the landscape level have been developed in a participatory way involving both policy makers and local population. First, the policies and local interventions currently planned in Tra Bui by the provincial Forestry department and the district office for Agriculture and Rural Development were assessed through interviews with relevant policy makers. The interviews revealed that forest protection is promoted by the national Decree 99, which is implemented at the local level through a compensation to local farmers of 180,000 VNDong (6.7 euros)/ha/year to protect assigned parcels of forests, and by the Program 134, which aims to reallocate land to households and to improve soil fertility (SocRepViet 2004). In addition, local scale land use strategies consist on the use of manure to improve soil fertility and encourage the introduction of fixed cultivation. Then, the local population was involved via a participatory rural appraisal (PRA) and via semi-structured expert interviews, which allowed local people and their leaders to express their knowledge of land management interventions and key landscape processes affecting the villages. Furthermore, socio-economic data and information on local land management practices were gathered through 56 interviews with local households.

This analysis highlighted that substantial deforestation is occurring on steep slopes to expand the cultivated land and satisfy the food demands of the expanding population. The main problem of local farmer consists on the low agricultural yields, caused by the topography (steep slopes), the poor soils and insufficient knowledge of soil conservation techniques. Low yields led farmers to clear forestland for cultivation, exacerbating land degradation and flood risk. Farmers indicated their need for new agricultural techniques to increase yields in the face of the increasing land scarcity. The household interviews also allowed to identify three main types of land users (the "agents"): farmers engaged in subsistence farming without Land Use Right Certificates (81 % of farming households), farmers engaged in subsistence farming with Land Use Right Certificates (14 %) and farmers engaged in cash crop production with Land Use Right Certificates (5 %) (Salvini et al. 2016).

## Policy Scenarios

This assessment was the input for a companion modelling (ComMod) process. The ComMod process consists of a role-playing game with local farmers and an Agent-Based Model. The role-playing game provided a participatory means to develop policy and climate change scenarios. Based on the interviews with policymakers, four categories of scenarios were included: (i) Business As Usual (BAU), with no policy intervention to protect the forest; (ii) REDD+, either with a payment of a subsidy to farmers to avoid plant acacia ("PES_AC") or where a stricter forest protection is implemented ("ForPro"); (iii) Climate-Smart Agriculture (CSA), with sustainable agricultural intensification and expanding Tephrosia fallow ("CSA_Tephrosia"); (iv) Climate-Smart Landscape (CSL), which implement simultaneously REDD+ and CSA interventions either via strict forest protection and Tephrosia fallow ("ForPro_CSA"), or with a PES for avoiding acacia plantations and Tephrosia fallow ("PES_CSA"). All the land use scenarios were assessed for two climate scenarios: under current climate (and associated rice yields), and under a climate change scenario that decreases rice yields of 5.9 % by 2030, as estimated by IFPRI (2010).

In each scenario, the farmers expressed their intention to improve soil fertility (by using manure and adopt Tephrosia fallow), plant acacia or deforest. The analysis also revealed that deforestation is mainly driven by the need of land for rice cultivation in order to satisfy family food needs, and to establish acacia stands as cash crop. Land ownership was found to play a major role in decision-making, limiting the willingness of farmers without Land Use Right Certificates to adopt Tephrosia fallow and their possibility to plant Acacia on public lands (Salvini et al. 2016).

### Carbon Emission Scenarios

These information were then used as inputs to the Agent-Based Model (ABM), a spatially explicit model to simulate landscape dynamics and the associated carbon emissions over decades. The ABM provides to policymakers ex-ante information on key processes driving change in the landscape and to explore the impact of the policy scenarios. The ABM was constructed by assigning decision rules to each of the three agent types (farmers) described above, using the GAMA simulation platform. Carbon emissions or removals were related to land use activities in each land cover class, which consisted of deforestation of forest land and acacia plantation, or rice cultivation or fallow practice on cropland or grassland areas. The carbon emissions or removeals were derived by combining the VGTB Land Cover (Schultz and Avitabile, 2012a) and Carbon stock maps for the year 2010, subtracting the carbon stocks of the classes before and after the change. Model simulations were run for each category of policy under each climate scenario to calculate and compare the resulting landscape dynamics and associated $CO_2$ emissions (Salvini et al. 2016).

The results of the ABM model indicated that in the BAU scenario rice and acacia cultivation leads to widespread deforestation, which occurs at progressively greater distances from the settlements. Since poor forest is located closer to the settlements, most emissions are first caused by deforestation of poor forest and later of medium forest. The avoided emissions (compared to the BAU scenario) calculated per policy category in each climate scenario are presented in Fig. 10.

PES_CSA is the scenario with lowest carbon emissions (hence, highest avoided emissions) because deforestation due to acacia plantation and rice production is minimized. There is a considerable increase in emissions avoided over time with avoided emissions increasing from 160 Gg (Gigagrams) in 2024 to 320 Gg in 2044. However, this scenario requires a major investment from the government to compensate farmers for the lost income derived from acacia plantations.

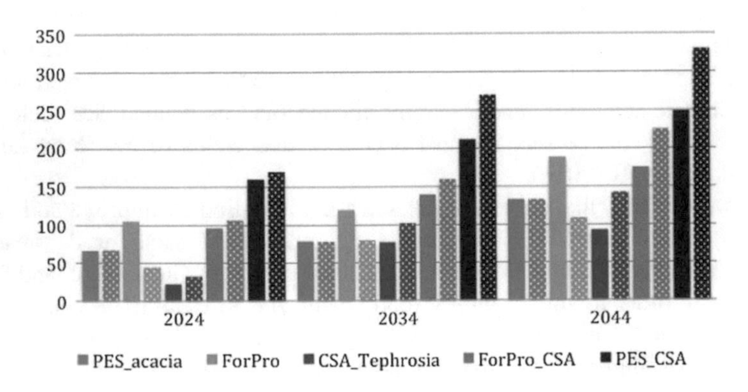


**Fig. 10** Avoided carbon emissions (in Gg) for different policy scenarios (PES_AC, ForPro, CSA_Tephrosia, ForPro_CSA and PES_CSA) in: (i) the current climate (*full color*), (ii) a climate change scenario (*striped color*) (Reproduced from Salvini et al. 2016)

The second-best scenario is ForPro_CSA, where all farmers adopt Tephrosia fallow, which increases yields while storing carbon on the fallow land and stricter forest protection reduces deforestation for rice cultivation. This shows that policies aimed at improving adaptation to climate change are more effective if implemented in synergy with mitigation policies (ForPro_CSA scenario rather than ForPro and CSA_Tephrosia separated).

The CSA_Tephrosia scenario presents higher emissions than ForPro_CSA because subsistence farmers without Land Use Right Certificates do not adopt the Tephrosia fallow technique and there is no policy to prevent their expansion of rice cultivation into forest areas. The PES_acacia scenario provides substantial emission reductions but, as for PES_CSA, its implementation requires a high investment related to the lost income from acacia planatations. ForPro has higher emission reductions than CSA_Tephrosia because it assumes reduced deforestation due to the establishment of acacia plantations, which contributes to more emissions than rice cultivation. However, this would result in less income for subsistence farmer without Land Use Right Certificates and represents an important trade-off between mitigation goals (lower emissions from deforestation) and improvement of local livelihoods (revenues from selling timber from acacia plantations). Instead, CSA_Tephrosia would be more effective on the long term because Tephrosia would contribute in maintaining rice yield higher than the BAU in a climate change scenario, hence reducing deforestation related to rice cultivation (Salvini et al. 2016).

## Land Management Strategies

Tra Bui Commune faces the dual challenge of protecting the forest while improving agricultural practices to adapt to climate change impacts. The conducted research found that the main driver of deforestation in Tra Bui is extensive agriculture practices for rice production and the expansion of acacia plantations. The participatory framework used in this study stimulated the active involvement of local stakeholders in land use scenario development and in the design of benefit-sharing mechanisms, which inform policymakers about how land use decisions are made at the local level.

Considering the policies and interventions planned by the government, two main land management strategies are recommended to lower reforestation and at the same time sustain local livelihoods. The first one is sustainable intensification of rice production via the introduction of soil fertility improving activities. A potential intensification technique is Tephrosia fallow, which is expected to increase the yield of up to 19 % and store more carbon in fallow land (up to 9.6 ton/ha). The increase of yield per unit area would lower pressure on forests. Moreover, its introduction is likely to lower forest degradation due to firewood collection since the woody fallow can be used for firewood. However, the provision of Land Use Right Certificates to all farmers is needed because this technique implies some investment in land which farmers are willing to make if there is land use ownership recognition.

The second strategy is Payment for Ecosystem Services (PES) to lower deforestation related with the establishment of acacia plantations. Acacia represents an important source of income for the local farmers, hence lowering deforestation related to acacia plantation could be achieved if there is an alternative income for farmers or a payment that covers the opportunity cost, an equivalent of about 7 millions VNDong/ha/year. This amount is much higher than the current payment that DARD is planning to provide to preserve the forest, which is of 180,000 VNDong/ha/year (Salvini et al. 2016).

This study highlights that forest protection and rural development planning should be coordinated. Currently, the policies aimed at agriculture management and forest protection are designed and implemented separately by two different government departments (MARD and MONRE). This sectorial approach is less effective in reducing emissions, reducing the cost effectiveness of project implementation. In particular, if agriculture intensification is done in synergy with forest protection, the latter would be much more effective since the introduction of a soil fertility improving technique would reduce their need of fertile land from forest stands. In fact, on one hand protecting the forest is not successful if adaptation needs are not satisfied (farmers will still deforesting illegally for crops and timber) and the compensation offered by the government to preserve the forest is considered insufficient by local farmers. On the other hand, letting local population adapt without providing alternatives to deforestation does not lead to benefit in terms of mitigation.

In addition, the involvement of local stakeholders is very important to improve the design of local interventions to better fit the local context. However, local communities in Vietnam are not currently actively involved in decision-making. Rather, the design of policies implemented by MARD and MONRE is currently top-down without consideration of the local setting, local knowledge and the goals and needs of local stakeholders. The land ownership also plays a central role in land use decisions and adoption of sustainable agriculture, since farmers without LURC are less willing to invest in agricultural management. Lastly, benefit sharing should be carefully designed: the current compensation of 180,000 dongs/year/ha to local farmers for forest protection may be too low to stimulate farmers to modify their land use, hindering the effectiveness of REDD+ policies (Salvini et al. 2016). The complete description of the data analysis and the results can be found in Salvini et al. (2016).

## References

Angelsen A, Kaimowitz D (2001) Agricultural technologies and tropical deforestation. CABI Publishing, Oxon, UK

Avitabile V, Schultz M, Herold N, de Bruin S, Pratihast A, Vu Quang H, Herold M (2016) Carbon emissions from land cover change in Central Vietnam. Carbon Management in press. http://dx.doi.org/10.1080/17583004.2016.1254009

BNDES, (2009) Amazon Fund Annual Report. Brazilian Development Bank, Brazil

Breiman L (2001) Random forests. Mach Learn 45:5–32

Chave J, Andalo C, Brown S, Cairns MA, Chambers JQ, Eamus D, Folster H, Fromard F, Higuchi N, Kira T, Lescure JP, Nelson BW, Ogawa H, Puig H, Riera B, Yamakura T (2005) Tree allometry and improved estimation of carbon stocks and balance in tropical forests. Oecologia 145:87–99

DANIDA (2008) Climate and disaster check of Danish sector programmes in Vietnam. Hanoi, Vietnam, 98 p

Danielsen F, Skutsch M, Burgess ND, Jensen PM, Andrianandrasana H, Karky B, Lewis R, Lovett JC, Massao J, Ngaga Y et al (2011) At the heart of REDD+: a role for local people in monitoring forests? Conserv Lett 4:158–167

FAO (2010) Global forest resources assessment 2010. Country report, Viet Nam. Available at www.fao.org/docrep/013/al664E/al664E.pdf. Accessed April 2016

Harris NL, Brown S, Hagen SC, Saatchi SS, Petrova S, Salas W, Hansen MC, Potapov PV, Lotsch A (2012) Baseline map of carbon emissions from deforestation in tropical regions. Science 336(6088):1573–1576

Herold N (2012) Soil organic carbon dynamics under forest change in the Vu Gia Thu Bon river basin, Central Vietnam. Project report, 2012. Vu Gia Thu Bon information system. http://leutra.geogr.uni-jena.de/vgtbRBIS/metadata/start.php

IFPRI (2010) Impacts of climate change on agriculture and policy options for adaptation. The Case of Vietnam. Discussion paper 01015 August 2010

IPCC (2006) IPCC guidelines for national greenhouse gas inventories. In: Eggleston HS, Buendia L, Miwa K, Ngara T, Tanabe K (eds). Prepared by the National Greenhouse Gas Inventories Programme, IGES, Japan

Lugo AE, Brown S (1993) Management of tropical soils as sinks or sources of atmospheric carbon. Plant Soil 149:27–41

MARD (2012) Decision No 2089/QĐ-BNN-TCLN on forest area 2011 of Vietnam. Vietnam

Meyfroidt P, Lambin EF (2008) The causes of the reforestation in Vietnam. Land Use Policy 25 (2):182–197

Mokany K, Raison R, Prokushkin AS (2006) Critical analysis of root: shoot ratios in terrestrial biomes. Glob Change Biol 12(1):84–96

MONRE (2009) Climate change, sea level rise scenarios for Vietnam. Ministry of Natural Resources and Environment, Hanoi, Vietnam, 34 p

National Strategy on Climate Change (2011) Decision No 2193/QD-Tg dated on 05 Oct 2011 about approval of National Strategy on Climate Change, Vietnam

National Strategy on Green Growth (2012) Decision No 1393/QĐ-Tg dated on 25 Sept 2012 on approval of National Strategy on Green Growth, Vietnam

Nielsen A (2007) The regularized iteratively reweighted MAD method for change detection in multi- and hyperspectral data. IEEE Trans Image Process 16:463–478

Olofsson P, Foody GM, Herold M, Stehman SV, Woodcock CE, Wulder MA (2014) Good practices for estimating area and assessing accuracy of land change. Remote Sens Environ 148:42–57

Pearson T, Walker S, Brown S (2005) Sourcebook for land use, land-use change and forestry projects. Bio Carbon Fund of the World Bank, Washington, DC

Penman J, Gytarsky M, Hiraishi T, Krug T, Kruger D, Pipatti R, Buendia L, Miwa K, Ngara T, Tanabe K, Wagner F (2003) Good practice guidance for land use, land-use change and forestry. IPCC National Greenhouse Gas Inventories Programme and Institute for Global Environmental Strategies, Kanagawa, Japan

Pratihast A, Herold M, Avitabile V, De Bruin S, Bartholomeus H, Souza C, Ribbe L (2012) Mobile devices for community-based REDD+ monitoring: a case study for Central Vietnam. Sensors 13(1):21–38

REDD Vietnam (2012) Vietnam's opinions on REDD+ emissions reduction in the carbon fund 4th meeting. Available at www.vietnam-redd.org. Accessed June 2013

Salvini G, Avitabile V (2013) Modeling of land change processes for 2010–2020 in the Vu Gia Thu Bon river basin, Central Vietnam. Project report, 2013. Vu Gia Thu Bon information system. http://leutra.geogr.uni-jena.de/vgtbRBIS/metadata/start.php

Salvini G, Ligtenberg A, van Paassen A, Bregt AK, Avitabile V, Herold M (2016) REDD+ and climate smart agriculture in landscapes: a case study in Vietnam using companion modelling. J Environ Manage 172:58–70

Sang PM, Lamb D, Bonner M, Schmidt S (2012) Carbon sequestration and soil fertility of tropical tree plantations and secondary forest established on degraded land. Plant Soil 1–14

Schultz M, Avitabile V (2012a) VGTB land cover 2010 v2—updated. Project report, 2012. Vu Gia Thu Bon information system. http://leutra.geogr.uni-jena.de/vgtbRBIS/metadata/start.php

Schultz M, Avitabile V (2012b) 2001–2005–2010 land cover change analysis of the Vu Gia Thu Bon river basin, Central Vietnam. Project report, 2012. Vu Gia Thu Bon information system. http://leutra.geogr.uni-jena.de/vgtbRBIS/metadata/start.php

SocRepViet (2004) Socialist Republic of Vietnam. Decision 134/2004/QD-TTg (Program 134), 2004. Available at: http://chuongtrinh135.vn/english/News/ActionsofOMP135/Program-134-135-The-essence-of-poverty-reduction-task-_115_397_6.aspx

Tranh D (2011) Good practices for hydropower development, 2011. Small publications series no. 39, CRBOM Center for River Basin Organizations and Management, Solo, Central Java, Indonesia. Available at: http://www.crbom.org/SPS/Docs/SPS39-HP-VN.pdf

UNITED NATIONS (UN) (2012) Climate change fact sheet: the effects of climate change in Vietnam and the UN's responses. Version of 10 August 2012

Zanne AE, Lopez-Gonzalez G, Coomes DA, Ilic J, Jansen S, Lewis SL, Miller RB, Swenson NG, Wiemann MC, Chave J (2009) Data from: towards a worldwide wood economics spectrum. Dryad Digital Repository. doi:10.5061/dryad.234

Zarin DJ, Harris NL, Baccini A, Aksenov D, Hansen MC, Azevedo-Ramos C, Azevedo T, Margono BA, Alencar AC, Gabris C, Allegretti A, Potapov P, Farina M, Walker WS, Shevade VS, Loboda TV, Turubanova S, Tyukavina A (2016) Can carbon emissions from tropical deforestation drop by 50 % in 5 years? Glob Change Biol 22:1336–1347. doi:10.1111/gcb.13153 Dataset accessed through global forest watch climate on 31/03/2016. climate.globalforestwatch.org

# Connectivity Conservation Management: A Biodiversity Corridor for Central Vietnam

**Claudia Raedig, Hoang Ho Dac Thai and Udo Nehren**

**Abstract** The Vu Gia Thu Bon (VGTB) watershed in Central Vietnam is enclosed by the foothills of the Annamite Mountains in the north, west and south. The Annamite Mountains are known for their elevated level of biodiversity, but the VGTB watershed faces increasing pressures from economic growth, urbanization, and agricultural expansion. This leads to an increasing loss and fragmentation of near-natural forest ecosystems by an extension of the road network and clearing of forests for agricultural land and plantations of exotic species. The main objective of this research is to identify priority areas for biodiversity conservation based on tree species distribution patterns, and to link these priority areas by a biodiversity corridor. Species distribution models show highest levels of potential tree species richness in the northwestern part of the watershed bordering Lao PDR, and in the highlands in the south at the border to Kon Tum Province. The protected area network covers a proportion of the areas of higher potential tree species richness, but leaves a gap in the northwestern part of the VGTB watershed. In order to close that gap, a new protected area should be established, and be connected to existing protected areas with stepping stones, small areas of elevated high potential tree species richness or with plantations of native tree species with high economic and ecological value. Awareness-building measures for both, biodiversity conservation as well as the use of native tree species for income generation have to be carried out to sensitize local communities.

**Keywords** Biodiversity corridor · Central Vietnam · Biodiversity conservation · MAXENT · Tree species diversity · Vu Gia Thu Bon watershed

C. Raedig (✉) · U. Nehren
Institute for Technology and Resources Management in the Tropics and Subtropics, TH Köln - University of Applied Sciences, Cologne, Germany
e-mail: claudia.raedig@th-koeln.de

U. Nehren
e-mail: udo.nehren@th-koeln.de

H. Ho Dac Thai
Hue University - Institute of Resources and Environment, Hue, Vietnam
e-mail: hoanghdt@hueuni.edu.vn



A. Nauditt and L. Ribbe (eds.), *Land Use and Climate Change Interactions in Central Vietnam*, Water Resources Development and Management,
DOI 10.1007/978-981-10-2624-9_5

## Introduction

Vietnam hosts 310 mammal species, 840 bird species, 286 reptile species, 162 amphibian species, 3170 fish species and 14000 plant species (World Bank 2005; VDR 2011). Moreover, Vietnam is one of the few countries in the world with steadily increasing forest coverage (FAO 2010, 2015). Seriously scarred in the years of the Vietnam War between 1965 and 1975, the vegetation has grown steadily since the early 1990s from 24.7% (24.6–31.1%) of the country's area in 1992 to 38.2 % (34.4–42.1%) in 2005 (Meyfroidt and Lambin 2008). However, little is known on species occurrences and species distribution patterns in the present-day Vietnamese forests.

In terms of biogeography, Central Vietnam represents a crucial area in Southeast Asia, since elements of the subtropical and the tropical flora encounter here. This becomes apparent in the composition of the natural vegetation which alternates across latitudes and across elevations, with frequent and often abrupt changes (Sterling et al. 2006). At elevations of up to 800 m a.s.l., forests in Central Vietnam are primarily broadleaf evergreen forests dominated by tropical taxa. Dipterocarps which represent an economically important source of timber occur only infrequently. With increasing elevation, the number of trees of tropical families decreases and trees from taxa of temperate forests such as *Lithocarpus* (Fagaceae) or *Nageia* (Podocarpaceae) become more common. At elevations above 1000 m a. s.l., wet, submontane and montane forests occur. One important component of this altitudinal range is conifers, particularly pines and cypresses, with some of the species being economically important (Sterling et al. 2006).

The Vu Gia Thu Bon (VGTB) watershed in Central Vietnam has its highest peaks in the south, the maximum elevation being the Ngoc Linh Mountain reaching almost 2600 m a.s.l. The foothills of the Annamite Mountains surround the VGTB watershed, with some areas of higher elevation traversing into the hilly midlands (Fig. 1). The midlands as well as the uplands host extensive forest areas, while most lowland forests have been converted to agricultural land. However, vast parts of the forested areas in the midlands and uplands are plantations, dominated by non-native *Acacia* species and hybrids thereof (Bueren 2004; ICEM 2008). Detailed spatial distribution patterns for native tree species of VGTB are only insufficiently known, impeding the identification of priority areas for conservation.

Figure 1 shows the network of protected areas in the VGTB watershed based on two decrees issued by the Vietnamese Ministry of Agriculture and Rural Development (MARD 2008) and by the Quang Nam Provincial People's Committee (PPC 2012). The protected areas located in the southwest of the watershed consist of the Song Thanh Nature Reserve, which is the largest nature reserve in the VGTB watershed, and the Ngoc Linh Nature Reserve, which continues into the adjacent Kon Tum Province towards the south. At the border to the northern neighboring province of Thua Thien-Hue lies the most recently established protected area along the famous Ho Chi Minh Trail, the Saola Reserve, as well as Ba Na-Nui Chua Reserve and the Ban Dao Son Tra Reserve, whereof the latter is

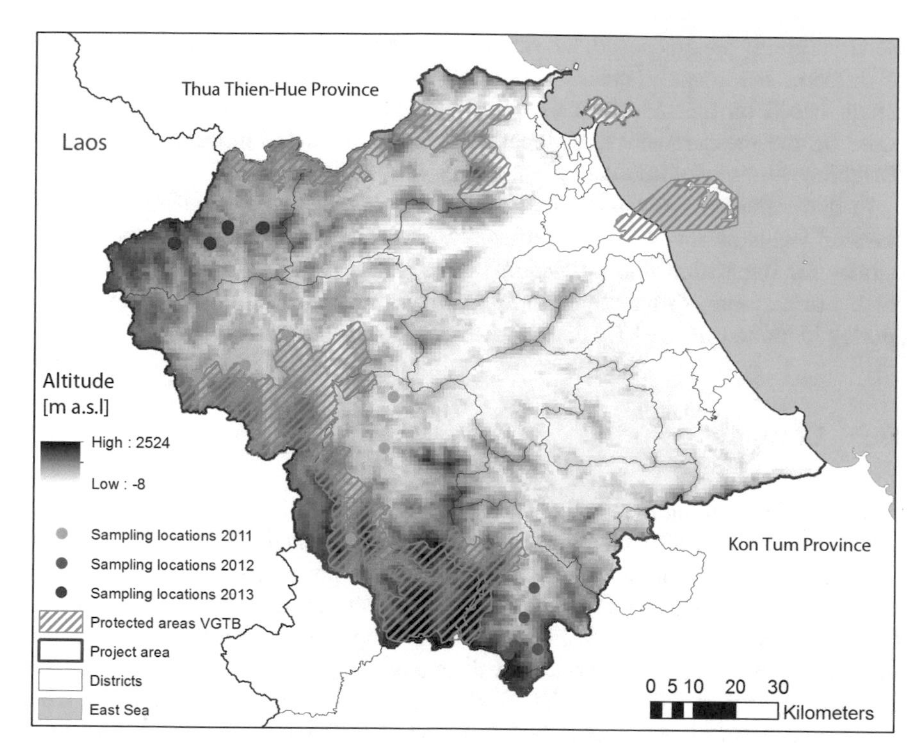


**Fig. 1** Overview on the three field campaigns carried out from 2011 to 2013 in the midlands and uplands of the VGTB watershed. In addition, elevations based on Hijmans et al. (2005) and protected areas based on MARD (2008) and Quang Nam PPC (2012) are shown

located in the province of Da Nang. The Cu Lao Cham Biosphere Reserve consists of the Cu Lao Cham islands as well as the adjacent terrestrial part of mainland Quang Nam Province.

Growing awareness of the importance of biodiversity conservation is on the one hand reflected by Vietnam's signature of internationally binding treaties such as the Convention on Biological Diversity (CBD) in 1993 and the Convention on International Trade in Endangered Species (CITES) in 1994 (Sterling et al. 2006), and by the entry into force of the Biodiversity Law in 2009. On the other hand, increasing pressures from economic growth, urbanization and agricultural expansion hamper the implementation of regulations for the protection of biodiversity. As a consequence, forest fragmentation induced by agricultural activities, by an extension of the road network, and particularly by plantations of non-native tree species lead to a decreasing cover of near-natural forest in the VGTB watershed.

This, in turn, leads to habitat shrinkage, to larger dispersal distances and thereby to biodiversity loss. Linking of forest remnants is one way to reconnect isolated forest patches and thus counteract fragmentation. Connectivity has two main aspects, the structural and the functional component. Whereas the latter refers to the behavioural response of species (or their individuals) and ecological processes to the structure of the landscape, the structural component refers to the spatial layout of different landscape elements (Wiens 2006). Functional connectivity is either actual, based on the measurement of actual movements of species, or potential, based on the measurement of indirect knowledge about the movement of species (Calabrese and Fagan 2006).

In this research, species distribution modelling will allow to identify areas of elevated levels of potential tree species richness. These areas are given the highest priority for integration into a functional biodiversity corridor in the VGTB watershed. Furthermore, this research aims at exploring the usefulness of native tree species identified in the VGTB watershed as plantation trees.

## Field Campaigns

Three field campaigns in subsequent years from 2011 to 2013 were carried out to collect tree species distribution data in the VGTB watershed (Fig. 1). Based on existing forest maps, areas with a size of 1 km × 1 km of continuous forest were selected as basic sampling sites. For each of these sites, nine subplots with a size of 10 m × 10 m were identified. First, a central subplot was assigned and sampled, and subsequently eight subplots in a distance of approximately 250 m in each direction (N, NE, E, SE, S, SW, W, NW) were sampled. For each subplot, all trees larger than 5 cm of diameter at breast height (DBH) were identified. Only trees which could be identified to species level were used for subsequent calculations. Species numbers for the three different campaigns were determined (Fig. 2).

In the next step, a set of ecologically and economically valuable species was composed based on the report of the project 'Lesser known timber species of Vietnam' jointly implemented by the WWF and the Vietnamese-German Forestry Programme (n.d.) and based on Hoang (2005). Species identified in the three field campaigns were then compared to the resulting set of 41 species and a list of ten ecologically and economically valuable species found in the VGTB watershed was assembled (Table 1).

## Species Distribution Modelling

MAXENT (Maximum Entropy Modeling) software version 3.3.3 k was chosen to model potential tree species richness based on the identified tree species and on environmental data (Phillips et al. 2006; Elith et al. 2011). The strength of MAXENT software is the capacity to model species occurrence data when dealing with small numbers of species occurrences (Pearson et al. 2007; Williams et al. 2009). Environmental predictors were taken from the World-Clim Global Climate GIS database (Hijmans et al. 2005). The most-correlated environmental predictors were omitted based on a Spearman rank correlation matrix to reduce collinearity and number of predictors, resulting in selection of the predictors

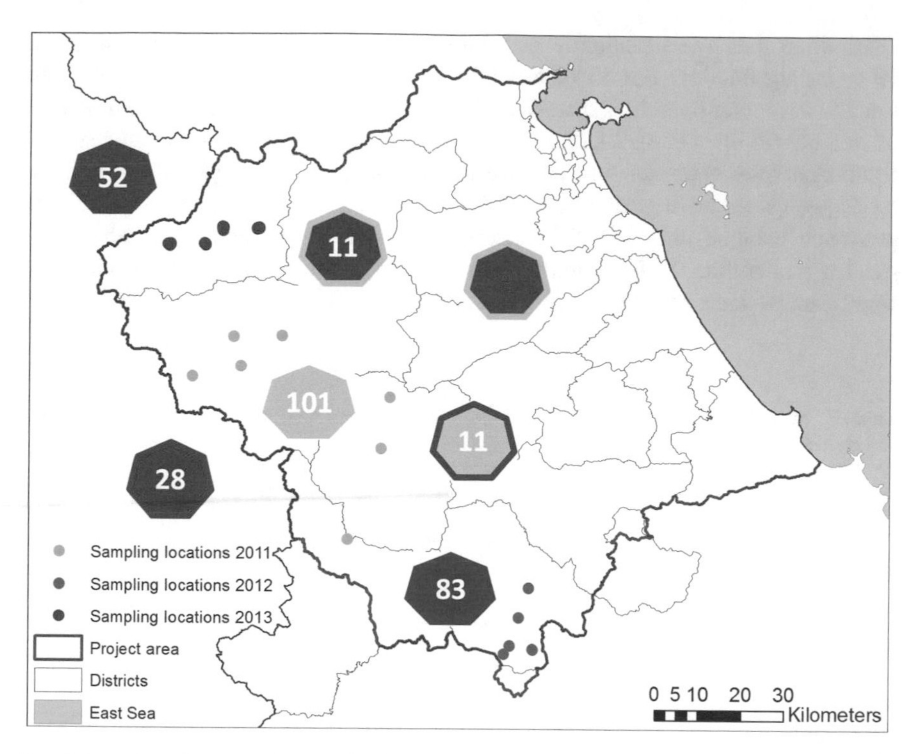


**Fig. 2** Number of tree species identified to species level in the three field campaigns (*highlighted in the colour of the corresponding year*). In addition, the number of species shared for two respective campaigns (*highlighted with the colours of the two corresponding years*) and the number of species shared in all campaigns (*all three colours*) are shown

**Table 1** List of ecologically and economically valuable species found in VGTB

| Scientific name | |
|---|---|
| *Aquilaria crassna* | Pierre |
| *Canarium album* | (Lour.) Räusch |
| *Canarium bengalense* | Roxb. |
| *Endospermum chinense* | Benth |
| *Lithocarpus ducampii* | (Hickel and A. Camus) A. Camus |
| *Madhuca pasquieri* | (Dubard) H.J. Lam |
| *Peltophorum dasyrhachis* | (Miq.) Kurz |
| *Shorea guiso* | (Blanco) Blume |
| *Shorea roxburghii* | G. Don |
| *Terminalia myriocarpa* | van Heurck and Müll. Arg. |

altitude, maximum temperature of the warmest month, precipitation seasonality, and precipitation of the warmest quarter (Raedig and Kreft 2011). The relevant steps of the modeling followed largely Raedig and Kreft (2011). The automatical features option to constrain predictor variables for training was chosen, and distribution data were randomly partitioned into a 50 % training fraction, and a 50 % testing fraction. For 45 species with more than five occurrences, a binomial test and a receiver operation characteristic analysis (yielding $AUC_{TEST}$ values) were carried out (Phillips et al. 2006). If the *p*-values calculated in the binomial test were significant ($p < 0.05$ for any threshold calculated, Pawar et al. 2007), and the $AUC_{TEST}$ values were greater than 0.7 (Thuiller et al. 2003), the species model was generated using a threshold *P* to determine the area predicted as suitable for a species greater than 0.5 (logarithmic output format). The list of 20 tree species with significant *p*-value and $AUC_{TEST}$ value is summarized in Table 2.

**Table 2** Species with more than six occurrences (in different plots) and with significant *p*-value and robust $AUC_{TEST}$ value

| Scientific name | | Occurrences | $AUC_{Train}$ | $AUC_{Test}$ |
|---|---|---|---|---|
| *Actinodaphne obovata* | (Nees) Bl. | 7 | 0.956 | 0.926 |
| *Aphanamixis polystachya* | (Wall.) R.N. Parker | 9 | 0.910 | 0.904 |
| *Antheroporum pierrei* | Gagnep. | 13 | 0.766 | 0.883 |
| *Canarium album* | (Lour.) Räusch. | 17 | 0.685 | 0.705 |
| *Castanopsis tonkinensis* | Seemen | 14 | 0.969 | 0.903 |
| *Clerodendrum japonicum* | (Thunb.) Sweet | 10 | 0.907 | 0.710 |
| *Dalbergia pinnata* | (Lour.) Prain | 10 | 0.845 | 0.941 |
| *Diospyros touranensis* | Lecomte | 8 | 0.843 | 0.964 |
| *Knema furfuracea* | (Hook. fil. and Thomson) Warb. | 19 | 0.852 | 0.732 |
| *Lithocarpus ducampii* | (Hickel and A. Camus) A. Camus | 21 | 0.954 | 0.835 |
| *Macaranga denticulata* | (Blume) Müll. Arg. | 11 | 0.941 | 0.914 |
| *Madhuca pasquieri* | (Dubard) H.J. Lam | 8 | 0.940 | 0.956 |
| *Neolamarckia cadamba* | (Roxb.) Bosser | 13 | 0.910 | 0.764 |
| *Ormosia pinnata* | (Lour.) Merr. | 9 | 0.832 | 0.818 |
| *Paranephelium spirei* | Lecomte | 9 | 0.941 | 0.928 |
| *Photinia arguta* | Wall. ex Lindl. | 7 | 0.774 | 0.935 |
| *Pterospermum lanceifolium* | Roxb. ex DC. | 10 | 0.811 | 0.815 |
| *Syzygium brachiatum* | (Roxb.) Miq. | 16 | 0.915 | 0.959 |
| *Syzygium lanceolatum* | (Lam.) Wight and Arn. | 9 | 0.850 | 0.766 |
| *Syzygium zeylanicum* | (L.) DC. | 29 | 0.801 | 0.823 |

## Identification of the Priority Areas for Conservation and Design of a Biodiversity Corridor

Potential tree species richness was calculated by summing up incidences for *P* greater than 0.5 per grid cell. The resulting map (Fig. 3) was then compared with the map of protected areas of the VGTB watershed to identify areas of elevated tree species richness that are not included in the protected area network. Furthermore, additional information about the road network of the VGTB watershed was overlaid to identify areas without fragmentation due to roads. Thereby identified areas with an elevated level of tree species richness and with a low fragmentation level were considered as priority areas for conservation in the VGTB watershed. It was further explored, whether these priority areas could be linked to form large scale corridors, creating continuous forested areas in the long run.

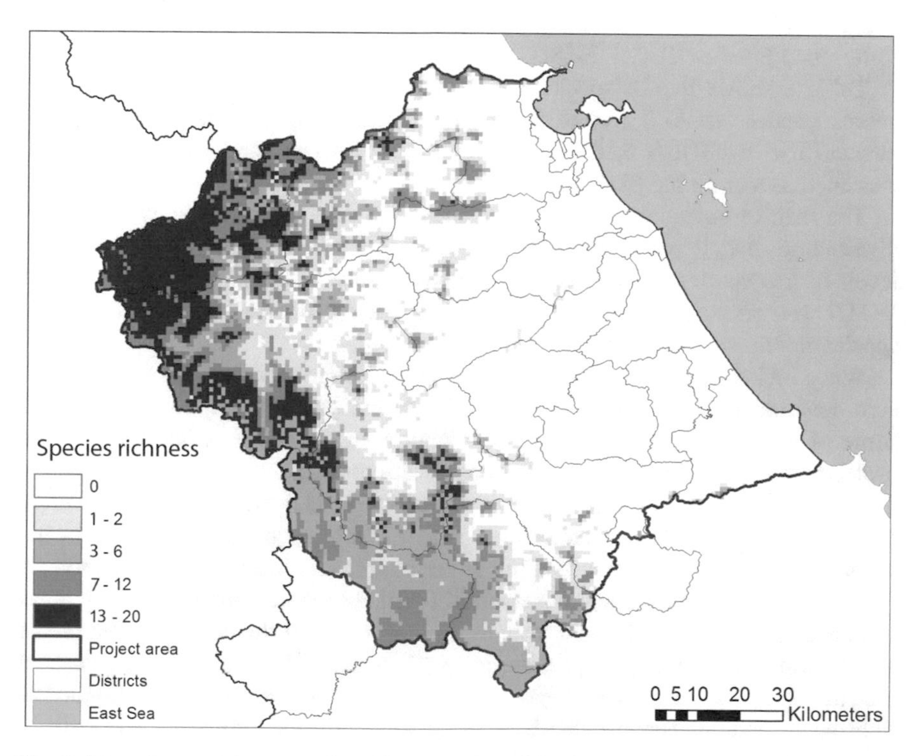


**Fig. 3** Map of potential tree species richness in the midlands and uplands of VGTB watershed based on the modelled distribution data of the species shown in Table 2

## Results

In the first field campaign which was carried out in 2012 in the Vu Gia area in the west of the VGTB watershed, 101 native tree species were identified. In the following field campaign in 2012 which was centred on the southern Thu Bon area of the watershed, 83 native tree species were found. In the last field campaign, 52 native tree species were identified in the northwestern region of the VGTB watershed (Fig. 2). Although the field campaigns in 2012 and 2013 took place with the largest distance between each other, they share 28 species common to both areas, whereas the adjacent field campaigns share only 11 common species, respectively. Only four species have been identified in all three campaigns. Altogether, 187 tree species were identified in the three field campaigns.

The species set of native tree species with a potential for plantation species for the VGTB watershed are shown in Table 1. All species except *Aquilaria crassna* were found in the wild. Because of its use for incense, this tree species is high in demand in Vietnam, and critically endangered according to the IUCN Red List of Threatened Species (IUCN 2015).

Table 2 shows the 20 species which had significant *p*-values and $AUC_{TEST}$ values greater than 0.7. Out of these 20 species, 18 species have not yet been evaluated in the IUCN Red List of Threatened Species, and two species were classified as vulnerable (IUCN 2015).

The map of potential tree species richness for the 20 species for which robust distribution models could be constructed is shown in Fig. 3. Areas with elevated levels of tree species richness are predominantly located in the western area of the VGTB watershed. In the southern part of the watershed, smaller areas of elevated species richness are found.

When including the protected areas (Fig. 4), the large spatial gap in the protected area network in the northwest of the VGTB watershed becomes visible. Whereas some of the smaller centres of species richness in the south are covered by the protected area network, other smaller centres lie outside.

Figure 5 incorporates the road network of the VGTB watershed. The overlay with this layer reveals one larger area which is not segmented by roads in the western region of the watershed. This larger continuous forest area could serve as the core area for biodiversity conservation (outlined in red). Smaller priority areas (outlined in orange) are located in both the northern and the southern area of the VGTB watershed. When the priority areas are connected to the existing protected area network, two main axes appear, one reaching from west to east (highlighted in light blue), and another reaching from north to south (highlighted in light red).

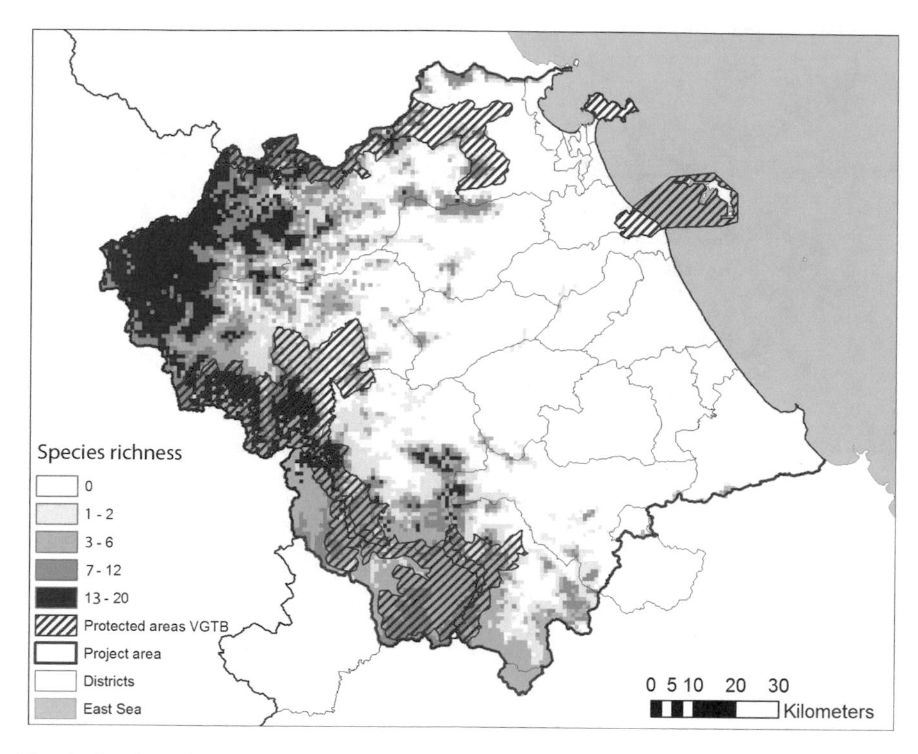


**Fig. 4** Overlay of the protected area network with the potential tree species richness map

## Discussion

Tree species data were collected in three consecutive years in a data-poor region, and served to develop species distribution models for the detection of priority areas for conservation. In the long run, these priority areas can act as stepping stones for the establishment of two biodiversity corridors in the VGTB watershed. The two corridors would be connected and the core priority area (outlined in red in Fig. 5) would have a crucial role for both corridors. First, the core priority area is part of the west–east corridor linking the Annamite Mountains with the sea. It further links the north–south corridor with protected areas in the northern neighbour province of Thua Thien-Hue and in the southern province of Kon Tum. Second, this priority area could also be linked with forest areas in Laos to promote transboundary biodiversity conservation.

Both corridors are separated by the National Road 14D, and in order to ensure connectivity between the corridors, suitable structures like wildlife crossings could be built to allow for exchange of larger mammals between the corridors. The main highway traversing the western region of Quang Nam, the National Road 14, intersects both corridors, and would represent the eastern limitation to the proposed protected area. The road partly following the border between Quang Nam and Thua

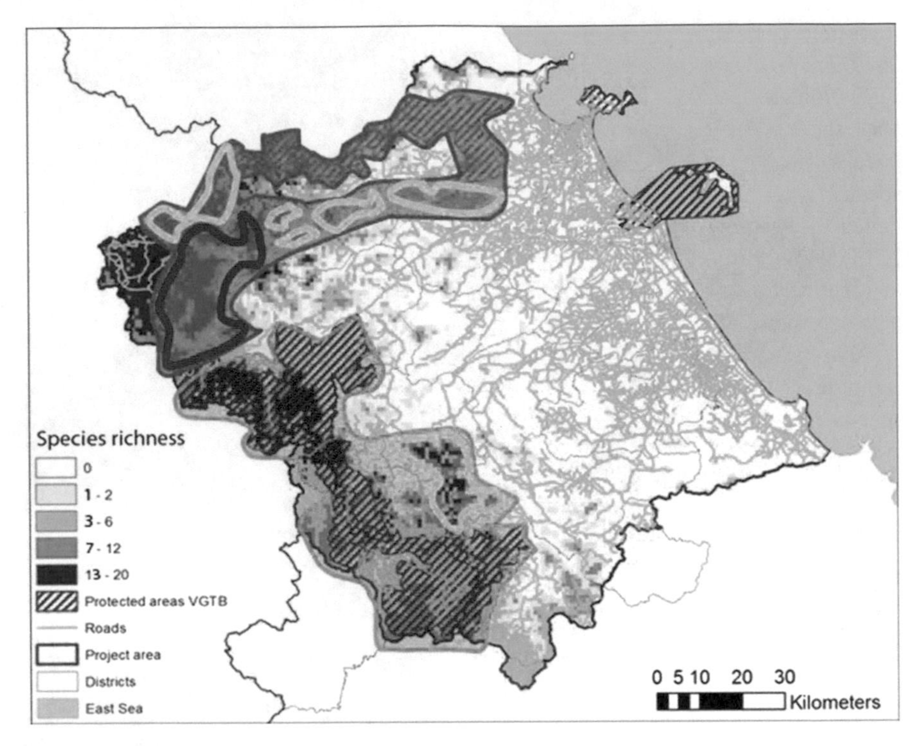


**Fig. 5** Linking priority areas for conservation for establishing forest corridors in the VGTB watershed. Priority areas for conservation are marked in *orange*, and the core priority area is outlined in *red*. Corridor areas are highlighted in *blue* (west–east axis;) and in *rose* (north–south axis). The road network is based on data from MONRE (2010)

Thien-Hue provinces, the National Road 14G, is the reason for the parallel structure of the west–east corridor; only at its eastern limit, the Ba Na-Nui Chua Reserve crosses this road; at all these intersections of corridor areas by major roads, wildlife crossings should be constructed.

The three field campaigns were carried out in a rapid assessment approach, since working conditions in the mountainous areas of the VGTB watershed allowed usually only the establishment and tree species assessment of one 1 km × 1 km plot in one day. It was important for the selection of the sampling design that the data collected could also be used by the Forest Inventory and Planning Institute Hue. Therefore, the rapid assessment approach was chosen, which allowed collecting tree species data in an area of continuous forest cover of 1 km$^2$, but distributed on nine subplots, which could be logistically managed in one day. Whereas in the 2011 field campaign, seven of such 1 km$^2$ plots could be found, in the 2012 field campaign five, and in the 2013 field campaign only four were found. Fragmentation levels particularly in the latter area were high, and it was difficult to find forest stands with the size of 1 km$^2$. This explains the lower species numbers identified in the 2013 field campaign. When comparing the species identified in each field campaign, the highest number of shared species was found between the 2012 and 2013 campaigns, although they are spatially more distant from each other. This indicates that for a complete assessment of species richness of the VGTB watershed, further data collection is indispensable, and that also smaller forest areas have to be considered. This is also corroborated by the absolute number of tree species identified in this research which roughly resembles one-sixth of the altogether 1129 plant species identified in former surveys in Quang Nam (DARD 2013). However, for the identification of species distribution patterns, with the aim to identify forest areas of elevated species richness and to establish connectivity between them, the focus on 1 km$^2$ plots of continuous forest is more suitable.

As a further outcome of this research, native tree species with high economic and ecological value were identified to replace exotic plantation trees. Two of the species identified in the present study belong to the family of Dipterocarpaceae that are known for their excellent timber (see Table 1). Kammesheidt (2011) revealed hilly degraded areas in elevations between 300 and 600 m a.s.l. as suitable for dipterocarp plantations, but also suggested the cultivation of other trees endemic to Southeast Asia. Whereas exotic tree plantations are often introduced with the argument to protect remaining natural forests, for most tropical forests rather the opposite is true, and the pressure on them is aggravated. When timber of high quality becomes scarce, illegal logging in near-natural forests and protected areas increases (Kammesheidt 2011). The cultivation of native tree species with high-quality timber—in contrast to cultivation of exotic trees with low-quality wood for the pulp and paper industry—could reverse that trend.

A further advantage of the cultivation of native tree species is the adapted provision of ecosystem services. Exotic species provide ecosystem services adapted to the ecosystem they originate from. Often, these services are not relevant for the new ecosystem or even harmful, e.g., when native animals cannot feed on the exotic trees' pollen or their fruits, or when the introduced species use more water than the native ones. On the opposite, native tree species provide ecosystem services adapted to their ecosystems, and the reintroduction of such species as plantation trees can contribute to ecosystem functioning and thus to maintaining of the ecosystem services, provisioning as well as regulating, supporting and cultural ecosystem services.

## Recommendations for Conservation Activities in VGTB Watershed

Centres of elevated potential tree species richness with a low degree of fragmentation by the existing road network yield a set of priority areas which—when linked—form two perpendicular biodiversity corridors within the VGTB watershed. The core priority area identified connecting the corridors as well as allowing for transboundary conservation to neighboring provinces and to Laos could serve as a backbone for the protected area network in the VGTB watershed. Because of its key importance, this core priority area should ideally be designated as a protected area.

Near-natural forest stands with elevated levels of species richness could act as stepping stones within the two corridors fostering connectivity. Furthermore, plantations of native tree species could act as stepping stones, since such tree species need less fertilizer inputs, pesticides and grooming in comparison to exotic species, allowing for more diverse understory vegetation, and thus promoting the provision of biodiversity-related ecosystem services. The concept of corridors and stepping stones should be integrated into the forest management of the VGTB watershed.

Awareness-building measures are suggested to sensitize local communities for biodiversity conservation. The protected area network resembles a top-down conservation strategy, which can only succeed when accompanied by bottom-up conservation measures. As long as local communities are not aware of the benefits resulting from forest conservation, an extension of the existing protected area network will not find their acceptance. This will also facilitate the work of the forest rangers, who often have to endure widespread resentment in the communities. Furthermore, the potential of the use of native tree species for cultivation has to be promoted, and the planting of native tree species should be facilitated and integrated into forest management.

**Acknowledgments** We would like to express our gratitude to the scientists from the Forest Inventory and Planning Institute Hue for species identification and tree height measurements carried out during the field campaigns, in particular to Dao Nguyen Sinh, Pham Van Nghiem, Nguyen Van Hung and the Director of Song Thanh Natural reserve, Tran Van Thu. We further thank the rangers from the Vietnamese Forest Department who supported the field campaigns by finding suitable plot locations.

## References

Bueren VM (2004) *Acacia* hybrids in Vietnam. Impact Assessment Series. ACIAR Australian Centre for International Agricultural Research

DARD Department of Agriculture and Rural Development (2013) Decision No 3277. Quang Nam Provincial People's Committee PPC

Elith J, Phillips SJ, Hastie T, Dudík M, Chee YE, Yates CJ (2011) A statistical explanation of MaxEnt for ecologists. Divers Distrib 17:43–57

Fagan WF, Calabrese JM (2006) Quantifying connectivity: balancing metric performance with data requirements. In: Crooks KR, Sanjayan M (eds) Connectivity conservation. Cambridge University Press, Cambridge, pp 297–317

FAO Food and Agriculture Organization of the United Nations (2010) Global Forest Resources Assessment 2010. Country Report Vietnam

FAO Food and Agriculture Organization of the United Nations (2015) Global forest resources assessment 2015. Desk reference

Hijmans RJ, Cameron SE, Parra JL, Jones PG, Jarvis A (2005) Very high resolution interpolated climate surfaces for global land areas. Int J Climatol 25:1965–1978

Hoang Ho DT (2005) Restoration of degraded natural forests in the North Central Coast region of Vietnam. Dissertation

ICEM International Centre for Environmental Management (2008) Strategic environmental assessment of the Quang Nam Province hydropower plan for the Vu Gia-Thu Bon River Basin. Hanoi, Viet Nam, pp 1–205

IUCN (2015) The IUCN red list of threatened species. Version 2015-4. http://www.iucnredlist.org. Accessed in April 2016

Kammesheidt L (2011) Guest editorial: planting native quality timber trees in South-East Asia: pipedream or lucrative business? J Tropical Forest Sci, 355–357

MARD Ministry of Agriculture and Rural Development (2008) Decree 2370/QĐ/BNN-KL dated 05/08/2008

Meyfroidt P, Lambin EF (2008) Forest transition in Vietnam and its environmental impacts. Glob Change Biol 14:1319–1336

MONRE Ministry of Natural Resources and Environment (2010) Road network of the provinces Quang Nam and Da Nang

Pawar S, Koo MS, Kelley C, Ahmed MF, Chaudhuri S, Sarkar S (2007) Conservation assessment and prioritization of areas in Northeast India: priorities for amphibians and reptiles. Biol Conserv 136:346–361

Pearson RG, Raxworthy CJ, Nakamura M, Townsend PA (2007) Predicting species distributions from small numbers of occurrence records: a test case using cryptic geckos in Madagascar. J Biogeogr 34:102–117

Phillips SJ, Anderson RP, Schapire RE (2006) Maximum entropy modeling of species geographic distributions. Ecol Model 190:231–259

PPC Provincial People's Committee (2012) Decision 2265/QĐ-UBND dated 23/7/2012. Establishment of the Sao La natural reserve. Quang Nam PPC

Raedig C, Kreft H (2011) Influence of different species range types on the perception of macroecological patterns. Syst Biodivers 9:159–170

Sterling JE, Hurley MM, Minh DL (2006) Vietnam: a natural history. Yale University Press, London

Thuiller W, Vayreda J, Pino J, Sabate S, Lavorel S, Gracia C (2003) Large-scale environmental correlates of forest tree distributions in Catalonia (NE Spain). Glob Ecol Biogeogr 12:313–325

VDR Viet Nam Development Report (2011) Natural resources management. A joint development partner report to the Viet Nam Consultative Group Meeting, Hanoi, 7–8, 2010

Wiens JA (2006) Connectivity research: what are the issues? In: Crooks KR, Sanjayan M (eds) Connectivity conservation. Cambridge University Press, Cambridge, pp 23–27

Williams JN, Seo C, Thorne J, Nelson JK, Erwin S, O'Brien JM, Schwartz MW (2009) Using species distribution models to predict new occurrences for rare plants. Divers Distrib 15:565–576

World Bank (2005) Vietnam environment monitor: biodiversity. World Bank, Washington, DC

WWF Worldwide Fund for Nature, Vietnamese-German Forestry Programme (n.d.) Lesser known timber species of Vietnam, project report

# Rice-Based Cropping Systems in the Delta of the Vu Gia Thu Bon River Basin in Central Vietnam

**Rui Pedroso, Dang Hoa Tran, Viet Quoc Trinh, Le Van An and Khac Phuc Le**

**Abstract** Despite high pressures for agricultural land conversion, increasing competition for water, and the relatively low net benefits of rice production, rice is still by far the predominant farm occupation in the Vu Gia Thu Bon basin in Central Vietnam. This study examined the reasons for such persistence by surveying and analyzing a comprehensive set of qualitative (planting and harvesting dates) and quantitative data (yields, labor and nonlabor inputs, prices) for all the crops present in the cropping systems of 113 farms in the region. The net benefit derived from rice production was on average 23 M VND $ha^{-1}$, with a relatively low labor input of 144 man-days per $ha^{-1}$. The net benefits generated by vegetable production are more than 9 times higher (ca. 215 M VND $ha^{-1}$) with a labor demand of ca. 928 man-days $ha^{-1}$. Despite the very high net benefits of vegetable production, in this region they do not translate into an equivalently high added value per ha and man-day. These values are 'only' nearly double than those for rice, and not much higher than those for watermelon, chili, and groundnut. The results indicate that farmers' decisions for not rushing in diversifying production to vegetables are wise when looking at the high risks of vegetable production, shortage of on-farm labor resources, and high opportunity costs of nonfarm labor opportunities. Rice is a robust crop and a pillar of families' food security, demanding low labor inputs. Under current conditions, farmers will most probably continue predominantly cropping rice. There is nevertheless the need to improve the rice system. Technical efficiency of rice production in the delta of the VGTB basin is 78 %, a low figure if compared to recent average estimations of 86 % for the Vietnamese Mekong and Red River deltas. The small scale of production, land fragmentation and irrigation challenges due to salinity intrusion are the main factors impacting on technical efficiency in the region.

---

R. Pedroso (✉) · V.Q. Trinh
Institute for Technology and Resources Management, Cologne, Germany
e-mail: rui.pedroso@th-koeln.de

D.H. Tran · L.V. An · K. Phuc Le
Faculty of Agronomy, College of Agriculture and Forestry,
Hue University, Hue, Vietnam



A. Nauditt and L. Ribbe (eds.), *Land Use and Climate Change Interactions in Central Vietnam*, Water Resources Development and Management,
DOI 10.1007/978-981-10-2624-9_6

## Introduction

Vietnam is one of the biggest rice exporters worldwide, a success due to extensive land and market reforms and introduction of new technologies over the last 30 years. Rice production was decentralized and markets liberalized inducing higher rice prices. Farm profits are now retained by farmers giving people the incentives to invest on the farm These changes have induced an enormous increase in total factor productivity (TFP) and Vietnam managed to strongly reduce rural poverty during this period (Hansen and Nguyen 2007; Kompas et al. 2012) Nevertheless these achievements, there are strong signs of a TFP slowdown in Vietnam since 2002. This can be witnessed in all rice producing regions except for the Mekong River Delta (Kompas et al. 2012). The latter authors speak in this context from restrictions on land use and market regulations that still call for further reforms. The pursue of land consolidation by abolishing restrictions on land size and the effective development of real estate markets and land property rights are recognized here as decisive for setting TFP back on track. Land fragmentation is seen as a major factor explaining efficiency and productivity of rice production in Vietnam. The country's agricultural land fragmentation was mainly caused by the land allocation process during the reform years in the course of the Doi Moi policy (World Bank 2003; Van Hung et al. 2007; Linh 2012). The process of land allocation was based on strong equity principles between households, which led to the distribution of rather small land plots to households with different locations within the communes and land qualities (Ravallion and van de Walle 2001; Van Hung et al. 2007; Linh 2012). Moreover, the rapidly growing population and strong urban developments have reclaimed a huge amount of agricultural land for urban and industrial development; between 2001 and 2010 ca. 1 million hectares (World Bank 2011). The disruption of Rice landscapes can impact on irrigation systems and on rice production efficiency. The new official socioeconomic development plans for the Quang Nam province indicate high current and future rates of land conversion from agriculture to other uses specially tourism along the coastal line (PPC Quang Nam 2012; Quang et al. 2014). Despite these evidences, the official statistics regarding rice production for the whole province show only small acreage decreases since 2000. The cultivated area of rice per year in the Quang Nam province shows a change from 94,360 ha in year 2000 to 87,904 ha in year 2013, a decrease of 6456 ha (ca. 7 %). The lowest yearly cultivated area registered during this time was in year 2006, with a minimum of 83, 631 ha, a reduction of about 11 % (Fig. 1). According to these statistics, total rice output has kept a continuous positive trend in the last years, where decreases in planted area could be compensated by yield improvements (Fig. 2). The other two most important staples in the region are cassava and maize, which are growing at very high rates. The planted area of Maize has grown ca. 47 % between 2000 and 2013. Cassava shows the second highest growth rate with ca. 31 % in the same time period. According to key experts' interviews the high growth rates of these two crops is due to the higher demand for animal fodder in the case of Maize and in the case of Cassava due to the strong

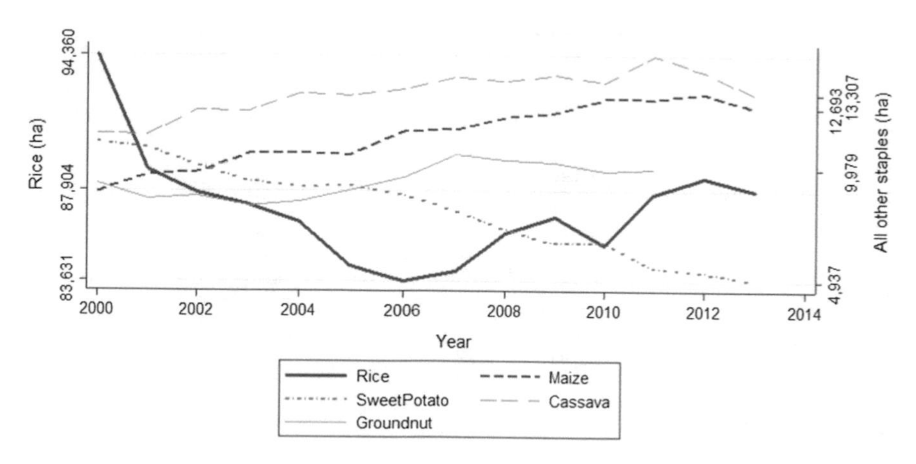


**Fig. 1** Planted area (ha) of main staples for whole Quang Nam

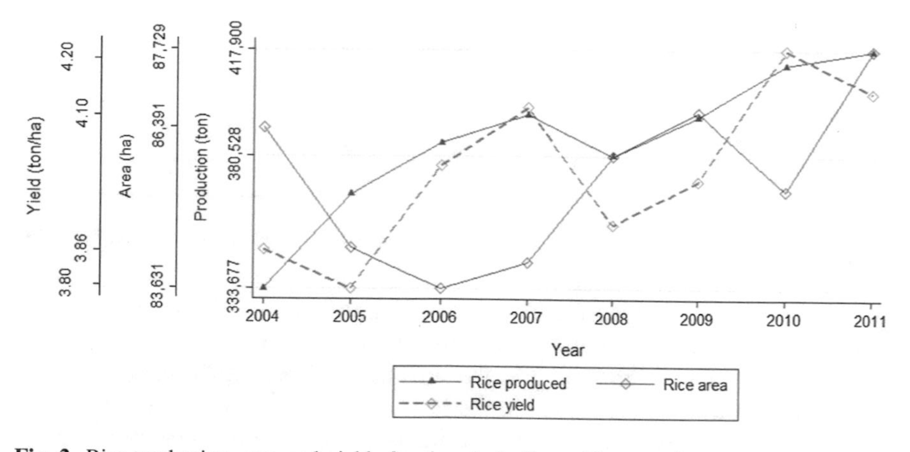


**Fig. 2** Rice production, area and yields for the whole Quang Nam province

export demand, especially to China for biofuel processing. Groundnut is another staple of the region showing an increasingly importance for peoples livelihoods (see Fig. 1).

The spatial production distribution can be seen in Fig. 3 with acreage figures for the year 2013. Accordingly, the largest rice producers are the districts Thang Binh with over 15,900 ha, followed by Dien Ban (11,412 ha), Dai Loc (8,707 ha), Duy Xuyen (7,735 ha), Que Son (6,800 ha), Nui Thanh (7,515 ha) and Phu Ninh (6,767 ha). All other districts in Quang Nam show production levels lower than 5000 ha. Cassava is the second most produced staple; the districts located on the coast are again the largest producers with Que Son (2,552 ha), Nui Thanh (2,350 ha) and Thang Binh (1,193 ha). Maize is the third staple in Quang Nam and mostly produced in Dien Ban (2,133 ha), Dai Loc (1,778 ha) and Nam Giang (1,145 ha). Groundnut is typically cultivated in Thang Binh (2,284 ha), Dien Ban

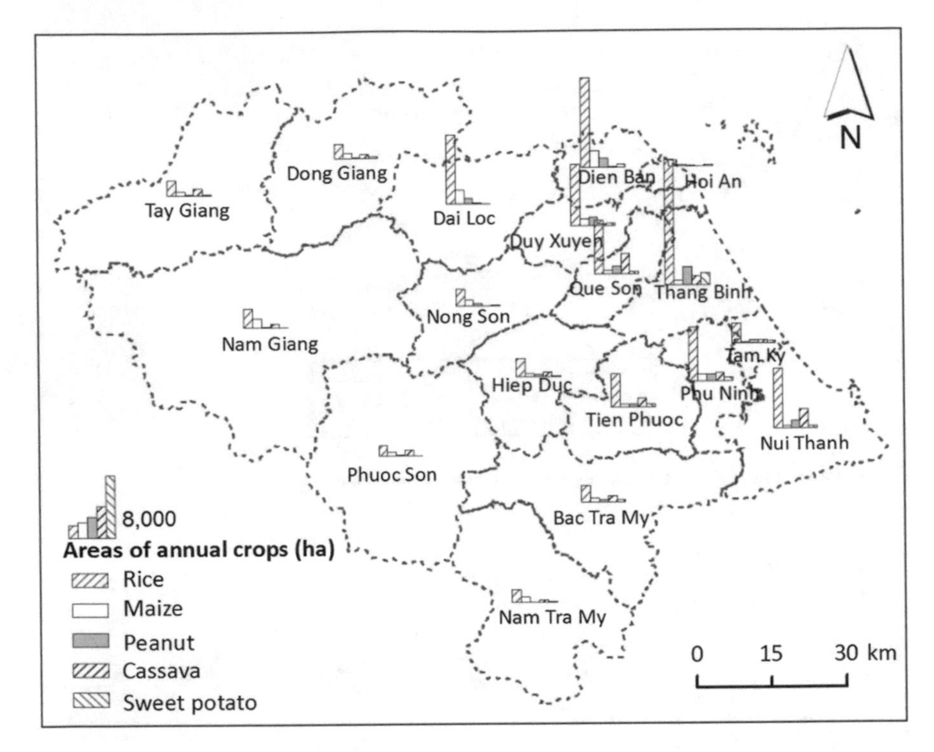


**Fig. 3** Planted areas and spatial distribution of most important staples in Quang Nam for 2013

(1,164 ha) and Duy Xuyen (1,102 ha). Sweet potato is mainly produced in the Thang Binh district (1,601 ha), all other districts show acreage values under 500 ha for 2013. Agricultural production in the upland districts is quite marginal when compared with the acreages found in the lowlands of the Vu Gia Thu Bon delta.

Besides land conversion and fragmentation, another important pressure on agricultural communities in this region is the impact of climate change and climate variability. Vietnam's coast line and deltasDelta, where most rice is grown, are very much exposed to flooding, extreme weather events, sea level rise and salt water intrusion. These factors increase the population's vulnerability to food insecurity and poverty, making agriculture a hazardous proposition (MONRE 2009; Chung et al. 2015).

This study focus on the delta of the Vu Gia Thu Bon river basin (VGTB) mainly located in the Quang Nam province in Central Vietnam. The delta can be characterized by 13 rice irrigation zones delineated according to their hydraulic connectivity as in Fig. 4 (Viet 2014). The delta of the VGTB river basin is where the most fertile soils for rice production are found and where most of rice production takes place. Rice production is by far the most important activity and mainly found in the alluvial plains of the delta (see also Fig. 3).

The main objective of this study is to characterize rice production in the delta of the VGTB river basinRiver basin in terms of yields, crop benefits, input demand,

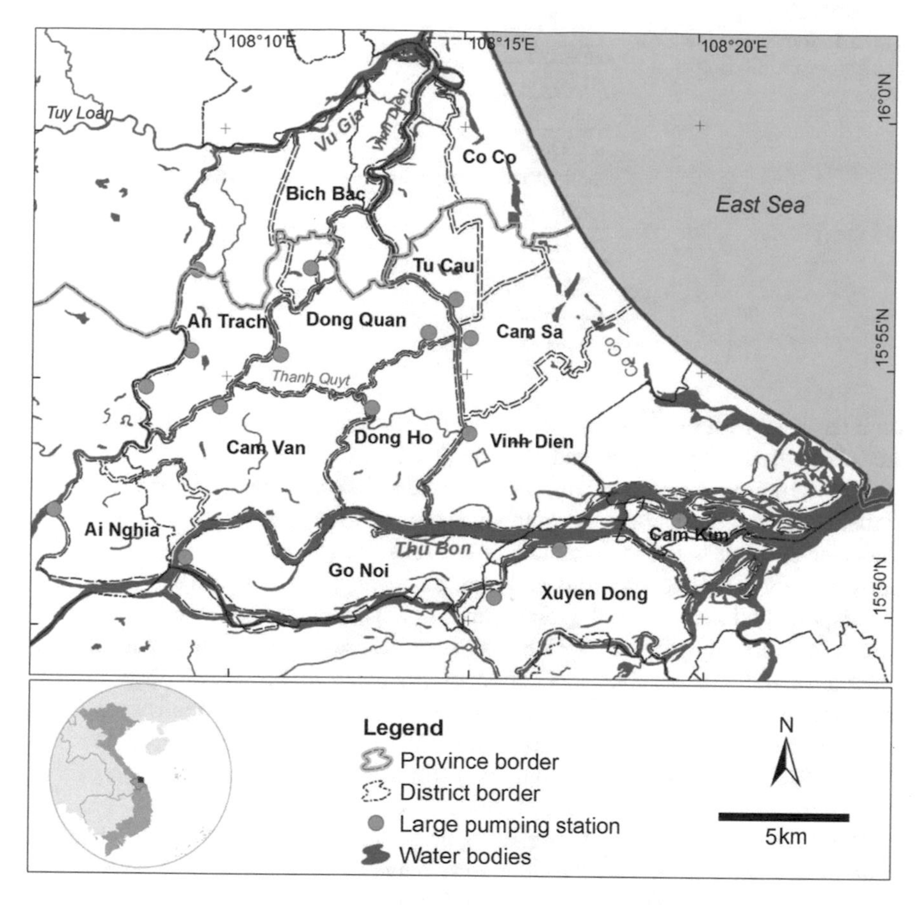


**Fig. 4** Study region in the delta of the VGTB river basin, Quang Nam province

profitability, and technical efficiency (TE). The analyses are divided in two main sections: a more descriptive one, where rice production is compared to the other most important crops in the delta, and a more strict section, where we apply stochastic frontier methods to analyze the average TE of rice production at different temporal and spatial scales, i.e., across the WSp and SA seasons, and also across the different irrigation zones.

## Descriptive Analysis of Crop Production in the Delta of the VGTB River Basin

A survey was performed in 2013 with a sample of 113 farmers from most representative communes for the different lowland irrigation zones (Fig. 4). The choice of representative communes from which farmers were randomly chosen was based

**Table 1** Representative communes for the irrigation zones

| Irrigation zone | Commune |
|---|---|
| Ai Nghia | Dai Hoa |
| Cam Van | Dien Tho |
| An Trach | Dien Tien |
| Dong Quang & Bich Bac | Dien Hoa |
| Dong Ho | Dien An |
| Go Noi | Dien Quang |
| Tu Cau & CoCo | Dien Ngoc |
| Cam Sa & Vienh Dien | Cam Ha |
| Xuyen Dong & Cam Kim | Duy Phuoc |

on interviews with key informants and local authorities in the lowlands districts Dai Loc, Dien Ban, Duy Xuyen, and Hoi An (Table 1).

The sampling procedure was a compromise between budget constraints and statistical representativeness. A comprehensive questionnaire was applied to each of the selected farm households including sections on: (1) household characteristics, (2) yields for all crops, rotations and crop calendar, (3) inputs of crop production, (4) disposal and marketing of agricultural products, and other relevant sections. All prices in this study refer to 2013 levels. One can say that the survey results will remain valid for a considerable period of time before becoming obsolete. This assumption is based on the fact that inflation in Vietnam is under control (ca. 4.1 % between 2011 and 2015, World-Bank 2016). The study fits in the five year plan (2010–2015) of the Provincial People's Committee (PPC) so that cropping patterns are not expected to change much in this period.

We have calculated farm household's crop net benefits per hectare, where net benefits are defined here as the difference between total gross benefits and total variable costs. This can also be called the gross benefits to a family's labor, management, land, and capital. The total gross benefit calculated for rice and groundnut included the opportunity value of home consumption. Gross benefit is, for these two crops, calculated as the sum of the annual production sold in the market, the amount disposed for paying wages, or feeding animals, and multiplied by the farm gate price, plus the opportunity value of the proportion of total production further processed and consumed by the family. For all other crops gross benefits equal only total production multiplied by the farm gate price. Rice and groundnut are two crops in the region for which more than 50 and 75 % of production, respectively, is consumed by the family. Not accounting for the opportunity value of the crops would lead to an undervaluation of the farm household crop benefit. The rice reserved for home consumption is milled in local mills after harvest and stored at home. We assume that only 62 % of this rice is actually consumed after accounting for whole processing losses (assumption based on interviews with local key informants). The rice consumed is valued at 13,000 VND $Kg^{-1}$. The farm gate price for rough rice is 7,000 VND $Kg^{-1}$ on average. Groundnut is basically produced in the WSp season under rain-fed conditions. Local households further process groundnut to cooking oil. We assume, based on key informants' interviews, that from 100 kg of harvested rough groundnut, 70 kg are grains (70 %). Furthermore and also based on experienced key informants, we assume that 3 kg of groundnut grains produce 1 L of cooking oil. The farm gate price for rough groundnut is on average 25,842 VND $Kg^{-1}$, the price for one liter groundnut oil was judged at 100,000 VND $L^{-1}$.

Subtracting total costs incurred for seeds, organic and chemical fertilizers, pesticides, hired labor, irrigation costs, and any other variable costs, we calculate the net benefits to families' labor, management, land, and capital. Added value per hectare and man-day was calculated by excluding hired labor costs from the total costs calculation (i.e., net benefits to the production factors labor, land and management) and divided by the labor demand.

The net benefits for vegetables were calculated for each individual family species and averaged. The vegetable labor demands were assessed for each field operation. For comparing different vegetable species, calculations should be done not only on a per unit of area basis but also on per unit of time in the field (Huong et al. 2013b). The vegetable values in this study were converted to units per hectare (ha) and per growing day (Gday) on the field. Household labor power supply was defined in terms of those individual members who participate in income activities. It included persons who are part of the family unit, reside at the household site and are actively involved in generating income. A man-day of work was defined here as the amount of work that can be carried out by an adult male in an 8-hour work day. Because the work productivity differs between males, females and people in different ages, we have calculated the household's on-farm labor supply in terms of man-day equivalent. This calculation is normally done by using standard conversion weights applied to males and females in different age groups (see Norman 1973).[1]

## *Cropping Pattern in the VGTB Delta Region*

The typical cropping pattern of the average farm household across the four cropping seasons can be seen in Table 2. The total sample area dedicated to each crop in the respective season was averaged across the 113 farmers found in the sample (the number of farmers in the sample cultivating a crop in the season can be seen in brackets). The two main rice cropping seasons are the Winter–Spring (WSp) season starting around December 25 and ending by mid-May and the Summer–Autumn (SA) season, starting by the end of May until mid-September (rice–rice system). The Spring–Summer season (SpS) starts in April, and the Autumn–Winter

[1]Labor supply calculations are presented in a submitted different article. Information can be provided on request.

**Table 2** Average area (sao) allocated per farm and season to the main annual crops

| Crop | WSp | | SpS | | SA | | AW | |
|---|---|---|---|---|---|---|---|---|
| Rice | 4.31 | (113) | | | 4.36 | (113) | | |
| Maize | 1.25 | (50) | 0.13 | (9) | 1.95 | (66) | 0.02 | (1) |
| Groundnut | 1.24 | (62) | | | 0.05 | (2) | | |
| Watermelon | 0.35 | (7) | 0.16 | (2) | | | | |
| Chili | 0.32 | (14) | | | | | | |
| Vegetables | 0.16 | (15) | 0.06 | (7) | 0.05 | (7) | 0.06 | (9) |
| Tobacco | 0.10 | (4) | | | | | | |
| Mung bean | 0.06 | (3) | 0.60 | (22) | 0.54 | (20) | | |
| Sweetpotato$^{+}$ | 0.06 | (4) | 0.01 | (2) | 0.08 | (6) | 0.11 | (8) |
| Sesame | 0.01 | (1) | 0.08 | (3) | 0.26 | (17) | | |
| Cassava | 0.01 | (2) | 0.03 | (2) | | | | |
| Kumquat | 0.06 | (3) | 0.06 | (3) | 0.06 | (3) | 0.06 | (3) |
| Banana | 0.01 | (1) | 0.01 | (1) | 0.01 | (1) | 0.01 | (1) |
| Fallow | 0.0 | | 6.8 | | 0.5 | | 7.6 | |
| Cultivated | 7.9 | | 1.1 | | 7.4 | | 0.3 | |

1 sao = 500 $m^2$
In brackets is the number of farmers cultivating the crop in the season
$^{+}$Sweetpotato for roots

(AW) starts around September, both seasons are not foreseen for rice. In the WSp and SA rice seasons nearly 100 % of the farm is cultivated, while in the SpS 14 % and in the AW season only 3 % of the farm is under cultivation. The sample average farm size is 7.9 sao, which is about 0.4 ha (1 sao equals 500 $m^2$). The main crops cultivated per farm in the WSp season are rice (4.31 sao), maize (1.25 sao) and groundnut (1.24 sao). They are followed by watermelon (0.35 sao), chili (0.32 sao), vegetables (0.12 sao) and tobacco (0.10 sao). All other crops have very small average acreages, e.g., cassava is of nearly no importance in the delta region, it is however the second most important staple in the whole Quang Nam district. Vegetables are basically grown all-year round, the species found in the sample are amaranth, bitter gourd, bottle gourd, chinese mustard, coriander, cucumber, lettuce, malabar spinach, sweet potato (vines), mint, okra, spring onion, water spinach, and wax gourd. Kumquat[2] and banana are typical perennial crops of the region but not grown by many farmers. In the SA season, rice is again the main crop (4.36 sao) followed by maize once more (1.95 sao). Groundnut is basically substituted by mung bean (0.54 sao) and sesame (0.26 sao). Sesame is a drought resistant crop and very much suitable for the SA season.

[2]Kumquat is an evergreen tree, producing edible golden-yellow fruits resembling small oranges. Mostly found in the coastal areas.

## Production

The study region is mostly characterized by fertile alluvial soils and good irrigation conditions (Viet 2014), we estimate rice yields of 5.5 t $ha^{-1}$ in the WSp season and 5.1 t $ha^{-1}$ in the SA season. Maize achieves average yields of ca 5.6 t $ha^{-1}$, which can be considered rather high for the region (Ha et al. 2004). Groundnut is the third most important staple achieving yields of 2.45 t $ha^{-1}$. The most important cash crops in the WSp season are watermelon, chili, vegetables, and tobacco. Vegetables are grown all year-round with a yearly average of ca. 27 t $ha^{-1}$. Tobacco is mostly grown in the WSp season and yields reach 2.7 t $ha^{-1}$ on average. In the SA season groundnut is no longer grown, instead of it mung bean (2.5 t $ha^{-1}$) and sesame (0.5 t $ha^{-1}$). Sweet potato is cultivated all-year round either for home consumption (vines as family vegetables, and roots for animal feeding) or as cash crop, i.e., for selling vines only. When cultivated for the family sweet potato shows little inputs, yields and net benefits. When cultivated for vines, sweet potato is followed by farmers as intensive vegetable cultivation, the level of inputs is very high, as well as yields and net benefits (detailed production data is presented in Pedroso et al. 2017).

## Inputs

Regarding the labor inputs, the total average labor use in rice is 144 man-days $ha^{-1}$, which is, after sesame with 119 man-days $ha^{-1}$, the lowest labor input of all crops. Maize shows an average labor input of 213 man-days $ha^{-1}$ and groundnut 215 man-days $ha^{-1}$. The vegetables labor input is calculated per hectare and per growing day in the field (Gday). The vegetables labor demand averaged over all the different species is 9.1 man-day $ha^{-1}$ $Gday^{-1}$. Taking the vegetables average growth duration of 102 Gday and multiplying both figures gives a 928 man-day $ha^{-1}$ average labor demand for vegetables (A detailed analysis of labor and nonlabor inputs is presented in Pedroso et al. 2017).

## Profitability of Crop Production

The average yearly net benefits for the most relevant crops in the lowlands can be seen for the two main seasons in Table 3. Rice and maize net benefits in the WSp season are relatively low, achieving ca. 24 M VND $ha^{-1}$ and 21 M VND $ha^{-1}$ respectively.[3] The average value for rice slightly declines in the SA season to ca. 22 M VND $ha^{-1}$, which is about the same value of maize in this season. The low

[3]The average exchange rate for the year 2013 was ca. 26,000 VND $EUR^{-1}$. This means that the 24 M VND $ha^{-1}$ net benefits for rice are equivalent to ca. 923 EUR $ha^{-1}$.

**Table 3** Average benefits, variable costs and added values ('00000 Dong $ha^{-1}$) for the crops of the two main seasons and year-round vegetables

| Gross benefit per hectare | | | | | | Variable costs per hectare | | | | Net benefits per hectare to family labor, land, management and capital | | | | Added value per man-day | | | |
|---|---|---|---|---|---|---|---|---|---|---|---|---|---|---|---|---|---|
| | *N* | Mean | Min | Max | CV | Mean | Min | Max | CV | Mean | Min | Max | CV | Mean | Min | Max | CV |
| *WSp season* | | | | | | | | | | | | | | | | | |
| Rice | 113 | 420 | 98 | 681 | 0.25 | 184 | 90 | 306 | 0.20 | 236 | −44 | 438 | 0.44 | 1.7 | −0.2 | 3 | 0.41 |
| Groundnut | 62 | 626 | 126 | 2022 | 0.59 | 85 | 35 | 185 | 0.36 | 541 | 57 | 1928 | 0.67 | 2.5 | 0.3 | 9 | 0.67 |
| Maize | 50 | 363 | 146 | 700 | 0.29 | 153 | 10 | 469 | 0.62 | 210 | −142 | 470 | 0.57 | 1.0 | −0.7 | 2 | 0.56 |
| Chili | 14 | 999 | 504 | 1733 | 0.36 | 317 | 92 | 674 | 0.51 | 682 | 188 | 1553 | 0.57 | 2.9 | 0.8 | 7 | 0.56 |
| Watermelon | 7 | 947 | 300 | 2400 | 0.71 | 136 | 71 | 216 | 0.46 | 810 | 91 | 2327 | 0.88 | 3.1 | 0.4 | 9 | 0.87 |
| Tobacco | 4 | 1004 | 570 | 1540 | 0.41 | 149 | 101 | 190 | 0.25 | 854 | 4 | 1390 | 0.48 | 2.2 | 1.1 | 4 | 0.47 |
| Sweet potato+ | 4 | 9.9 | 5.3 | 15 | 0.44 | 6.5 | 2.5 | 15 | 0.89 | 3.4 | −3.2 | 12 | 1.95 | 0.01 | −0.01 | 0.04 | 1.95 |
| *SA season* | | | | | | | | | | | | | | | | | |
| Rice | 113 | 401 | 53 | 2236 | 0.53 | 180 | 85 | 417 | 0.26 | 221 | −62 | 1925 | 0.89 | 1.6 | −0.2 | 9 | 0.84 |
| Maize | 66 | 335 | 131 | 786 | 0.32 | 115 | 5 | 278 | 0.42 | 220 | −47 | 653 | 0.49 | 1.0 | −0.2 | 3 | 0.48 |
| Mung bean | 20 | 382 | 10 | 846 | 0.45 | 69 | 34 | 166 | 0.55 | 313 | −32 | 800 | 0.56 | 1.4 | −0.1 | 4 | 0.56 |
| Sesame | 17 | 149 | 32 | 480 | 0.72 | 36 | 10 | 108 | 0.77 | 112 | 18 | 372 | 0.84 | 0.9 | 0.2 | 3 | 0.83 |
| Sweet potato+ | 6 | 13 | 1.3 | 41 | 1.10 | 27 | 6.8 | 60 | 0.73 | −14 | −33 | 4.5 | −0.86 | −0.1 | −0.1 | 0.0 | −0.86 |
| *Year-round* | | | | | | | | | | | | | | | | | |
| Vegetables* | 23 | 2332 | 540 | 5760 | 0.70 | 247 | 21 | 812 | 0.67 | 2084 | 227 | 5483 | 0.75 | 3.2 | 0.5 | 11 | 0.69 |

+Cultivation for home consumption only (vines for the family, roots for livestock feeding). Values for vines for the market are presented in Pedroso et al. (2017)

*Detailed statistics for each vegetable species and planting dates are found in Pedroso et al. (2017)

values are mostly connected with the low farm gate prices of ca. 7,000 VND $Kg^{-1}$ for rice and ca. 6,200 VND $Kg^{-1}$ for maize. Groundnut shows the highest net benefit of the three staple crops with 54 M VND $ha^{-1}$. The average added values per man-day are 165 K VND man-$day^{-1}$ for rice, 102 K VND man-$day^{-1}$ for maize, and 252 K VND man-$day^{-1}$ for groundnut. Added value man-$day^{-1}$ for rice and maize are low and below the region's average wage rate of ca. 200 K VND man-$day^{-1}$. As a result, the added value per man-day of groundnut cultivation is 2.5 times that of maize and ca. 1.5 times higher than that for rice. This is mainly because of the high groundnut farm gate price of ca. 26,000 VND $Kg^{-1}$, and the further processing of groundnut to cooking oil.

The cash crops sesame and mung beans are grown mainly in the SA season and present modest net benefits of ca. 11 M and 31 M VND $ha^{-1}$. The other cash crops chili, watermelon, tobacco, and vegetables, present much higher net benefits per hectare, achieving nearly 70 M VND $ha^{-1}$ for chili, to over 215 M VND $ha^{-1}$ for vegetables (see Table 3). The vegetable net benefits are nearly 9 times higher than those of rice. Because of the very high labor requirements of vegetables, the relatively very high net benefits of vegetables crops do not translate into equivalently high added value per man-day. Vegetables have an added value of ca. 320 K VND man-$day^{-1}$, which although nearly double that for rice, is only slightly higher than watermelon and chili, and not much higher than groundnut. The relatively very low labor demand of the rice crop translates into an added value per man-day that is higher than maize, mung bean and sesame, and not much lower than tobacco.

## Technical Efficiency in Rice Production in the VGTB River Basin

We follow an output oriented specification of the stochastic frontier production function (Kumbhakar and Lovell 2000)

$$\ln y_i = f(\boldsymbol{x}_i; \beta) + v_i - u_i \tag{1}$$

In Eq. (1) the $i$ subscript denotes observations (farms), $y_i$ is a scalar of observed output, $f(\boldsymbol{x}_i; \beta)$ is a log-linear Cobb–Douglas production function, where $x_i$ is a $1 \times J$ vector of input variables (factors of production), $\beta$ is the corresponding $J \times 1$ vector of coefficients. The random error $v_i$ is assumed as having zero-mean and normal distributed, and the production inefficiency error term $u_i$ is assumed to have a half-normal distribution. We account for heteroscedasticity in the modelModel. Ignoring it in $v_i$ would produce biased estimates of technical inefficiency and a downward biased estimate of the intercept in β. Ignoring heteroscedasticity in $u_i$ causes biased estimates of the $\beta$ parameters and of technical inefficiency estimates (Kumbhakar and Lovell 2000; Wang and Schmidt 2002). Because we have assumed a half-normal distribution of $u_i$, heteroscedasticity is dealt with in this study by parametrizing $\sigma_u^2$ (for more details Caudill et al. 1993; Hadri 1999; Battese and Coelli 1995; Wang and Schmidt 2002). The Cobb–Douglas specification in Eq. (1) was chosen including originally in $x$ the variables[4]: *land* (sao), *labor* (man-day sao$^{-1}$), *capital* (VND) and variable costs *varc* (VND). The variable *land* (sao) representing the acreage for rice in each farm was dropped from the Cobb–Douglas function in the final specification due to collinearity problems. This variable was however included in the heteroscedasticity equation, where several other variables possibly related to technical inefficiency were also included. These variables are related to land fragmentation (the total number of plots in the farm, the average plot area, or the distance from the plots to the homestead and to the nearest market), variables related to education and experience (school attendance and age), or variables related to environmental factors (perceived soil fertility and a proxy for salinity water intrusion in the SA season). The model parameters are estimated through maximization of the log likelihood function (see Kumbhakar and Lovell 2000).

## *Stochastic Frontier and Inefficiency Model Estimates for Rice Production*

The technical efficiency in rice production is compared for the WSp and SA seasons by using the same specification in both models. The variables in the inefficiency models related to education, age, the total number of plots in the farm, the average plot area, or the distance from the plots to the homestead and to the nearest market, as well as soil fertility, were found not significant in both WSp and SA models and dropped from the specifications.[5] The estimates are presented in Table 4 (least squares models indicated as OLS and stochastic frontier models as SF).

The signs of the elasticity coefficients of the SF functions are as expected and are highly significant in both season models. The sum of elasticities for the input variables gives 1.24 in the WSp and 1.18 in the SA model. The hypotheses of constant returns to scale (CRTS) cannot be rejected in any of the models of Table 4 as given by the $\chi_1^2$ statistic. The likelihood ratio test (LR) against the null hypothesis of no technical inefficiency (OLS) (three degrees of freedom), has a value of 12.672 in the WSp model and a value of 17.459 in the SA model. The critical value at the 0.01 significance level is 10.501 (see Kodde and Palm 1986). It can be understood hereafter that the null hypotheses of no technical inefficiency can be rejected for both WSp and SA models.

In the inefficiency model, the estimated coefficients of *land* (rice acreage) and *rplots* (number of rice plots on the farm) are both significant for the WSp season, *rplots* appears slightly nonsignificant in the SA model.

[4]Descriptive statistics on these variables are presented in a forthcoming study by the authors. Information can nevertheless be delivered upon request.

[5]Descriptive statistics on these variables are presented in a forthcoming study by the authors.

**Table 4** Maximum likelihood and OLS estimates

| | WSp | | SA | | |
|---|---|---|---|---|---|
| Independent variable ln prod | OLS | SF | OLS | SF1 | SF2 |
| ln labor | 0.382***<br>(4.46) | 0.323***<br>(4.07) | 0.482***<br>(5.18) | 0.374***<br>(4.66) | 0.378***<br>(5.08) |
| ln capital | 0.323**<br>(2.11) | 0.291**<br>(2.08) | 0.221<br>(1.06) | 0.306**<br>(2.03) | 0.272<br>(1.88) |
| ln varc | 0.643***<br>(7.64) | 0.622***<br>(8.30) | 0.537***<br>(5.76) | 0.497***<br>(5.59) | 0.506<br>(6.56) |
| Intercept | −1.152<br>(−1.40) | −0.398<br>(−0.52) | −3.924***<br>(−2.73) | −3.03***<br>(−2.49) | −3.094***<br>(−2.82) |
| Sum of coefficients<br>$\chi_1^2$ statistic for CRTS | | 1.24<br>2.37 | | 1.18<br>1.38 | 1.16<br>1.17 |
| ln $\sigma_\upsilon^2$ | | −3.140***<br>(−10.92) | | −3.544***<br>(−10.17) | −3.374***<br>(3.46) |
| Inefficiency model | | | | | |
| land | | −0.509**<br>(−2.15) | | −0.387**<br>(−2.33) | −0.490**<br>(−2.34) |
| rplots | | 0.490*<br>(1.81) | | 0.204<br>(1.13) | 0.362<br>(1.54) |
| pumping | | | | | −0.318***<br>(−4.08) |
| intercept | | −1.678***<br>(−2.80) | | −0.688<br>(−1.46) | 5.47***<br>(3.46) |
| $R^2$ | 0.81 | | 0.80 | | |
| Log likelihood | | −11.99 | | −21.27 | −6.725 |
| LR statistic+ | | 12.67 | | 17.46 | 46.54 |
| Observations 113 | | | | | |

*t* Statistics in parentheses
*$p < 0.10$, **$p < 0.05$, ***$p < 0.01$
+The likelihood ratio test statistic $LR = -2log[Likelihood(H_0)] - log[Likelihood(H_1)]$, has a chi-squared distribution with degrees of freedom (*df*) equal to the number of parameters assumed to be zero in the null hypotheses, $H_0$, provided $H_0$ is true. The critical values of the distribution for three *df* 10.501 for significance level of 1 %
Kodde and Palm (1986)

For the WSp model, the calculated marginal effects of *land* on the unconditional mean of $u_i$ is negative and equals −0.060. The level of inefficiency can be reduced, on average, by 6 percent for every 1 percentage increase in land acreage. The marginal effects of *rplots* are positive and equal 0.05. This means that the level of inefficiency rises, on average, by 5 percent for every 1 percentage increase in *rplots*.

For the SA model, the calculated marginal effects of *land* and *rplots* on the unconditional mean of $u_i$, are similar to the WSp. The calculated marginal effect of

**Table 5** Technical efficiency estimations for the delta irrigation zones

| | WSp (%) | SA (%) | Rice area (ha) in 2010 |
|---|---|---|---|
| Ai Nghia | 84.1 | 82.3 | 876.3 |
| Cam Van | 86.2 | 79.5 | 2037.5 |
| An Trach | 84.3 | 72.8 | 1553.1 |
| Dong Quang & Bich Bac | 72.6 | 71.1 | 1724.9 |
| Dong Ho | 84.1 | 79.4 | 1265.8 |
| Go Noi | 81.0 | 77.6 | 941.9 |
| Tu Cau & Co Co | 76.1 | 50.5 | 808.8 |
| Cam Sa & Vienh Dien | 77.6 | 70.8 | 2130.7 |
| Xuyen Dong & Cam Kim | 82.2 | 78.9 | 1627.7 |
| | 81 | 74 | 12,966.9 |

the variable pumping is in the SA negative and equals -0.041. The level of inefficiency can be reduced, on average, by 4.1 percent for every 1 percentage increase in the available time for pumping (salinity reduction). The efficiency scores were calculated for each farm individually and averaged for each irrigation zone (Table 5). The temporal and spatial distribution of efficiencies is quite interesting and corresponds to expectations. We calculate an average TE of 81 % for the WSp and 74 % for the SA season (weighted according to the 2010 areas of the irrigation zones). As expected the efficiencies are lower in the SA season. This season presents a greater challenge to farmers in terms of irrigation management (water availability and salinity intrusion events). This seems to have severe impacts on TE in the affected irrigation zones (see Fig. 4).

This is most evident in the Tu Cau and Co Co irrigation zones, where we have a dramatic drop on efficiency from 76.1 % in the WSp to 50.5 % in the SA season. These irrigation zones are very much affected by salinity intrusion in the SA season. Salinity intrusion reduces water availability and difficulties irrigation management, which is expected to impact on yields and efficiencies. The Tu Cau irrigationIrrigation zone receives water from the Vinh Dien river, which is extremely affected by salinity intrusion. The Co Co irrigation water comes from small lakes along coastal line, which are feed by the return flows of Tu Cau zone. Thus, the salinity intrusion in the Co Co is highly correlated with salinity intrusion in the Tu Cau irrigation zone (see Fig. 4). Technical efficiency in the Cam Sa and Vienh Dien irrigation zones is also very much impacted by salinity intrusion in the SA season although not as much as in Tu Cao and Co Co. Salinity intrusion in the affected irrigation zones is controlled at the pumping station level by constraining the available pumping hours to the irrigation zone. During events of salinity intrusion irrigation is stopped, i.e., the available water (available pumping time) for the respective irrigation zone is reduced.[6] Based on several years of operation data for the individual pumping stations, we constructed a risk index, the variable

[6]Events with salinity levels over the local threshold of 1.0 g $L^{-1}$.

*pumping*, expressing the available pumping hours under the impacts of saltwater intrusion (see Table 6). As the duration of available pumping hours is less than 3, pumping stations cannot operate to provide water for irrigation. As the duration of "fresh" water ranges from 4 to 12 h, it is unable to provide sufficient water. As the duration of "fresh" water ranges from 12 to 24 h, pumping stations can operate to provide sufficient water with care on unpredictable high saltwater intrusion.

We tested the hypotheses regarding the negative impact of salinity intrusion on technical efficiency by including the proxy variable *pumping* in the SA stochastic frontier model specification (see SF2 in Table 4). As it can be seen, the variable *pumping* is highly significant in the SA season and also negative, meaning that an increase in the available pumping hours (salinity reduction) would reduce technical inefficiency. Dong Quang and Bich Bac show very low efficiency levels as well. For Dong Quang these low levels seem to be related to salinity intrusion (see the east part of the irrigation zone along the Vinh Dien river in Fig. 4). For Bich Bac the low efficiency levels are most probably a consequence of the strong urbanization in the area and consequent land fragmentation. An analysis on the rates of land conversion in the last 15 years sheds some light on the causes for Bich Bac (see Table 7).

High-resolution land use maps (scale of 1:2000 and 1:10,000) in 2000 and 2010 were used to detect spatial changes in paddy rice and other annual crops during this period. The conversion of planted areas of paddy rice and other annual crops to

**Table 6** Available pumping time for underthreshold salinity intrusion

| Level | Hours |
|---|---|
| High | <3 |
| Medium | 4–12 |
| Low | 12–24 |
| No | >24 |

**Table 7** Land conversion changes (%) between 2000 and 2015

| | Paddy | Other annuals |
|---|---|---|
| Ai Nghia | −2 | 0 |
| An Trach | −3 | −3 |
| Bich Bac | −64 | −45 |
| Cam Kim | 0 | 0 |
| Cam Sa | −8 | −5 |
| Cam Van | −2 | −2 |
| Co Co | −48 | −64 |
| Dong Ho | −1 | 0 |
| Dong Quang | −9 | −4 |
| Go Noi | −1 | 0 |
| Tu Cau | −26 | −10 |
| Vinh Dien | −1 | −3 |
| Xuyen Dong | −2 | −5 |

nonagricultural land uses for the 2010–2015 period is determined through updating urban development maps of both Quang Nam and Da Nang. In Bich Bac there were major land conversions to urban uses and road construction in the last 15 years, where ca. 64 % of paddy fields and 45 % of land for annual crops were lost. The urban development and road construction strongly impact on the irrigation system by disrupting connectivity, fragmenting and isolating large areas of paddy fields and making irrigation management extremely difficult (personal communication with the Irrigation Management Company in Quang Nam and own observations). The high land conversion rates impact on efficiencies as seen in the importance of factors like the scale of operations and fragmentation delivered in Tables 4 and 5. The Tu Cau and Co Co irrigation zones are other areas very much affected by land conversion and in this case also affected by salinity. Both factors impact on efficiency with dramatic consequences for the affected communes as can be seen in Tables 4 and 5 (Fig. 5).

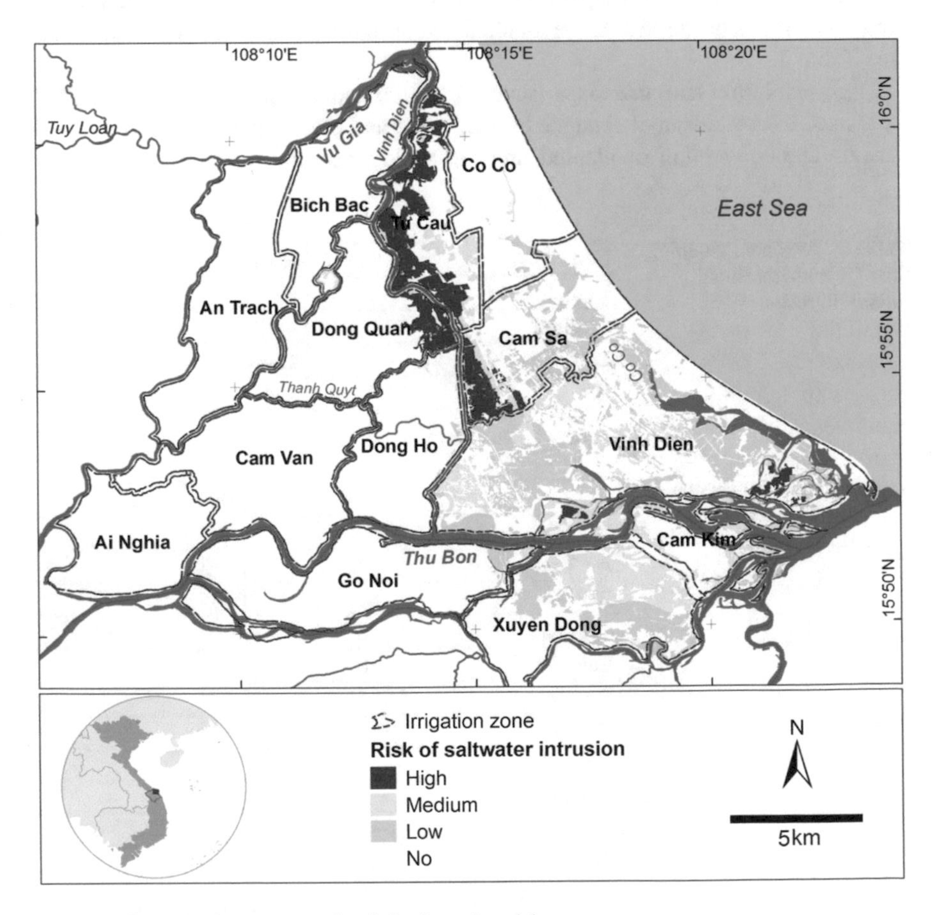


**Fig. 5** Delta irrigation zones and salinity intrusion risk

## Discussion

The results show that despite the low net benefits and added values of rice production, rice is still by far the dominant crop of the VGTB delta region. Rice plays a crucial role in families' food security and has an invaluable cultural role in the region. Furthermore, it is also a very robust crop, and by far the crop that demands the lowest labor inputs. The temporal distribution of labor demands is another advantage of rice production. Most labor is required at the beginning of the season for land preparation and sowing and again at the end of the season for harvesting and postharvest activities. During the season, the rice crop demands almost no labor from the farm household. This is a great advantage because the household can engage in nonfarm income activities. Maize is the second most important crop but shows very low net benefits and added value per man-day. Maize cultivation seems to be justifiable only because of the major role that maize plays in livestock feeding in the region (more than 40 % of own production is destined for feeding own animals). The market situation for maize is nevertheless changing rapidly in the region. The main driving factor is the growing demand for livestock fodder and the settlement of an international maize fodder processing company in the Dien Ban district (personal communication). These developments will surely bring some changes in the actual prices paid for maize. Groundnut is a very attractive crop not only because of its stable market demand and price, but also because farmers can process it into cooking oil if prices are not attractive enough. Despite the high net benefits of cash crops (especially vegetables) and increasing pressure on farmers for changing cropping patterns toward high value crops, farmers are reluctant to do so and areas used for high value crops are still very low. The main cause of this wise behavior of farmers lies in the very high labor demands and an added value per man-day that fall behind expectations. The households' labor force available for on-farm work is ca. 1.9 persons (for details see Pedroso et al. 2017). This labor force after conversion to equivalent man-days translates to only 1.6 man-days.[7] The typical farm size in the region is 0.4 ha, or 4000 $m^2$. Let us imagine a farm-household growing year-round vegetable on 1000 $m^2$, with an average net benefit of ca. 210 M VND $ha^{-1}$, and a labor requirement of 9.1 man-days $ha^{-1}$ $Gday^{-1}$. This means that the family would need 0.91 man-day/1000 $m^2$/Gday. The vegetable labor demand would require about 60 % of the total available on-farm labor power (See Jansen et al. 1996; Huong et al. 2013a, b). If farmers increase the acreage of vegetables to 2000 $m^2$, they would need 1.8 man-days/2000 $m^2$/Gday, which now clearly exceeds the available labor power of 1.6 man-days equivalent available. Even though farmers would have enough labor power for cultivating vegetables on 1000 $m^2$, those growing vegetables only use 700 $m^2$ (1.4 sao) on average (for details see: Pedroso et al. 2017). The reasons for farmers' reluctance to diversify toward cash crops, in particular vegetables are nevertheless not exclusively related to the very high labor inputs. Many other reasons discourage farmers, e.g., high biotic and abiotic risks like high disease

[7]See Norman (1973) for an explanation of the weighting scheme.

and pest incidences, climatic conditions, lack of knowledge about vegetables technologies, or price risks. Moreover, the local value chains are rudimentary and do not ensure a reliable and fast marketing of highly perishable products, in particular to Da Nang city. The very small scale of vegetable production is another problem that causes lack of marketing control by the farmers.

The production of rice in the delta of the VGTB shows TE levels of 81 % in the WSp season and 74 % in the SA season, a yearly average of 78 %. This level of TE is quite lower than the value of 86 % currently found for the Mekong River Delta (MRD) and Red River Delta (RRD) (see Kompas et al. 2012). These authors also used stochastic production frontier (SFA) methods with a Cobb–Douglas (CD) production function specification. The authors use an own farm survey from 2004 for the Mekong River Delta (MRD) and Red River Delta (RRD). The main hypothesis of the study was that land consolidation in Vietnam is a challenge for increasing productivity and efficiency. The authors isolated the effects of farm size and average plot size as representatives for land fragmentation, as well as soil quality, irrigation quality and education level rankings. The total farm size and the average plot size of rice farmers are found to have a significant positive impact on efficiency, i.e., increasing farm and plot average size will increase farm efficiency. Our results for TE in the delta of the VGTB river basin also show similar results regarding the scale of production (the rice acreage) and the land fragmentation problem (the number of rice plots on the farm).

## Conclusions

There is an increasing pressure on rice cultivation in the region, given high rates of land conversion for urbanization and industrial purposes, and also high competing demand for water in the SA season. There is also a broad discussion in favor of reducing rice acreages and intensifying agriculture toward high value crops like vegetables. The high net benefits achieved with vegetable production are known to the general public and decision makers in the region. Information on net benefits alone is however asymmetric, and in this way the public debate is biased. Current vegetable cultivation in the region, when analyzed in terms of added value per man-day$^{-1}$, is no longer as attractive as it seems at first glance. The very high labor demands for vegetables in the region do not allow the very high net benefits to translate into high added value per ha and man-day. The present study shows that farmers keep on holding to rice as the main crop and this decision is wise, and can be understood against the background of lack of on-farm labor force in the family, aging, increasing nonfarm income opportunities, and the lower than expected added value per man-day in respect of vegetables.

We think that before farmers are willing to take the high risks and increase their vegetable acreages, an effort must be made to improve labor efficiencies, improve vegetable technologies, reduce marketing risks, and to solve the problem of small scales of production. Regarding the latter, farmers could organize themselves into producer organizations for better bargaining positions in input and output markets, and also allow a better development of integrated production and marketing value chains. We see the need for a more gradual diversification toward high value vegetables.

Against this background and regarding rice, it is inevitable that there must be an increase in TE if rice should further play a determinant role in food security and conservation of cultural values of rural populations. The TE of rice production in the delta of the VGTB river basin is with 78 % quite lower than the 86 % found for the MRD and RRD regions. Efficiency can greatly be improved in the region if policy measures are undertaken for increasing the scale of production and also for land consolidation. The intrusion of salinity is a major problem for some of the irrigation zones and innovative adaptation methods are needed in rice irrigation scheduling and irrigation engineering for increasing technical efficiency of rice production in the region.

## References


Battese GE, Coelli TJ (1995) A model for technical inefficiency effects in a stochastic frontier production function for panel data. Empirical Econ 20:325–332

Caudill SB, Ford JM (1993) Biases in Frontier estimation due to heteroscedasticity. Econ Lett 41 (1):17–20

Chung NT, Jintrawet A, Promburom P (2015) Impacts of seasonal climate variability on rice production in the central highlands of Vietnam. Agric Agric Sci Procedia 5:83–88

Ha DT, Thao TD, Khiem NT, Trieu MX, Gerpacio RV, Pingali PL (2004) Maize in Vietnam: production systems, constraints, and research priorities. CIMMYT, Mexico, D.F

Hadri K (1999) Estimation of a doubly heteroscedastic stochastic frontier cost function. J Bus Econ Stat 17:359–363

Hansen H, Nguyen T (eds) (2007) Market policy and poverty reduction in vietnam. Vietnam academy of social sciences. Hanoi, Vietnam Culture and Information Publishing House

Huong PTT, Everaarts AP, Neeteson JJ, Struik PC (2013a) Vegetable production in the Red River Delta of Vietnam. I. Profitability, labour requirement and pesticide use. NJAS-Wageningen J Life Sci 67:27–36

Huong PTT, Everaarts AP, Neeteson JJ, Struik PC (2013b) Vegetable production in the Red River Delta of Vietnam. II. Profitability, labour requirement and pesticide use. NJAS-Wageningen J Life Sci 67:37–46

Jansen HGP, Midmore DJ, Binh PT, Valasayya S, Tru LC (1996) Profitability and sustainability of peri-urban vegetable production systems in Vietnam. Neth J Agric Sci 44:125–143

Kodde DA, Palm F (1986) Wald criteria for jointly testing equality and inequality restrictions. Econometrica 54(5):1243–1248

Kompas T, Che NC, Nguyen HTM, Nguyen HQ (2012) Productivity net returns, and efficiency: land and market reform in vietnamese rice production. Land Econ 88(3):478–495

Kumbhakar SC, Lovell KCA (2000) stochastic frontier analysis. Cambridge University Press, Cambridge

Linh Vu Hoang (2012) Efficiency of rice farming households in Vietnam. Int J Dev Issues 11 (1):66–73

MONRE (Ministry of Natural Resources and Environment) (2009) Climate change, sea level rise scenarios for Vietnam, Hanoi. Available at: http://vgbc.org.vn/vi/nc/169-climate-change-sea-level-rise-scenarios-for-vietnam-2009

Norman DW (1973) Methodology and problems of farm management investigations: experiences from northern Nigeria African Rural Employment Paper 8. Department of Agricultural Economics, Michigan State University, East Lansing, Michigan, USA 47 pp

Pedroso R et al (2017) Cropping systems in the Vu Ghia Thu Bon river basin, Central Vietnam: On farmers' stubborn persistence in predominantly cultivating rice, NJAS—Wageningen J Life Sci http://dx.doi.org/10.1016/j.njas.2016.11.001

PPC Da Nang (People's Committee of Da Nang City) (2011) Report on master plan of socio-economic development of Da Nang City in 2020 (Vietnamese language), Da Nang, Vietnam

PPC Quang Nam (Quang Nam Provincial People's Committee) (2012) Socio-economic development of Quang Nam Province in the last 15 years (1997–2011) (Vietnamese language). Tam Ky, Vietnam, p 97

Quang PN, van Westen ACM, Zoomers A (2014) Agricultural land for urban development: the process of land conversion in Central Vietnam. Habitat Int 41:1–7

Ravallion M, van de Walle D (2001) Breaking up the collective farm: welfare outcomes of Vietnam's massive land privatization. World Bank Policy Research Working Paper 2710, The World Bank, Washington, DC

Van Hung P, MacAulay TG, Marsh SP (2007) The economics of land fragmentation in the north of Vietnam. Aust J Agric Resour Econ 51(2):195–211

Viet TQ (2014) Estimating the impact of climate change induced salinity intrusion on agriculture in estuaries—the case of Vu Gia Thu Bon, Vietnam. Doctoral dissertation. RUHR University Bochum, Germany, p 175

Wang H, Schmidt P (2002) One-step and two-step estimation of the effects of exogenous variables on technical efficiency levels'. J Prod Anal 18:129–144 (Kluwer Academic Publishers)

World Bank (2003) Vietnam: delivering on its promise: vietnam development report, 2003. Hanoi

World Bank in collaboration with the Asian Development Bank (2011) Recognizing and reducing corruption risks in land management in Vietnam. Hanoi

World-Bank, Inflation, consumer prices (annual %) (2016). http://data.worldbank.org/indicator/FP.CPI.TOTL.ZG

The National Publishing House—Su That (2015) MapVietnam. At www.worldbank.org/mapvietnam/ (2016) Inflation, consumer prices (annual %). At: http://data.worldbank.org/indicator/FP.CPI.TOTL

# Measuring GHG Emissions from Rice Production in Quang Nam Province (Central Vietnam): Emission Factors for Different Landscapes and Water Management Practices

**Agnes Tirol-Padre, Dang Hoa Tran, Trong Nghia Hoang, Duong Van Hau, Tran Thi Ngan, Le Van An, Ngo Duc Minh, Reiner Wassmann and Bjoern Ole Sander**

**Abstract** This study comprises greenhouse gas (GHG) emission measurements on rice fields in the Vu Gia/Thu Bon Basin in Central Vietnam, as part of an interdisciplinary research project. The experiments were conducted in the delta lowland (DL) and hilly midland (HM), over three seasons (summer–autumn in 2011 and 2012; winter–spring season in 2012) with two water management treatments namely continuous flooding (CF) and alternate wetting and drying (AWD). GHG emissions were dominated by methane ($CH_4$) emissions showing large difference among seasons, whereas nitrous oxide ($N_2O$) emissions were negligible and irrelevant in the overall carbon footprint. However, temporal patterns were not conclusive over the entire observation period. The observed seasonal $CH_4$ emission rates ranging from 83 to 696 kg $CH_4$ per ha were relatively higher compared to other field studies and can, at least in part, be attributed to organic amendments applied in accordance to farmers' practice. The practice of AWD reduced $CH_4$ emissions significantly ($P < 0.0001$) in all seasons, corresponding in average to 71 % of the emissions from CF. On the other hand, AWD had no significant effect on $N_2O$ emissions. The average seasonal $CH_4$ emission in the DL (420 kg $ha^{-1}$) was also significantly higher than in the HM (206 kg $ha^{-1}$). Compared with IPCC

A. Tirol-Padre (✉) · N.D. Minh · R. Wassmann · B.O. Sander
International Rice Research Institute (IRRI), Los Baños, Philippines
e-mail: a.padre@irri.org

D.H. Tran · T.N. Hoang · D.V. Hau · T.T. Ngan · L.V. An
College of Agriculture and Forestry, Hue University, Hue, Vietnam

N.D. Minh
Soil and Fertilizers Research Institute, Duc Thang Ward, North Tu Liem District, Hanoi, Vietnam



A. Nauditt and L. Ribbe (eds.), *Land Use and Climate Change Interactions in Central Vietnam*, Water Resources Development and Management,
DOI 10.1007/978-981-10-2624-9_7

default values, this data set from Vietnamese rice fields indicates a higher emission level and Scaling factor for AWD. The average grain yield across all sites and seasons increased by 4 % ($P < 0.0002$) relative to CF with the practice of AWD.

## Introduction

Rice production is the second largest anthropogenic source of GHG methane (IPCC 2007), with global warming potential (GWP) that is 25 times higher than carbon dioxide ($CO_2$). Therefore, sustainable rice farming must aim to reduce $CH_4$ emissions, especially against the backdrop of increasing food production to keep pace with population growth. Carbon dioxide, methane, and nitrous oxide are the key GHG that contribute to global warming at 60, 15, and 5 %, respectively (IPCC 2007). Global and regional estimates of GHG emissions from rice paddy fields vary greatly with assumptions made on the importance of different factors affecting emissions.

Rice is Vietnam's main food product, accounting for about 50 % of the gross production of all food crops. Vietnam being the world's fifth largest rice producer and one of the top-three rice exporters has now become a sustainable rice supplier. However, rice cultivation is also the biggest source of agricultural $CH_4$ emission in Vietnam. According to the 2nd National Communication to the UNFCCC (MONRE 2010), emission from rice cultivation accounts for 57.5 % of agricultural GHGs or 26.1 % of national source strength of all GHGs. However, the First Biennial Update Report (MONRE 2014) estimated the national $CH_4$ emission from paddy rice fields at 50.5 % of agriculture and 18.1 % of all GHG emissions in Vietnam.

While these figures corroborate the significance of rice production within the overall GHG budget in Vietnam, they also illustrate the range of uncertainty inherent in these estimates. Due to lack of actual field observations at that time, these estimates have been computed using global default values as Emission Factors (IPCC Tier 1 approach). In the meantime, several Vietnamese groups have conducted field experiments (Vu et al. 2015; Pandey et al. 2014), but the existing data base on GHG emissions remain scarce. As one of the first published experiments, Minh et al. (2015) presented some data from the same field experiment shown here for a comparison of simulated data and field observations. To the best of our knowledge, this study represents the first published data set on GHG emission from Vietnamese rice production outside the Red River Delta that covers a multi-season period and that incorporates different management techniques.

This study on GHG emissions from rice fields was part of a larger research project titled "Land use and climate change interactions in Central Vietnam" (LUCCi). This project deployed interdisciplinary research methods to consider both natural and social science aspects for optimized land use and water resources management strategies in the Vu Gia/Thu Bon (VGTB) River Basin. This extended basin is the key economic and agricultural regions in the Central Coast region of Vietnam.

The basin accounts for 220,040 ha of agricultural land. Rice is by far the most important crop with 120,000 ha (55 % of agriculture land) of cultivated area within the basin (GSO-VN 2016). At national scale, this corresponds to 1.7 % of the total rice area of Vietnam. Rice is typically grown in two croppings per year, i.e. winter–spring (WS), grown from January to April and summer–autumn (SA), from June to September. Rice yields in these seasons account for 5.17 t/ha (WS) and 3.98 t/ha (SA), respectively (data from GSO-VN 2016 for Quang Nam Prov.). This is about 15 % (WS) and 11 % (SA) less than the national yield average of the respective seasons.

The objectives of this study were fully aligned with the LUCCi project's primary goal which was, "*to provide a sound future land use management framework that links climate change mitigation—through the reduction of GHG emissions—with adaptation strategies to secure food supply in a changing environment*". Anchoring on this goal, this study has analyzed the predominant land-use practices such as rice production, with emphasis on the use of water resources. Moreover, the water-saving technologies studied in these experiments covered another thrust of the LUCCi project, which is to render rice production more adaptive to possible climate change impacts. This in-depth analysis of GHG emissions from rice production forms a key component for the envisioned carbon-optimized land and water use strategies for Central Vietnam. At the same time, the results of this study will allow major conclusions for similar land use systems in Vietnam as well as Southeast Asia. Through this LUCCi sub-project, improved irrigation practices were assessed in terms of optimizing mitigation alongside adaptation potential.

As rice is absent in the mountainous sections of the basin, we selected two locations for our experiment representing hilly midland (HM in Dai Quang) and

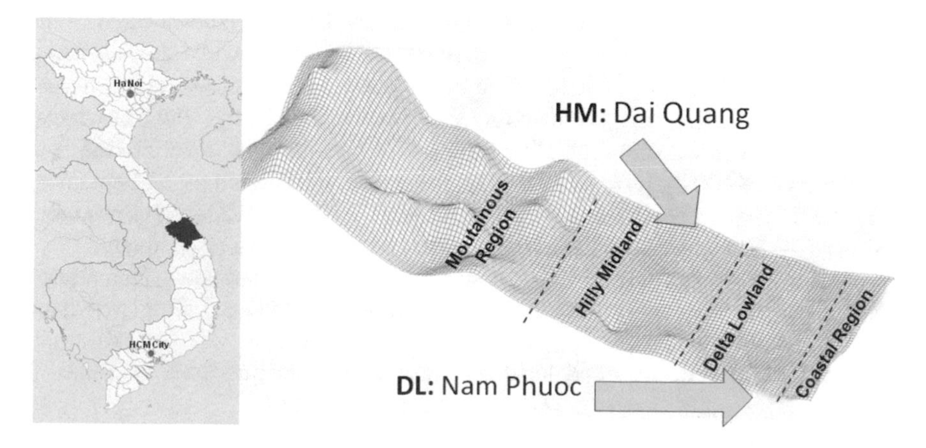


**Fig. 1** Location, geography and topography of the study area

delta lowland (DL in Nam Phuoc), respectively (Fig. 1). The range of emissions for the rice growing environments that can be found in the southern part of Vietnam was the first empirical data. While Central Vietnam has distinctive features, the rice growing environments comprise many commonalities with the Mekong Delta. Apart from the climatic situation, the VGTB basin comprises a delta that is quite similar to the Mekong in terms of bio-physical conditions. However, the Mekong Delta has been going through intensive development of irrigation infrastructure over recent decades while such developments are less pronounced in Central Vietnam. In turn, large sections of the VGTB coastal area are affected by tidal dynamics in terms of salinity and flooding risks. Although the two experimental sites of this study are too far from the coastline for immediate impacts at this point, the effects of sea level rise will undoubtedly lead to an aggravating water scarcity in the region that will trigger a need for water-saving irrigation practices.

## Site and Experimental Description

The experiments were conducted on two research stations in Quang Nam province (South Central Coast region) in the two districts Duy Xuyen (Nam Phuoc commune, delta lowland area) and Dai Loc (Dai Quang commune, hilly midland area) in three consecutive cropping seasons from June 2011 to August 2012.

The station in Nam Phuoc (15°50′ 56.3″N, 108°16′ 37.1″E) is located at the center of the VGTB delta, which is only about 8 km from the coast while the station in Dai Quang (15°53′ 20.9″N, 108°03′ 23.2″E) is located more upstream the Vu Gia about 25 km from the coast. Annual rainfall in the province averages 2000–2500 mm. The major part of annual precipitation is observed between October and February, which characterizes the rainy season. The soil at Nam Phuoc and Dai Quang station can be characterized as Fluvisol. Detailed soil parameters can be obtained from Table 1.

At both stations, experiments were set-up in randomized block designs. The main factor (block) was the water management (CF or AWD) while the sub-factor comprises three fertilizer treatments (farmers practice (FP), site-specific nutrient management (SSNM) or fertilization as recommended by the provincial Department of Agriculture). The Rice variety HT1 was directly sown into the field at a seed rate of 100 kg $ha^{-1}$. Accordingly, the rice was harvested on 14 September 2011 and 15 September 2011, 19 April 2012 and 20 April 2012, and 9 September 2012 and 10 September 2012 for Nam Phuoc and Dai Quang, respectively.

Stubbles were incorporated into the soil during land preparation which started about 11 days before sowing with a first plowing. The final puddling was done two days before sowing.

All details of soil, crop and fertilizer management are presented in Tables 1 and 2.

**Table 1** Site characterization and timings of field operations

| **Site Location** | | | | | | | |
|---|---|---|---|---|---|---|---|
| Commune | Nam Phuoc | | | Dai Quang | | | |
| District | Duy Xuyen | | | Dai Loc | | | |
| Latitude | 150 50' 56.3"N | | | 150 53' 20.9"N | | | |
| Longitude | 1080 16' 37.1"E | | | 108003' 23.2"E | | | |
| **Soil Properties:** | | | | | | | |
| %Clay | 22.47±1.13 | | | 27.19±3.82 | | | |
| %Sand | 22.10±1.36 | | | 19.10±0,82 | | | |
| %Silt | 54.42±1.54 | | | 53.69±3.87 | | | |
| Bulk density | 2.46 | | | 2.48 | | | |
| pH | 4.48±0.18 | | | 4.59±0.15 | | | |
| OC (%) | 1.15±0.07 | | | 1.19±0.13 | | | |
| Total N (%) | 0.10±0.004 | | | 0.11±0.009 | | | |
| Available P mg $P_2O_5$ 100g$^{-1}$ | 1.50±0.22 | | | 1.46±0.19 | | | |
| Available K mg $K_2O$ 100g$^{-1}$ | 1.51±0.25 | | | 1.54±0.27 | | | |
| **Timing of Field Operations** | | | | | | | |
| **Season** | **SA 2011** | **WS 2012** | **SA 2012** | **SA 2011** | **WS 2012** | **SA 2012** | **Remarks** |
| Plowing dates | May 28 | Dec 15 | May 20 | May 30 | Dec 15 | May 25 | 2x per season; plow depth =20cm |
| | June 6 | Dec 30 | May 29 | Jun 7 | Dec 30 | Jun 1 | |
| Sowing date | Jun 8 | Jan 1 | Jun 1 | Jun 9 | Jan 2 | Jun 2 | |
| Herbicide spraying | Jun 10 | Jan 4 | Jun 2 | Jun 11 | Jan 3 | Jun 3 | Sofit 300EC (Pretilachlor) |
| Harvest date | Sept 14 | Apr 19 | Sept 9 | Sept 5 | Apr 20 | Sept 10 | |

## *Greenhouse Gas Sampling and Analysis*

Fluxes of $CH_4$ and $N_2O$ were measured weekly during mid-morning using the static chamber method. The first sample was taken about two weeks after sowing. The last sampling was conducted about two weeks before harvest.

A plastic base with a diameter of 50 cm was inserted about 10 cm into the soil at a location representing the average plant density of a given plot. These bases were installed at least a day before sample collection and were left in the field throughout the growth period. The base height and water depth inside the frames were measured at each gas sampling time.

**Table 2** Management details; for acronyms of sites and treatments see in section "Site and Experimental Description"; acronyms of applied fertilizers: NPK = fertilizer mix; U = urea; KCl = potassium chloride; $P_2O_5$ = phosphorus pentoxide; SP = super phosphate

| Applied Fertilizers | Season/ Treatment | Site: DL | | | Site: HM | | |
|---|---|---|---|---|---|---|---|
| | | FP | Recommended | SSNM | FP | Recommended | SSNM |
| N fertilizer (kg N ha$^{-1}$) | SA2011 | NPK: 5.9, 5.9, 2.95<br>U: 18.4, 18.4 | NPK: 4.4, 4.4<br>U: 36.8, 18.4, 18.4 | NPK: 4.4<br>U: 36.8, 29.9, 29.9 | NPK: 7.4, 8.8, 5.9<br>U: 18.4, 18.4 | NPK: 4.4, 4.4<br>U: 36.8, 18.4, 18.4 | NPK: 4.4<br>U: 36.8, 27.6, 23 |
| | WS2012 | NPK: 5.9, 5.9, 2.95<br>U: 18.4, 18.4 | NPK: 4.4, 4.4<br>U: 36.8, 18.4, 18.4 | NPK: 4.4<br>U: 36.8, 29.9, 29.9 | NPK: 7.4, 8.8, 5.9<br>U: 40, 40 | NPK: 4.4, 4.4<br>U: 80, 40, 40 | NPK: 4.4<br>U: 80, 60, 50 |
| | SA2012 | NPK: 5.9, 8.5, 4.4<br>U: 23, 23, 36.8, 18.4 | | NPK: 5.9, 8.5, 2.9 | NPK: 7.4, 2.9<br>U: 36.8, 18.4, 18.4 | | NPK: 4.4<br>U: 36.8, 18.4, 18.4 |
| K fertilizer (kg K2O ha$^{-1}$) | SA2011 | KCl: 24, 24<br>NPK: 3.8, 3.8, 1.9 | KCl: 36, 36<br>NPK: 2.9,2.9 | KCl: 36, 36<br>NPK: 2.9 | KCl: 24<br>NPK: 4.8, 5.8, 3.8 | KCl: 36, 36<br>NPK: 2.9, 2.9 | KCl: 36, 36<br>NPK: 2.9 |
| | WS2012 | KCl: 24, 24<br>NPK: 3.8, 3.8, 1.9 | KCl: 36, 36<br>NPK: 2.9, 2.9 | KCl: 36, 36<br>NPK: 2.9 | KCl: 40<br>NPK: 4.8, 5.8, 3.8 | KCl: 60, 60<br>NPK: 2.9, 2.9 | KCl: 60, 60<br>NPK: 2.9 |
| | SA2012 | KCl: 2.8, 33.6<br>NPK: 3.6, 5.5, 2.7 | | KCl: 2.8, 33.6<br>NPK: 3.6, 5.5, 2.7 | KCl: 22.4,22.4<br>NPK: 4.5, 1.8 | | KCl: 44.8, 33.6<br>NPK: 2.7 |
| P fertilizer (kg P2O5 ha$^{-1}$) | SA2011 & WS2012 | NPK: 2.0, 2.0, 1.02 | NPK: 1.5, 1.5<br>SP: 64 | NPK: 1.5<br>SP: 64 | NPK: 2.6, 3.1, 2.0 | NPK: 1.5, 1.5<br>SP: 64 | NPK: 1.5<br>SP: 64 |
| | SA2012 | NPK: 2.3, 3.3, 1.7 | | NPK: 2.3, 3.3, 1.7<br>SP: 90 | NPK: 2.9, 1.1 | | NPK: 2.7<br>SP: 90 |

(continued)

**Table 2** (continued)

| | | | | | | | |
|---|---|---|---|---|---|---|---|
| Lime (kg $ha^{-1}$) | SA2011 & WS2012 | 0 | 400 | 400 | 0 | 400 | 400 |
| | SA2012 | 0 | | 500 | 0 | | 0 |
| Microbial organic fert.[1] (kg $ha^{-1}$) | SA2011 & WS2012 | 0 | 650 | 650 | 0 | 1000 | 1000 |
| | SA2012 | 400 | | 0 | 250 | | 500 |
| Straw management | SA 2011 | Burning[2] | | | Burning[2] | | |
| | WS2012 & SA2012 | Straw from previous crop removed<br>Stubbles[3] from previous crop incorporated | | | Straw from previous crop removed<br>Stubbles[3] from previous crop incorporated | | |

[1] Commercial name of the microbial organic fertilizer is Song Gianh with Organic C = 15%, Humic acid = 2.50%, and $P_2O_5$ = 1.50%
[2] Burned rice straw from preceding crop was buried 1 month before the start of the SA 2011 cropping season
[3] 10–15 cm rice stubbles from SA 2011 crop were left in the field during four months of wet fallow and incorporated during land preparation for WS2012 crop; 40cm of rice stubbles from WS 2012 crop were left in the field during 1-2 months of dry fallow and incorporated during land preparation for SA 2012 crop

The gas collection chambers, fabricated from a plastic pail with a height of 70 cm (120 L volume) were equipped with a thermometer and a sampling port and a battery-operated fan was installed inside.

At the time of sampling, the gas collection chambers were placed on the trough of the bases and sealed with water. Gas samples inside the chambers were collected at 0, 10, 20, and 30 min after chamber closure using a 60-ml syringe fitted with a stopcock. A 6 mL glass vial was washed with ca. 40 mL of sample gas before the gas was stored in the vial under pressure.

The samples were analyzed with an SRI 8610C gas chromatograph (GC) at Hue University within one week. The GC was equipped with 63Ni electron capture detector (ECD) for analysis of $N_2O$ gas and a flame ionization detector (FID) for analysis of CH4. The columns for the analysis of $CH_4$ and $N_2O$ were packed with Porapak Q (50–80 mesh) and the carrier gas used for both, FID and ECD, was nitrogen ($N_2$). $CH_4$ and $N_2O$ standard gases were purchased from the company "Messer" in Ho Chi Minh City. Sampling modalities in this experiment follow common practices (Sander et al. 2014) and fulfilled the minimum requirements of good GHG sampling (Butterbach-Bahl et al. 2015).

Flux rates were computed based on the ideal gas law, using chamber air temperature values measured at the time of sampling. For calculating the total $CH_4$ and $N_2O$ amounts emitted for a sampling interval, the measurements taken in the morning were assumed to represent daily average flux rates. It was further assumed that flux changes between two consecutive sampling days as well as between sowing and the first sampling and the last sampling and harvest are linear. The emission rates on the day of sowing and harvest were set to "0". The emission factor (EF) was calculated based on the seasonal cumulative emission divided by the season length in days. The GWP for the cropping season was calculated using GWP ($CH_4$) = 25 and GWP ($N_2O$) = 298 for a 100-year time horizon (Forster et al. 2007).

## *Methane Emissions Under Continuous Flooding (CF) and Alternate Wetting and Drying (AWD) in Delta Lowlands Results*

The SAS Mixed Procedure with Tukey–Kramer test (SAS Institute 2004) was used to analyze the variances in the average $CH_4$ and $N_2O$ emission rates due to day of sampling and treatment (water and fertilizer) for each location per season, and also the variances in cumulative $CH_4$ and $N_2O$ emissions due to site, season, and treatment. Treatment (water) means for each site per season were compared using pairwise mean comparison (pdiff) and adjusted probabilities were obtained using the Tukey–Kramer test.

The seasonal emission patterns are displayed in Figs. 2 and 3 for $CH_4$ and Figs. 4 and 5 for $N_2O$. In the first instance, this data base can be used to compare

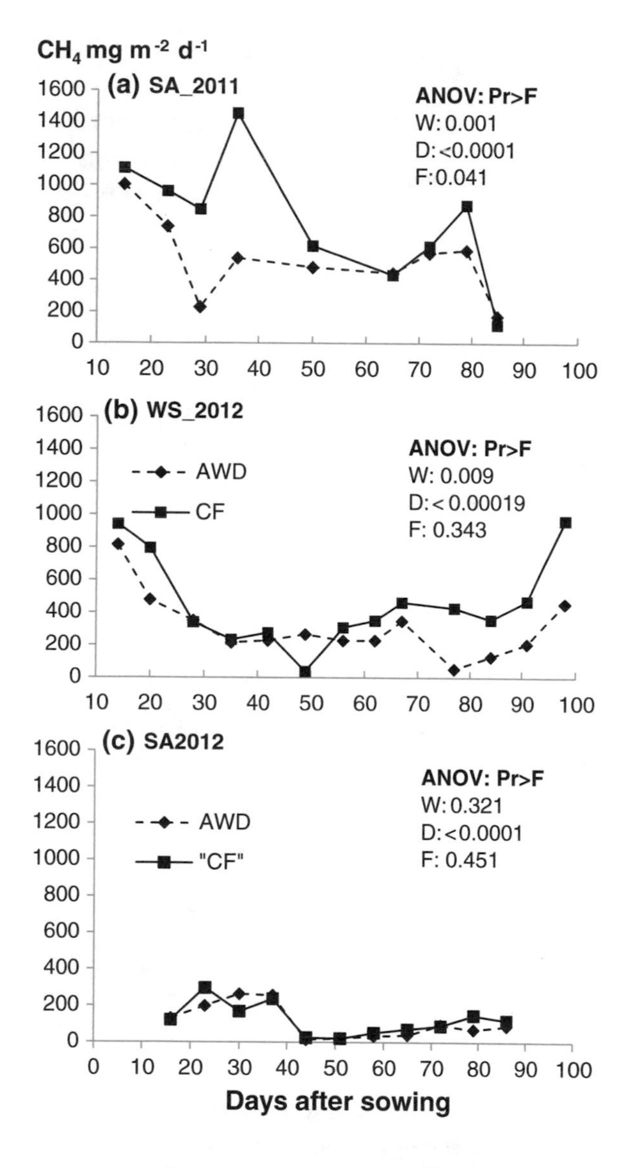


**Fig. 2** Daily $CH_4$ emissions under CF and AWD in DL during rice cultivation in **a** SA 2011 **b** WS 2012 and **c** SA 2012. Under ANOV, the significance probability value associated with the F Value (Pr > F) is shown for W (water management), D (day), and F (fertilizer). "CF" in SA 2012 denotes that CF was not strictly implemented in SA 2012

baseline emissions (under CF) at both sites. In all three seasons recorded for $CH_4$, the rice production in the DL was associated with higher levels of $CH_4$ emissions. This finding is also illustrated in Fig. 6 showing GWP per season. As for $N_2O$, emissions are generally at a very low level so there are no discernable differences among the sites. Even though our data set comprises only two seasons for $N_2O$ (in contrast to three for $CH_4$), the GWP diagram (Fig. 6) clearly shows the small contribution of this GHG to the overall carbon footprint of rice in the region.

In the next step, the data set can be used to define seasonal differences at a given site. However, the season-specific emission rates are rather heterogeneous. The

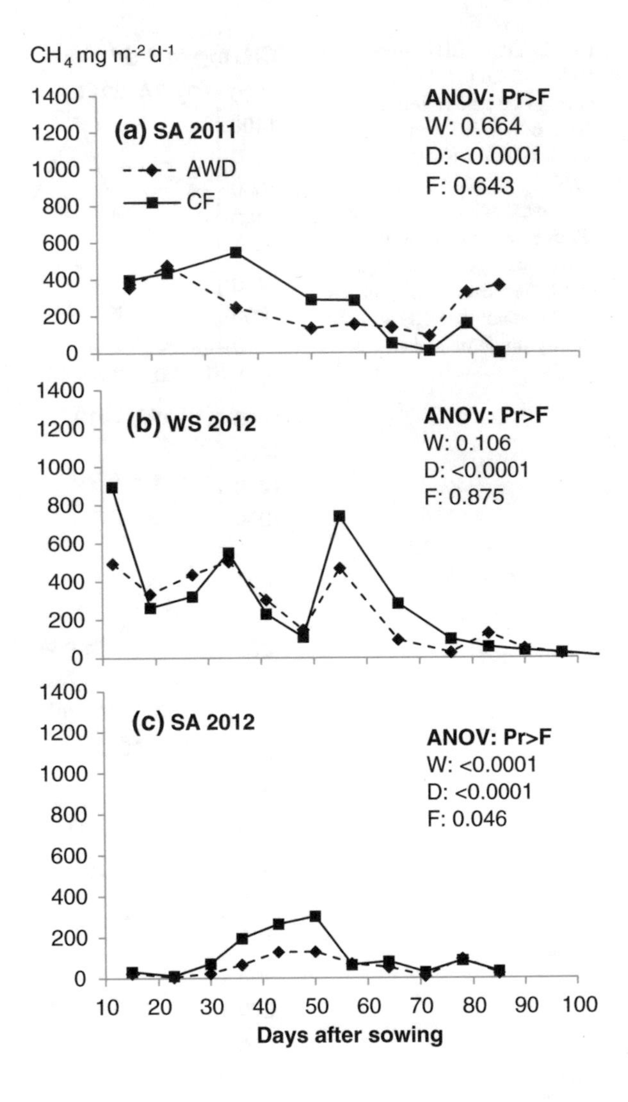


**Fig. 3** Daily $CH_4$ emissions under CF and AWD in HM during rice cultivation in **a** SA 2011, **b** WS 2012, and **c** SA 2012. Under ANOV, the significance probability value associated with the F Value (Pr > F) is shown for W (water management), D (day), and F (fertilizer)

initial summer–autumn season of 2011 yielded very high emissions at DL due to full incorporation of incompletely-burned rice straw by the farmer a month before sowing of the SA 2011 rice crop, which is the conventional practice of farmers in the area. Emissions were lower in the ensuing WS 2012, and SA 2012 when the rice straw was removed from the field. Therefore, SA 2012 could represent the baseline (under conventional practice). Emission Factor (EF) for this site and the SA 2012 would show the emission reduction without straw incorporation.

All field experiments encompassed a comparative assessment of CF and AWD. Practicing AWD consistently reduced emissions, which can also be corroborated by the GWP diagram (Fig. 6). Percentages ranged from small impacts corresponding to 87.3, 79.1 and 77.1 % of CF (HM in SA 2011, DL in SA 2012, and HM in WS

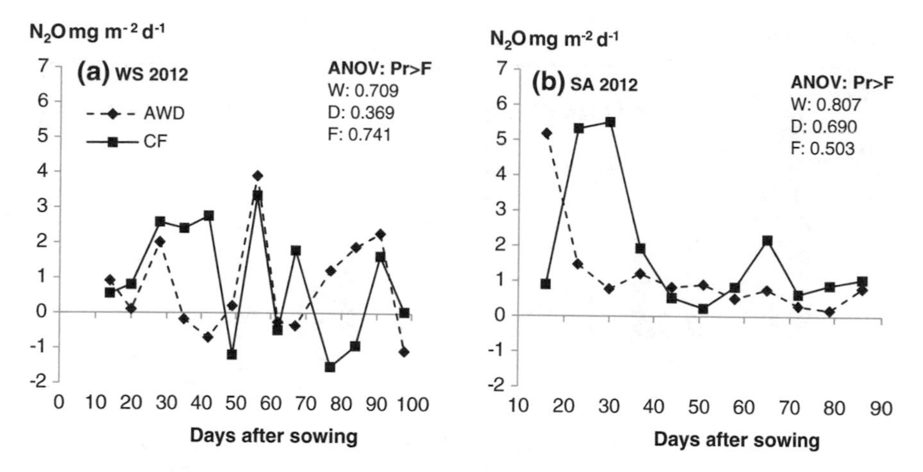


**Fig. 4** Daily $N_2O$ emissions under CF and AWD in DL during rice cultivation in **a** WS 2012, and **b** SA 2012. $N_2O$ emissions were not measured during SA 2011. Under ANOV, the significance probability value associated with the F Value (Pr > F) is shown for W (water management), D (day), and F (fertilizer)

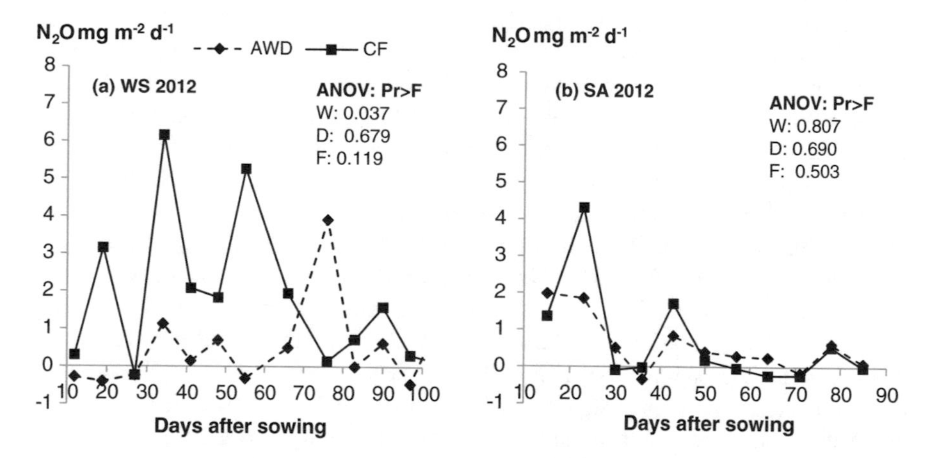


**Fig. 5** Daily $N_2O$ emissions under CF and AWD in HM during rice cultivation in **a** WS 2012, and **b** SA 2012. $N_2O$ emissions were not measured during SA 2011. Under ANOV, the significance probability value associated with the F Value (Pr > F) is shown for W (water management), D (day), and F (fertilizer)

2012 respectively) to significantly reduced levels corresponding to 54.5, 62.0, and 64.3 % (HM in SA 2012, DL in WS 2012, and DL in SA 2011, respectively). Higher $CH_4$ emissions under CF were observed in DL than in HM in all seasons. Greater emission reductions were also observed in DL as compared to HM in SA 2011 and WS 2012 but not in SA 2012 when the CF and AWD treatments were not effectively implemented in DL. The soil flood-water levels (measured every 2 days)

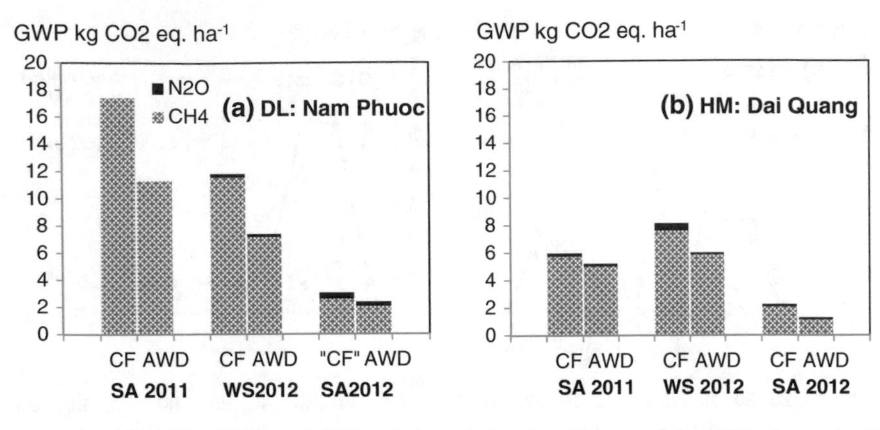


**Fig. 6** Total GWP from $CH_4$ and $N_2O$ emissions during three seasons of rice cultivation under CF and AWD in **a** DL and **b** HM. "CF" in SA 2012 denotes that CF was not strictly implemented in SA 2012

in DL did not vary between CF and AWD and indicated one effective drainage around the middle of the season in both treatments. The means of the measured $CH_4$ emission rates under CF (farmers' practice) ranged from 123 to 782 mg $m^{-2}$ $d^{-1}$ in DL (Fig. 2), and only 106–278 mg $m^{-2}$ $d^{-1}$ in HM (Fig. 3), during three consecutive seasons in 2011–2012. Lower measured values were obtained under AWD with means ranging from 110 to 530 mg $m^{-2}$ $d^{-1}$ in DL (Fig. 2) and 57 to 254 mg $m^{-2}$ $d^{-1}$ in HM (Fig. 3). Analyses of variance showed a significant effect of water management on the average $CH_4$ emission rates in DL during SA 2011 ($P = 0.001$) and WS 2012 ($P = 0.009$) but not during SA 2012 ($P = 0.321$). Methane emission reductions through AWD were observed during the early to mid-season stage (20–50 DAS) in SA 2011, and during the later rice growth stages (70–100 DAS) in WS 2012 (Fig. 2). In HM, water management significantly affected the $CH_4$ emission rate during SA 2012 ($P < 0.0001$) but not during SA 2011 ($P = 0.664$) and WS 2012 ($P = 0.106$). Day variations on the average $CH_4$ emission rates were highly significant in both sites ($P < 0.0001$). Fertilizer management significantly affected the $CH_4$ emission rate only in SA 2011 in DL ($P =$ 0.041) and in SA 2012 in HM ($P = 0.046$).

The means of measured $N_2O$ emission rates ranged from 0.5 to 1.8 mg $m^{-2}$ $d^{-1}$ overall sites, seasons, and water management. Day-to-day variations in measured $N_2O$ emission rates were not statistically significant but water management showed a significant effect during WS 2012 in HM, where $N_2O$ emission rates during the early to mid-season stages were higher under CF than AWD (Fig. 5).

Seasonal $CH_4$ emissions were consistently higher in DL during all 3 seasons ranging from 104 to 696 kg $ha^{-1}$ under CF and 82–450 kg $ha^{-1}$ under AWD), as compared to that in HM ranging from 83 to 305 kg $ha^{-1}$ under CF and 45–201 kg $ha^{-1}$ under AWD). Analyses of variance showed a highly significant ($P < 0.0001$) effect of site, season, and water management on seasonal $CH_4$ emissions, but fertilizer management had no significant effect ($P = 0.878$) (Table 3). AWD reduced

**Table 3** Seasonal $CH_4$ and $N_2O$ emissions from rice cultivation under CF and AWD during three seasons at two sites in Quang Nam Province, Central Vietnam

| Season | $CH_4$ (kg $ha^{-1}$ $season^{-1}$) | | | $N_2O$ (kg $ha^{-1}$ $season^{-1}$) | |
|---|---|---|---|---|---|
| | CF | AWD | % reduction by AWD | CF | AWD |
| *DL* | | | | | |
| SA 2011 | 696.3 | 450.0 | 35.3** | – | – |
| WS 2012 | 461.0 | 286.4 | 37.8* | 0.84 | 0.70 |
| SA 2012 | 103.7 | 82.0 | 20.9 | 1.50 | 1.07 |
| *HM* | | | | | |
| SA 2011 | 231.0 | 201.6 | 12.7 | – | – |
| WS 2012 | 304.6 | 234.8 | 22.9 | 1.83 | 0.50 |
| SA 2012 | 83.4 | 45.5 | 45.5 | 0.61 | 0.54 |
| *Analysis of variance* | | | | | |
| Source of variation | $CH_4$ | | | $N_2O$ | |
| | *F* value | | Pr > F | *F* value | Pr > *F* |
| Site (Si) | 39.74 | | <0.0001 | 0.41 | 0.524 |
| Season (S) | 47.75 | | <0.0001 | 0.29 | 0.596 |
| Water (W) | 17.83 | | <0.0001 | 3.39 | 0.073 |
| Fertilizer (F) | 0.13 | | 0.878 | 1.02 | 0.368 |
| Si × W | 4.22 | | 0.043 | 0.01 | 0.910 |
| S × W | 0.81 | | 0.447 | 0.30 | 0.587 |
| Si × S × W | 9.11 | | <0.0001 | 3.34 | 0.045 |

**, *Significant at the 1 and 5 % level, respectively, by Tukey test

**Table 4** Grain yields for three consecutive seasons under CF and AWD in DL and HM

| Season | Grain yield t $ha^{-1}$ | | | |
|---|---|---|---|---|
| | CF | | AWD | |
| | DL | HM | DL | HM |
| SA 2011 | 5.73 | 6.14 | 5.84 | 5.82 |
| WS 2012 | 5.48 | 6.12 | 5.97 | 6.27 |
| SA 2012 | 5.52 | 5.79 | 6.07 | 6.14 |
| Mean | 5.58 | 6.02 | 5.96 | 6.08 |
| CF Mean | 5.80 | | 6.02 | |
| *Analysis of variance* | | | | |
| Source of variation | | | *F* value | Pr > *F* |
| Water | | | 15.68 | 0.0002 |
| Site | | | 15.57 | 0.0002 |
| Season | | | 1.34 | 0.267 |

$CH_4$ emission by 13–45 % relative to CF. However, due to large variabilities among replicate measurements, statistically significant differences in seasonal $CH_4$ emissions between CF and AWD were obtained only in DL during SA 2011 and WS 2012 (Table 4). On the other hand, site, season, and water management showed no significant effects on seasonal $N_2O$ emissions (Table 3).

Grain yields were consistently higher under AWD than under CF except for one season in HM (SA 2011). Analysis of variance overall sites and seasons showed a highly significant ($P = 0.0002$) effect of water management and site on grain yield (Table 4). The average yields across sites and seasons showed a 4 % increase of grain yield under AWD relative to CF.

## Discussion

As indicated earlier, the database on GHG emission from rice fields in Vietnam is very scarce, hence the limited scope to set our data into the context of other comparable results. Pandey et al. (2014) and Vu et al. (2015) conducted experiments in northern Vietnam focusing on organic amendments.

In totality, the observed $CH_4$ emission rates in those studies were in a similar range as in our study, although the deviations among season and treatments were rather high in any given study. Vu et al. (2015) observed low emission rates in the spring rice season accounting to 70–140 kg $CH_4$/ha depending on the amendment. Emission rates in the summer season were about twice as high ranging from 160 to 290 kg $CH_4$/ha. In our study, we also recorded large differences among seasons. Emissions were generally high in the winter–spring season whereas the summer–autumn season showed large variations. As elaborated earlier, we assume that the high value for this season in 2011 may have been due to a transition in the soil after starting the field experiment. The study by Pandey et al. (2014) encompassed only one season. Emissions with farm yard manure were within the mid range of our records whereas emissions without organic amendments (control) were in the lower range of our records. Both treatments, however, showed higher $CH_4$ reductions under AWD management of around 70 %.

Figure 7 comprises a visual comparison of the emission data from this study with a total of 16 other field studies adapted from Sander et al. (2015). Each column represents the emission rates of AWD displayed by the lower dark section (see discussion below) as well as the emission rates of CF displayed by the total height. While the main criteria for selecting these studies was for illustration of AWD effects, the diagram also illustrates that the baseline emissions (CF) recorded in this study overall were in the range of those recorded in other East or Southeast Asian countries. Emissions from this study were high compared with field experiments conducted in India, although the very low level of emission rates in Indian rice systems have been previously documented and discussed (Wassmann et al. 2000; Pathak et al. 2003).

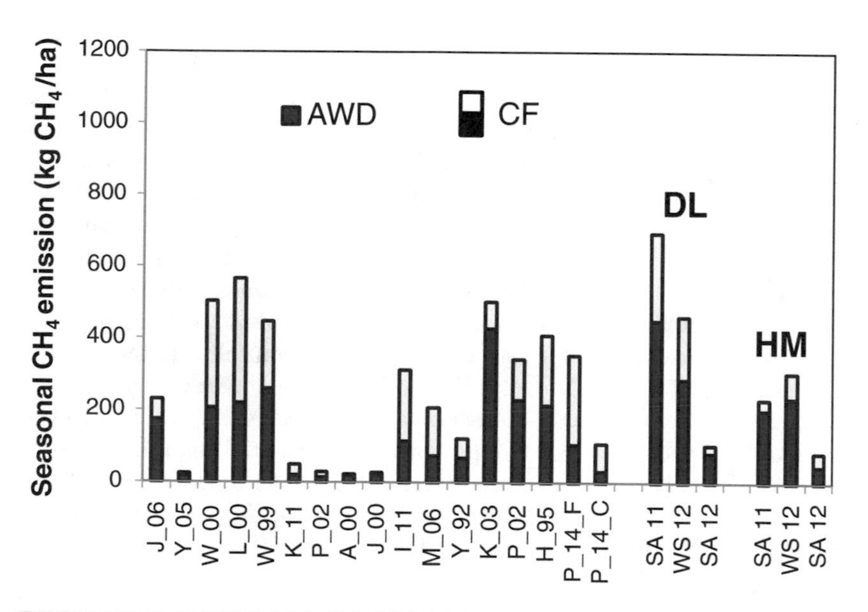


| | | | | | |
|---|---|---|---|---|---|
| J_06 | Jiao_2006 [CHI] | P_02 | Pathak_2002/2003 [IND] | K_03 | Kwun_2003 [KOR] |
| Y_05 | Yue 2005 [CHI] | A_00 | Adhya_2000 [IND] | P_02 | Park_2002 [KOR] |
| W_00 | Wang_2000 [CHI] | J_00 | Jain_2000 [IND] | H_95 | Husin_1995 [IDO] |
| L_00 | Lu_2000 [CHI] | I_11 | Itoh_2011 [JAP] | P_14_F | Pandey_2014_FYM [VIE] |
| W_99 | Wang_1999 [CHI] | M_06 | Minamikawa_2006 [JAP] | P_14_C | Pandey_2014_Control [VIE] |
| K_11 | Khosa_2011 [IND] | Y_92 | Yagi_1992 [JAP] | | |

**Fig. 7** Seasonal $CH_4$ emissions under AWD and CF (shown as differentials) of this study (SA + WS season; DL + HM) in comparison to published data; acronyms in *x*-axis denote the following publication (first author_year [country abbreviation]) while the data set from Vietnam (Pandey et al. 2014) is shown for two treatments

Looking at the results from this study in view of the IPCC methodology, the findings indicate values for Vietnam in terms of both, EF under CF conditions as well as scaling factor (SF) for multiple aeration. The average EFs obtained from this study over three seasons are 4.1 and 2.0 kg $CH_4$/ha/d for the DL and HM environment, respectively. These EFs are higher than the IPCC default value (i.e. 1.3 kg $CH_4$/ha/d) but comparable to results from other studies conducted in China (Lu et al. 2000; Wang et al. 2012), Korea (Kwun et al. 2003), and Indonesia (Husin et al. 1995). However, these EFs do not represent baseline factors as used in the IPCC guidelines but EFs for location-specific conditions, such as relatively high straw incorporation rates.

The findings of this study would translate into a slightly higher SF for multiple aeration (0.71) than the global default value (0.52). However, the experiments described here represent the first of its kind in both of the sites and accurate water management turned out to be difficult at times, which explains the slightly lower mitigation potential of AWD as compared to the global average.

The $N_2O$ records confirmed the low emission level in rice production. $N_2O$ emission rates were basically close to detection limit—irrespective of site, season, or treatment. Even though the $N_2O$ records showed some pronounced differences in relative terms, this can clearly be attributed to the extremely low level of emission rates. In turn this finding corroborates previous assessments on the negligible role of $N_2O$ in the overall GWP assessment of rice fields. In this field study, the viability of AWD as mitigation option was not impaired by changes in $N_2O$ emissions.

Finally, the field study draws major conclusions on the efficiency of AWD as a mitigation option in the context of Vietnamese rice production. In relative terms, the impact of AWD on $CH_4$ emissions was within the range observed in previous studies published so far (Fig. 7). To emphasize the only comparable study from Vietnam (Pandey et al. 2014), we have displayed these results for one treatment with farm yard manure (P_14_F) as well as for the control treatment without organic amendment (P_14_C). Collectively, these five records from Vietnam shown in Fig. 7 correspond to the current data base on AWD impacts in Vietnam. Thus, it can be deduced that AWD represents a viable mitigation option for this country. In this study, emission savings ranged from 12.7 to 45.4 % with an average of 29.2 %. Again, potential impacts of $N_2O$ can be deemed negligible with the overall carbon footprint.

## Conclusion

This study comprises the first set of GHG emissions data from rice production in two different environments in Central Vietnam. $CH_4$ emissions in the first two seasons were very high but average EFs under location-specific flooded conditions were 4.1 kg $CH_4$/ha/d for the DL and 2.0 kg $CH_4$/ha/d for HM environment. The $CH_4$ reduction under AWD) was lower as compared to other studies and thus the SF for multiple aeration as estimated from this measurement campaign would be $SF_{m.a.} = 0.71$.

Emissions of $N_2O$ appear irrelevant to the overall computations of the carbon footprint of rice production under both water management types, CF and AWD. Furthermore, no significant difference in $N_2O$ emissions was observed between the two water management treatments.

Results from this measurement campaign underline the suitability of AWD) as mitigation option in Vietnam with clear reduction in $CH_4$ emission, no overall change in $N_2O$ emission and an increase in grain yield. Other impacts and potential benefits of AWD, such as increased lodging resistance or higher pest and/or disease tolerance, will need further investigation.

This study was conducted at only two sites in the South Central Coast region of Vietnam, which is of course insufficient to claim validity at the national scale. However, it represents the best available data for Central Vietnam to date and can be seen as a step of moving toward region-specific EF for Vietnam.

In fact, this dataset will allow the defining of a management-specific EF for AWD that may alternatively be used in a GHG inventory instead of SF. Several ongoing field experiments in Vietnam and other countries are now including AWD trials because this practice is seen as the most promising mitigation option in rice production. The underlying assumption of the IPCC Tier 2 methodology was that EF will only be available for CF. This assumption may not be valid any more due to the strong interest that AWD has received in recent years and the growing data base of emission rates for this treatment. It seems odd to use a less accurate approach based on a SF for CF as long as direct measurements are available.

**Acknowledgments** This work was supported by the project "Land use and climate change interactions in Central Vietnam" (LUCCi) funded by the Federal Ministry of Education and Research, Germany, and coordinated by the Institute for Technology and Resources Management in the Tropics and Subtropics of Cologne University of Applied Sciences, Germany.

The position of B.O. Sander at IRRI was funded by the Federal Ministry for Economic Cooperation and Development, Germany. The positions of R. Wassmann and A. Padre were in part funded by the CGIAR Research Programs on "Climate Change, Agriculture and Food Security" and the "Global Rice science Partnership."

## References

Adhya TK, Bharati K, Mohanty SR, Ramakrishnan B, Rao VR, Sethunathan N, Wassmann R (2000) Methane emission from rice fields at Cuttack, India. Nutr Cycl Agroecosyst 58:95–105

Butterbach-Bahl K, Sander BO, Pelster D, Diaz Pines E (2015) Quantifying greenhouse gas emissions from soils and manure management In: Rosenstock TS, Rufino MC, Butterbach-Bahl K, Wollenberg E, Richards MB (eds) Methods for measuring greenhouse gas balances and evaluating mitigation options in smallholder agriculture, Springer (in press)

Forster P, Ramaswamy V, Artaxo P, Berntsen T, Betts R, Fahey DW, Haywood J, Lean J, Lowe DC, Myhre G, Nganga J, Prinn R, Raga G, Schulz M, Van Dorland R (2007) Changes in atmospheric constituents and in radiative forcing. In: Solomon S, Qin D, Manning M, Chen Z, Marquis M, Averyt KB, Tignor M, Miller HL (eds) Climate change 2007: the physical science basis. Contribution of Working Group I to the Fourth Assessment Report of the

Intergovernmental Panel on Climate Change. Cambridge University Press, Cambridge, United Kingdom and New York, NY, USA, pp 130–234

GSO-VN (2016) General Statistics Office of Vietnam (http://www.gso.gov.vn)

Husin YA, Murdiyarso D, Khalil MAK, Rasmussen RA, Shearer MJ, Sabiham S, Sunar A, Adijuwana H (1995) Methane flux from indonesian wetland rice: the effects of water management and rice variety. Chemosphere 31:3153–3180

IPCC (Intergovernmental Panel on Climate Change) (2006) IPCC guidelines for national greenhouse gas inventories vol. 4. Institute for Global Environmental Strategies (IGES), Hayama, Japan

IPCC (Intergovernmental Panel on Climate Change) (2007) Climate change 2007. Cambridge University Press, Cambridge, UK and New York, NY (USA)

Jain MC, Kumar S, Wassmann R, Mitra S, Singh SD, Singh JP, Singh R, Yadav AK, Gupta S (2000) Methane emissions from irrigated rice fields in northern India (New Delhi). Nutr Cycl Agroecosyst 58:75–83

Jiao Z, Hou A, Shi Y, Huang G, Wang Y, Chen X (2006) Water management influencing methane and nitrous oxide emissions from rice field in relation to soil redox and microbial community. Commun Soil Sci Plant Anal 37:1889–1903

Khosa MK, Sidhu BS, Benbi DK (2011) Methane emission from rice fields in relation to management of irrigation water. J Environ Biol/Acad Environ Biol India 32:169–172

Kwun S-K, Shin YK, Eom K (2003) Estimation of methane emission from rice cultivation in Korea. J Environ Sci Health Part A 38:2549–2563

Lu WF, Chen W, Duan BW, Guo WM, Lu Y, Lantin RS, Wassmann R, Neue HU (2000) Methane emissions and mitigation options in irrigated rice fields in southeast China. Nutr Cycl Agroecosyst 58:65–73

LUCCi–Land Use and Climate Change Interactions in Central Vietnam. Retrieved April 2016 from http://www.lucci-vietnam.info/projectregion/natural-environment

Minamikawa K, Sakai N (2006) The practical use of water management based on soil redox potential for decreasing methane emission from a paddy field in Japan. Agric Ecosyst Environ 116:181–188

Minh ND, Trinh MV, Wassmann R, Sander BO, Hoa TD, Trang NL, Khai MN (2015) Simulation of methane emission from rice paddy fields in Vu Gia-Thu Bon River Basin of Vietnam Using the DNDC model: field validation and sensitivity analysis. VNU J Sci: Earth Environ Sci 31 (1):36–48

MONRE—Ministry of Natural Resources and Environment (2010) VIET NAM'S Second National Communication of Viet Nam to the United Nations Framework Convention on Climate Change, Hanoi 2010, 152 pp

MONRE—Ministry of Natural Resources and Environment (2014) Initial Biennial updated report of Viet Nam to the UNFCCC, Hanoi 2014, 94 pp

Pandey A, Mai VT, Vu DQ, Bui TPL, Mai TLA, Jensen LS, de Neergaard A (2014) Organic matter and water management strategies to reduce methane and nitrous oxide emissions from rice fields in Vietnam. Agric Ecosyst Environ 196:137–146

Park M-E, Yun S-H (2002) Scientific basis for establishing country $CH_4$ emission estimates for rice-based agriculture: A Korea (south) case study. Nutr Cycl Agroecosyst 64:11–17

Pathak H, Bhatia A, Prasad S, Singh S, Kumar S, Jain MC, Kumar U (2002) Emission of nitrous oxide from rice-wheat systems of Indo-Gangetic plains of India. Environ Monit Assess 77:163–178

Pathak H, Prasad S, Bhatia A, Singh S, Kumar S, Singh J, Jain MC (2003) Methane emission from rice–wheat cropping system in the Indo-Gangetic plain in relation to irrigation, farmyard manure and dicyandiamide application. Agric Ecosyst Environ 97:309–316

Sander BO, Wassmann R (2014) Common practices for manual greenhouse gas sampling in rice production: a literature study on sampling modalities of the closed chamber method. Greenhouse Gas Meas Manage 4(1):1–13. doi:10.1080/20430779.2014.892807

Sander BO, Wassmann R, Siopongco JDLC (2015) Water-saving techniques: potential, adoption and empirical evidence for mitigating greenhouse gas emissions from rice production. In:

Hoanh CT, Smakhtin V, Johnston T (eds) Climate change and agricultural water management in developing countries. CABI Climate Change Series, pp 193–207
SAS Institute Inc (2004) SAS/STAT® 9.1 Users Guide. Cary, SAS Institute Inc, NC
Vu QD, de Neergaard A, Toan DT, Quan QH, P Ly, Tien MT, Stoumann Jensen L (2015) Manure, biogas digestate and crop residue management affects methane gas emissions from rice paddy fields on Vietnamese smallholder livestock farms. Nutr Cycl Agroecosyst 103:329–346
VuItoh M, Sudo S, Mori S, Saito H, Yoshida T, Shiratori Y, Suga S, Yoshikawa N, Suzue Y, Mizukami H, Mochida T, Yagi K (2011) Mitigation of methane emissions from paddy fields by prolonging midseason drainage. Agric Ecosyst Environ 141:359–372
Wang B, Xu Y, Wang Z, Li Z, Guo Y, Shao K, Chen Z (1999) Methane emissions from ricefields as affected by organic amendment, water regime, crop establishment, and rice cultivar. Environ Monit Assess 213–228
Wang ZY, Xu YC, Li Z, Guo YX, Wassmann R, Neue HU, Lantin RS, Buendia LV, Ding YP, Wang ZZ (2000) A four-year record of methane emissions from irrigated rice fields in the Beijing region of China. Nutr Cycl Agroecosyst 58:55–63
Wang J, Zhang X, Xiong Z, Khalil MAK, Zhao X, Xie Y, Xing G (2012) Methane emissions from a rice agroecosystem in South China: effects of water regime, straw incorporation and nitrogen fertilizer. Nutr Cycl Agroecosyst 93:103–112
Wassmann R, Neue HU, Lantin RS, Buendia LV, Rennenberg H (2000) Characterization of methane emissions from rice fields in Asia. I. Comparison among field sites in five countries. Nutr Cycl Agroecosyst 58:1–12
Yagi K, Tsuruta H, Kanda K-I, Minami K (1996) Effect of water management on methane emission from a Japanese rice paddy field: Automated methane monitoring. Global Biogeochem Cycles 10:255–267
Yue J, Shi Y, Liang W, Wu J, Wang C, Huang G (2005) Methane and nitrous oxide emissions from rice field and related microorganism in black soil, Northeastern China. Nutr Cycl Agroecosyst 73:293–301

# Hydrological and Agricultural Impacts of Climate Change in the Vu Gia-Thu Bon River Basin in Central Vietnam

**Patrick Laux, Manfred Fink, Moussa Waongo, Rui Pedroso, Giulia Salvini, Dang Hoa Tran, Dang Quang Thinh, Johannes Cullmann, Wolfgang-Albert Flügel and Harald Kunstmann**

**Abstract** This paper summarizes some of the climate (change) impact modeling activities conducted in the *Land use and Climate Change interactions in Central Vietnam* (LUCCi) project. The study area is the Vu Gia-Thu Bon (VGTB) river basin in Central Vietnam, which is characterized by recurrent floods during the rainy season, but also water shortages during the dry season. The impact modeling activities, such as the validation of the models are hindered by the scarcity of hydrometeorological data and an unfavorable distribution of the observation

P. Laux (✉) · D.Q. Thinh · H. Kunstmann
Institute of Meteorology and Climate Research, Karlsruhe Institute of Technology, Kreuzeckbahnstrasse 19, 82467 Garmisch-Partenkirchen, Germany
e-mail: patrick.laux@kit.edu

M. Fink · W.-A. Flügel
Department of Geography, Friedrich-Schiller-University Jena, Grietgasse 6, 07743 Jena, Germany
e-mail: manfred.fink@uni-jena.de

M. Waongo
Direction Générale de La Météorologie, 01 BP 576 Ouagadougou, Burkina Faso

R. Pedroso
Institute for Technology and Resources Management in the Tropics and Subtropics, Technology Arts Sciences, University of Applied Sciences, Betzdorfer Straße 2, 50679 Cologne, Germany
e-mail: rui.pedroso@th-koeln.de

G. Salvini · D.H. Tran
College of Agriculture and Forestry, Hue University, 102 Phung Hung Street, Hue City, Vietnam

D.Q. Thinh
Institute of Meteorology, Hydrology and Climate Change, 23/62 Nguyen Chi Thanh, Dong Da, Hanoi, Vietnam

J. Cullmann
World Meteorological Organization, 7 bis, Avenue de la Paix, CH 1211 Geneva 2, Switzerland



A. Nauditt and L. Ribbe (eds.), *Land Use and Climate Change Interactions in Central Vietnam*, Water Resources Development and Management,
DOI 10.1007/978-981-10-2624-9_8

network, i.e., station data is available only for the lowlands. In total, two different process-based and distributed hydrological models are applied in concert with climate change and land use projections. Based on that, the magnitudes and return periods of extreme flows are estimated. The modeling results suggest increases of extreme high flows due to climate change. A multi-objective agro-economical model was developed for a typical irrigation scheme in the region in order to optimize the area for cropping, irrigation-techniques and schedules. The model results suggest the irrigation technique *Alternate Wetting and Drying*, which has the potential to increase the benefits for the farmers and help to mitigate greenhouse gases at the same time. In addition, the regional-scale crop model GLAM is applied for groundnut under rainfed conditions, which is capable to identify regions suitable for cropping in the future. The paper further synthesizes recommendations for local stakeholders in Central Vietnam.

## Introduction

Vietnam will belong to the most severely affected countries worldwide by climate change: the average temperature in the region is expected to increase approximately 1.5 °C by 2020, seasonal rainfall could increase approximately by 20% by 2070, whereas dry season rainfall will decrease (UNDP 2008; MONRE 2009). Based on that, it is likely that the frequency of agricultural droughts will increase during the copping period, and the occurrence probability of floods will be increasing at national scale. By that time, however, it is not yet fully clear how climate change will impact on regional and local scales.

Recently, trend analyses based on the observation data in Central Vietnam revealed a significant increase in winter rainfall during the past decades (Souvignet et al. 2014), however, it is not clear how this trend will impact on the occurrence and magnitude of floods in the future. In addition, the network of hydrometeorological information is sparse and the analysis of spatial patterns is not feasible. The BMBF-funded *Land Use and Climate Change interactions in Central Vietnam* (LUCCi) project aims at improving decision support, mainly for hydrological and agricultural purposes in the Vu Gia-Thu Bon (VGTB) river basin of Central Vietnam.

Predominantly paddy rice is grown in the region, but the agricultural productivity is severely limited by water stress during the Summer–Autumn (SA) season (Nguyen 2015, personal communication). The prevailing irrigation technique is *Continuous Flooding* (CF). However, it is known that *Alternate Wetting and Drying* (AWD) has the potential to save water (e.g., Lampayan et al. 2015) and thus to increase the yields by extended cropping areas. At the same time AWD may reduce the GHG emission (e.g., Siopongco et al. 2013; Pandey et al. 2014). In dialog with agricultural stakeholders in Central Vietnam, it turned out that the performance of the AWD is tested in small field trials only, but until now no systematic and consolidated efforts have been undertaken. Farmer are reluctant to apply AWD, because AWD requires more water at the beginning of cropping season, which may lead to short-term water shortages if all the fields are prepared in the same period. Farmers also reported about problems with rats when the soil is not flooded.

Farmers are becoming increasingly interested in growing other crops than rice to spread their financial risk due to volatile market prices. One suitable option is groundnut, which is predominantly grown during the winter–spring season under rainfed conditions. It is one of the major staples in Central Vietnam and processed by the farmers to cooking oil.

During the LUCCi project, we developed and calibrated a suite of impact studies to address the aforementioned issues in the basin. For Central Vietnam, studies on the impacts of climate (change) on agricultural productivity are rare due to scarce data such as climate, soil, and crop management. We decided to apply a suite of different impact models in the LUCCi project rather than one integrated approach for water resources management, because of the higher flexibility in their usage (Haberland 2010). First, regional climate simulations and land use scenarios are conducted and locally refined (Laux et al. 2012, 2013, 2016), and existing regional climate information is collected and analyzed for the VGTB basin. Using the regional climate and land use information as input, two different hydrological models and one regional-scale crop-growth model are driven to derive spatially explicit climate change mitigation and adaptation options. Complex interactions between plants and climate need process-based approaches, particularly if nonlinear feedbacks are considered as induced by climate variability and climate change. Based on the calibrated hydrological models JAMS/J2000 (Fink et al. 2013) and WaSiM (Dang et al. 2016a) (submitted to Hydrological Science Journal)., estimates of future peak flows and low flows are conducted. While the focus of JAMS/J2000 was on land use change, WaSiM was driven with a suite of regional climate projections to derive climate model related uncertainties for the basin. Based on the spatially distributed simulation results, an **agro-economic optimization model** (AGRO-ECO) is being developed for a prototype rice irrigation scheme in the basin, in order to investigate the impact of different irrigation techniques on their economic value. The model is capable to maximizing the irrigated rice growing area under prevailing water limitations, simulated by WaSiM.

## Specific Objectives

Specific objectives of this study are to:

1. Setup and calibrate two fully distributed hydrological models to model the water balance in the VGTB basin;
2. Estimate future hydrological extreme events by using the calibrated hydrological models and regional climate projections (and land use scenarios);

3. Develop an economic optimization model with respect to farmers' profit by considering irrigated area for cropping, irrigation technology, and irrigation schedules for summer–autumn rice cultivation; and
4. Estimate future groundnut productivity for the VGTB basin.

# Methods and Data

## *Process-Based Hydrological Modeling Using JAMS/J2000*

In order to assess the impacts of climate and land use changes, the distributed hydrological model J2000 was used to simulate the water balance of this catchment. Furthermore, the model results were provided for subsequent hydraulic models for flood (Chapter "Integrated River Basin Management in the Vu Gia Thu Bon Basin") and saltwater intrusion analyses (Chapter "Bio-Physical and Socioeconomic Features of the LUCCi—Project Region: The Vu Gia Thu Bon River Basin") and models to represent hydrological infrastructure (Fink et al. 2013).

J2000 can be classified as a distributed, process-oriented hydrological model for hydrological simulations of mesoscale and macroscale catchments. It is implemented in the Jena Adaptable Modeling System (JAMS) framework (Kralisch and Krause 2006; Kralisch et al. 2007), which is a software framework for component-based development and application of environmental models. The model describes the hydrological processes as encapsulated or independent process modules. These modules describe, for example, input data regionalization and correction, calculation of potential and actual evapotranspiration, canopy interception, soil moisture and groundwater processes. The spatial representation of the model is utilizing the Hydrological Response Units (HRU) approach (Flügel 1996). This approach is enhanced by individual polygons (c.f. Fig. 2, detail window) with a multidimensional routing scheme (Pfennig et al. 2009). The HRUs are the result of overlay analyses of the Digital Elevation Model (DEM), the Topographical Wetness Index, the Mass Balance Index, the Annual Solar Radiation Index, and the individual soil classes. The land use and geology maps were used to assign the according properties to the resulting polygons by using a maximum membership function to reduce the number of HRUs for modeling. The HRU delineation process resulted in 477888 HRUs and 24192 reach segments for test site (Fink et al. 2013). For the representation of the irrigation in the model irrigated HRUs were selected. The assumptions are that only arable land with lower slopes is irrigated. The selection procedure is carried out for each separate modeled sub-catchment (Fig. 1) individually. The amount of rice fields reported in the according district statistics is used to calibrate the slope threshold. This procedure leads to higher thresholds in the upland than in the lowlands and ensures a realistic proportion of irrigated areas in the model.

The irrigation modules of J2000 are based on the assumption that every irrigated field has access to an irrigation source. Moreover, irrigation will be applied on

**Fig. 1** Modeled sub-catchments and irrigated HRUs (cyan)

demand, which is defined as quotient of actual and potential evapotranspiration. The irrigation water is taken from the subcatchment outlet. Beside the use of the downscaled climate data for two different GHG emission scenarios (Laux et al. 2013), the impact of land use change scenarios is analyzed using the J2000. The baseline scenario is based on a classification of remote sensing images from 2010. The future scenario consists of a combination of a VGTB land cover map (2020) at 30 m spatial resolution, obtained by the projection of the past land cover change dynamics (2001–2010) to the reference year 2020, and the master plan of the Department of Natural Resources DONRE for 2030. Since the projection has been created only for the mid and highlands, the master plan is used to represent the lowlands (Salvini and Avitabile 2013). Graphics representing the land use scenarios are shown in Fig. 2. The proportion of urban settlements is increasing while especially the forest is decreasing. The growth of the settlement basically takes place in the lowlands and leads there to a decreasing of cropland. This loss of cropland is compensated in the uplands, where forest is converted to cropland.

## *Process-Based Hydrological Modeling and Extreme Value Theory to Derive Extreme Discharge Events*

To investigate possible impacts of climate change on hydrological extreme events in VGTB basin, the following modeling chain approach has been employed. We base our analysis on regional climate projections based on different combinations of global and regional climate models in combination with different emission scenarios conducted in the LUCCi project and from other external sources (Ngo et al. 2014). Different bias correction techniques are applied to the available RCM

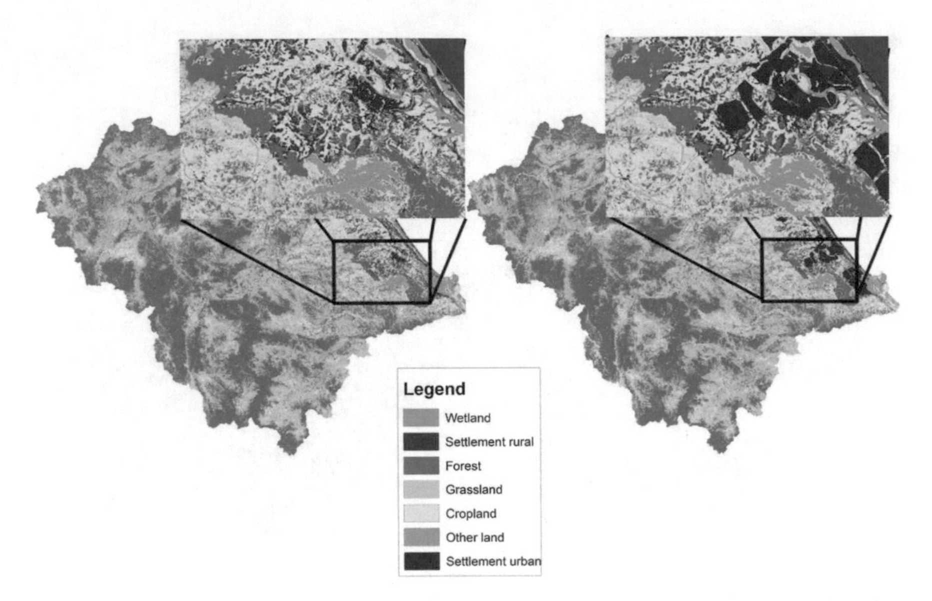


**Fig. 2** Land use scenario (*left* base line 2010, *right* near future 2020, from: Salvini and Avitabile 2013)

projections to correct for the biases in precipitation and temperature with respect to the observational data available in the LUCCi project (Souvignet et al. 2014). Downscaled and bias-corrected climate variables are then used as input for fully distributed hydrological water balance model WaSiM (Schulla 2012) to generate daily discharge series (Dang et al. 2016a). WaSiM is a fully distributed hydrological model dealing with simulations of the surface and subsurface hydrological cycle. It supports both continuous and event-based rainfall-runoff simulations. In this study, WaSiM was applied with the Penman–Monteith approach to calculate the potential evapotranspiration. Its soil module calculates the infiltration of water and the surface runoff generation using the Green and Ampt approach. A topographic analysis is done as a preprocessing step (TANALYS tool). The WaSiM model is set up for VGTB river basin with daily time step and 500 × 500 m spatial resolution. Required input data for WaSiM are a DEM, land use data, soil data and daily temperature, precipitation, relative humidity, wind speed, and global radiation. The WaSiM model was successfully calibrated and validated using the split sampling technique (Dang et al. 2016a).

The performance of the different bias correction methods for precipitation to reproduce the extreme value frequencies was analyzed. Annual maximum and minimum discharges are subsequently extracted from the corrected modeled series for the control period (1980–1999) as well as the future periods 2011–2030, 2031–2050, and 2080–2099. The generalized extreme value (GEV) distribution function is fitted to the data to produce flood frequency curves and to derive flood return periods.

## *Agro-Economic Model Based on GAMS (AGRO-ECO)*

The agricultural optimization model developed for this study is written in *General Algebraic Modeling System* (GAMS). During a field studies in 2015, we identified the QueTrung rice growing area as suitable case study targeting the research demands. The research demands have been elaborated jointly with the local stakeholders. In addition, we collected the required input data to initialize and constrain the model such as e.g., reservoir parameters, rice productivity rates, and the structure of the irrigation system.

Thus, we designed our model that resembles very closely the situation prevailing in the QueTrung rice growing area (Fig. 3). The irrigation system includes a reservoir and three irrigated rice cultivation areas, each of them divided into 4 blocks. The area of each block is 12 ha. One assumption, which is not fulfilled in reality is, that the irrigation system consists of concrete material canals only, and the irrigation system contains all the necessary structures for water control and distribution, cross-regulators.

AGRO-ECO accounts for the structure of the irrigation network, the reservoir by including its water balance (reservoir inflow and outflow, efficient reservoir storage, and evaporation), the irrigation techniques and schedules, and the objective function, which is maximizing the profit of the farmers. Thus, the allocation of land for cropping, the irrigation technology, and the irrigation schedules for each block are important decision variables, to be optimized with respect to the objective function.

Possible modeled irrigation options include AWD and CF (Fig. 4). For the AWD, the field is flooded with water at a level of about 5 cm above the soil surface and allowed to drop to 15 cm below the soil surface before being re-flooded. During early growth of rice (about 2 weeks after planting), the ponded water is kept about 2–3 cm above the soil surface. During the flowering period, the field is continuously kept submerged at around 5 cm above surface. For CF, the standing water is always kept at the depth of around 5 cm above surface during the whole vegetation period.

The model simulates the cropping period of 110 days including land preparation using a daily time step. Water requirement for land preparation is about 200 mm for both technologies.

AGRO-ECO uses the simulated discharges at the QueTrung sub-catchment as reservoir inflow, provided by WaSiM. WaSiM is driven by the interpolated meteorological data from the available observations for temperature and precipitation. More information of the AGRO-ECO model is given in Dang et al. (2016b).

## *The Regional Crop Model GLAM*

The process-based regional crop model GLAM is applied to model the complex crop-climate interactions. It has been used to model the impacts of the regional

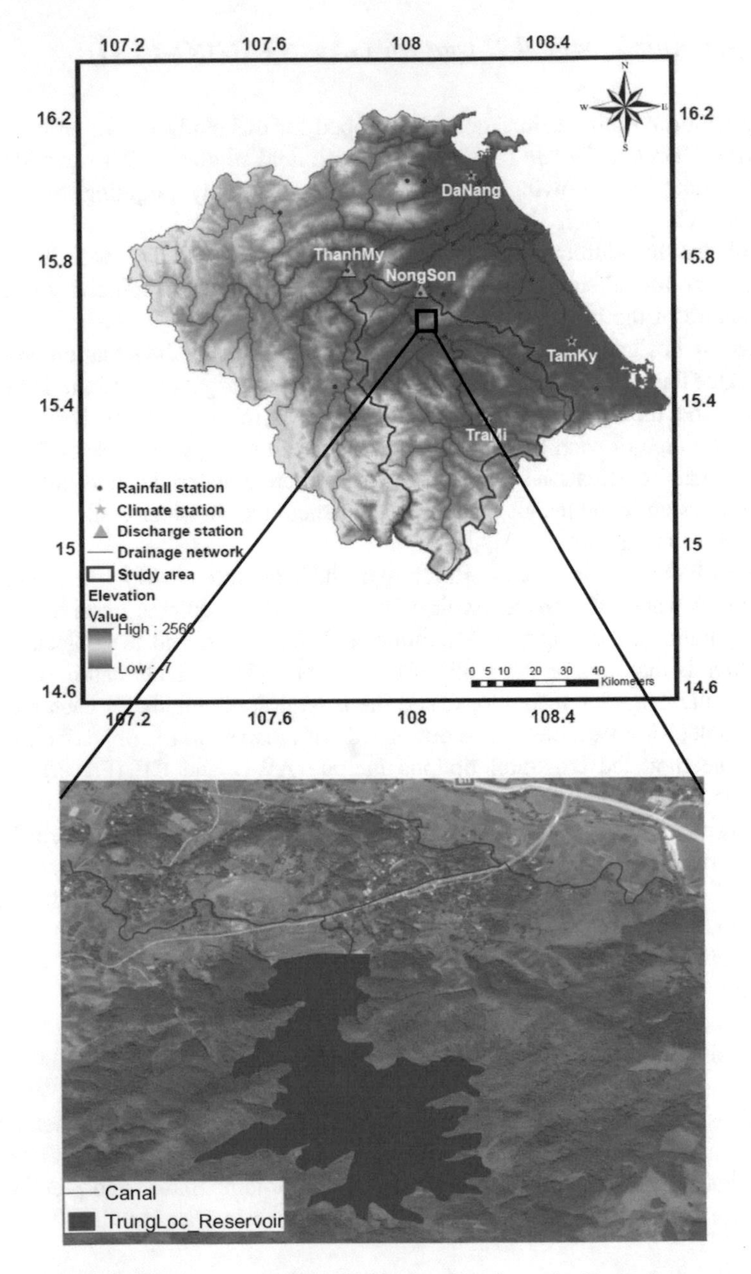


**Fig. 3** Que Trung rice irrigation system. Water is taken from the Trung Loc reservoir. The reservoir has an effective volume capacity of about $1.9 \times 10^6$ m$^3$ and consists of a dam with 331 m in length and 27.4 m in height. Water is diverted to the canal scheme via one submerged conduit running beneath the dam and controlled by a valve located on the downstream toe of the dam

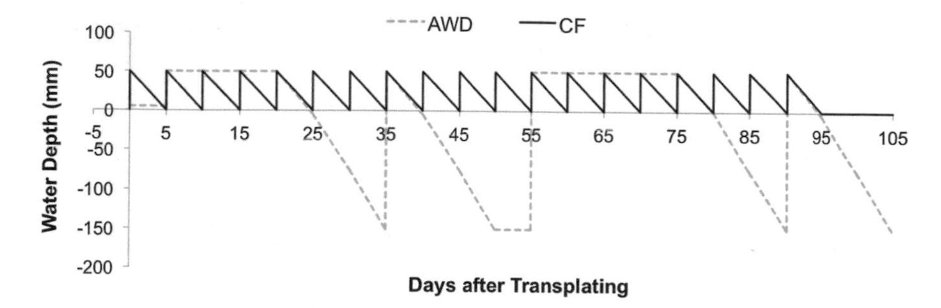


**Fig. 4** Schematic representation of irrigation schedule for both irrigation options, i.e., AWD and CF for the autumn-summer rice cropping season in the study region

climate on groundnut productivity. For calibration, GLAM was driven by WRF-ERA Interim for the period 2004–2012, using solar radiation, maximum and minimum temperature, precipitation, soil texture from the *Harmonized World Soil Database* (HWSD), yield data disaggregated, and discretized on the WRF grid (Fig. 5) based on observed yields for the districts in Quang Nam (Statistical Yearbooks of Quang Nam), i.e., the district yield was assigned to each grid cell within the district. The planting date was estimated using an intrinsic approach. It was planted if the soil water content exceeds half of soil water holding capacity.

Two different sets of crop specific parameters, including thermal durations, rate of change of harvest index, transpiration efficiency, max. rate change of LAI, are

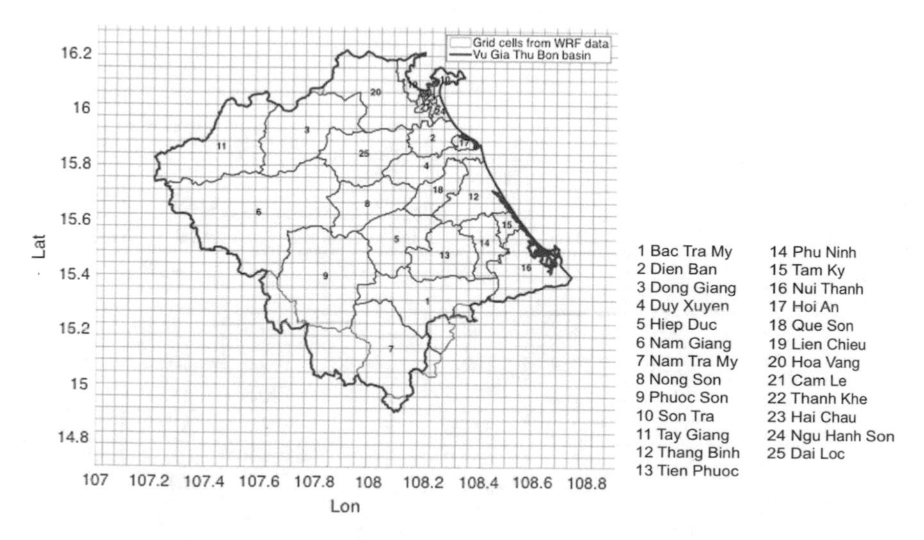


**Fig. 5** Districts considered for groundnut cropping. Based on observed yields for the districts in Quang Nam, the yield data were disaggregated and discretized onto the WRF raster

applied: a parameter set identified for India (subsequently referred to as P1), and a parameter set identified for China (subsequently referred to as P2).

# Results

## *Impacts of Land Use and Climate Scenarios on the Hydrology*

The baseline scenario results of the J2000 model and the model performance are described in Fink et al. (2013). In this paper, we focus on the scenario (c.f. Chapter “Vu Gia Thu Bon River Basin Information System (VGTB RBIS)—Managing Data for Assessing Land Use and Climate Change Interactions in Central Vietnam”) results. In Table 1 the quantiles of daily runoff from the entire study area for the different land use scenarios partitioned into two periods from 1972 to 2002 and from 2002 to 2032. The values indicate that the results are relatively heterogeneous according to future and nonfuture land use and future and non-future climate. In terms of the climate scenarios the low flows (<1%) are in three out of fourcases reduced, indicating a slight tendency for increasing drought conditions in the future. On average (50%) the runoff increases slightly in future. The same holds for the “normal” (<95%) high flow rates, whereas the extreme floods show no tendency. The impact of the land use change show a tendency of higher runoff rates except of the floods (>90%) where a tendency is difficult to identify. The given minimum and maximum values are shown to provide a comprehensive view. Because they are just bases on single modeled events, they should not be interpreted due to the high uncertainties.

**Table 1** Quantiles of runoff

| Runoff ($m^3/s$) | | | | | | | | |
|---|---|---|---|---|---|---|---|---|
| | Land use 2010 (base line) | | | | Land use 2020 (scenario) | | | |
| | 1972–2002 | 2002–2032 | 1972–2002 | 2002–2032 | 1972–2002 | 2002–2032 | 1972–2002 | 2002–2032 |
| | B1 | B1 | A1B | A1B | B1 | B1 | A1B | A1B |
| min (%) | 80.2 | 62.6 | 81.5 | 72.3 | 76.6 | 62.6 | 80.8 | 87.9 |
| 1 | 94.3 | 90.4 | 95.6 | 90.4 | 93.4 | 89.5 | 94.8 | 101.5 |
| 5 | 105.4 | 110.6 | 106.8 | 108.8 | 104.4 | 109.5 | 105.6 | 116.5 |
| 10 | 114.9 | 121.6 | 116.5 | 122.2 | 113.5 | 120.8 | 115.2 | 126.5 |
| 50 | 370.8 | 376.9 | 374.3 | 371.6 | 372.5 | 378.5 | 377.3 | 424.9 |
| 90 | 1655.9 | 1739.5 | 1654.2 | 1907.8 | 1658.2 | 1742.6 | 1660.8 | 1957.4 |
| 95 | 2460.6 | 2588.7 | 2422.0 | 2907.6 | 2461.0 | 2603.1 | 2432.6 | 2718.7 |
| 99 | 6389.4 | 6275.0 | 6404.5 | 7047.8 | 6385.8 | 6302.3 | 6438.6 | 5329.6 |
| max | 28307.2 | 21376.8 | 28431.5 | 36203.4 | 28354.5 | 21407.3 | 28478.5 | 28546.0 |

**Table 2** Trends Man Kendall test for A1B and B1 scenarios (future land use); *(***) indicates significant trends tested at α = 5 (1) % significance level

| | A1B 1971–2033 | | B1 1971–2050 | |
|---|---|---|---|---|
| | Significance | Trend slope | Significance | Trend slope |
| Runoff ($m^3/s$) | * | 2.68 | | 1.08 |
| Precipitation (mm) | * | 0.0193 | | 0.00659 |
| Mean temperature (°C) | *** | 0.0115 | *** | 0.00963 |
| Actual ET (mm) | | −1.19 | | −1.15 |

To look closer at the climate scenarios a Mann-Kendall trend test (Salmi et al. 2002) of yearly values has been performed (cf. Table 2). For the A1b scenario the test revealed positive trends for both precipitation and runoff (tested at significance level of 5%). Temperature indicates positive trends at 1% significance level. The less intensive scenario B1 shows the same tendencies than the trends of the A1b scenario, but only the temperature features a significant result. The simulated actual evapotranspiration (ET) show a nonsignificant decreasing tendency, which is difficult to explain looking at annual values because temperature and precipitation are increasing. One possible reason for this is a different seasonality in the precipitation and temperature between the first and the second period.

## *Impacts of Climate Change on Extreme Flows: A Multi-model Assessment*

It is found that none of the applied statistical bias corrections methods on precipitation led to improved representations of the peak flows after hydrological modeling (Dang et al. 2016a). Since this would be the prerequisite to derive return periods of extremes, the delta change method has been employed, i.e., the expected climate change signals are impressed on the observed discharges, and peak flows of return periods ranging from 1 to 50 years are estimated. Please note that the uncertainties of the results are increasing with increasing return periods since the underlying data is 20 years only. Figure 6 shows exemplarily the estimated peak flows for return periods ranging from 1 to 50 years for the period 2031–2050. It can be seen that the majority of adjusted RCM projections indicates increased peak flow values compared to the observations.

Overall, the results show a tendency toward increased high flows in the future while low flows has decreased tendency under both emission scenarios A1B and A2. However, it should be mentioned that the adjusted high and low flows of future climate scenarios varies considerably compared to that of the control period. For further results, the reader is referred to the study of Dang et al. (2016a). The large uncertainty among the projected peak flows, especially for high return periods can

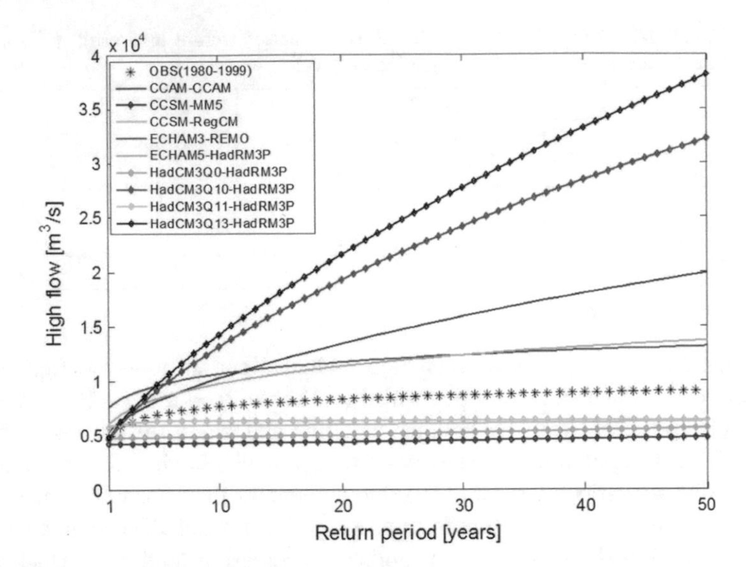


**Fig. 6** Peak flow for return periods ranging from 1 to 50 years after application of the delta change approach at NongSon station, for observed (control period) and period 2031–2050, and different GCM-RCM combinations based on the A1B emission scenario (Dang et al. 2016a *submitted to Hydrological Science Journal*)

be partly attributed to the relatively small sample size in this study. A clear attribution, however, is difficult due to the non-linear behavior of the hydrological runoff processes.

## *Optimized Management Strategies for Rice Cropping: A Case Study for the QueTrung Rice Growing Area (NongSon District)*

The mathematical model AGRO-ECO has been applied as a case study for the year 2012 for the irrigation demand site located in the Que Trung commune, (district: NongSon, province: QuangNam). Since water is a limiting factor for the summer-autumn season, the model optimizes the area and technique for irrigation with respect to the monetary benefit. Two different scenarios are considered: **scenario 1** assumes that the reservoir is full at the beginning of the cropping season, and **scenario 2** assumes a volume of 75% (Nguyen 2015, personal communication). Both scenarios are reasonable with respect to water volumes at the beginning of the summer-autumn season. The actual outflow from the reservoir to the irrigation system is designed for 0.3 $m^3/s$ in order to potentially irrigate 130–145 ha paddy rice. It is analyzed if the maximum outflow volume is sufficient to irrigate the whole area, and how the outflow volume may impact on the optimization results.

**Table 3** Summary of irrigated area and irrigation times for **scenario 1**

| Outflow ($m^3/s$) | Total irrigated blocks | Number of blocks applied | | Average irrigation events | | Water productivity ($kg/m^3$) | |
|---|---|---|---|---|---|---|---|
| | | AWD | CF | AWD | CF | AWD | CF |
| 0.3 | 9 | 1 | 8 | 20 | 31 | 0.50 | 0.36 |
| 0.4 | 10 | 1 | 9 | 21 | 30 | 0.52 | 0.36 |
| 0.5 | 10 | 9 | 1 | 20 | 35 | 0.50 | 0.35 |
| 0.6 | 10 | 6 | 4 | 20 | 34 | 0.52 | 0.36 |
| 0.9 | 10 | 8 | 2 | 21 | 31 | 0.51 | 0.36 |

Tables 3 and 4 show the results of the optimization results following the scenario 1. It can be seen that under the outflow design, not all the blocks can be irrigated. Only 9 blocks are irrigated, mainly under CF (8 out of 9). The water productivity is 0.5 and 0.36 $kg/m^3$ for AWD and CF, respectively. Increasing the maximum outlet would allow for an irrigation of more blocks, and a shift of the optimum toward the AWD technique.

If less water is available from the beginning of the cropping season (**scenario 2**), fewer blocks can be irrigated (Tables 5 and 6). Likewise, for scenario 1, increasing the potential outflow leads to a shift of the irrigation technique toward AWD.

## *Impacts of Climate Change on Attainable Groundnut Yields Under Rainfed Conditions*

First, the Yield Gap Parameter (YPG) is estimated to check how much the modeled yield deviates from the observed yield (2004–2011). THE YGP represents effects, which cannot be explicitly modeled by GLAM, such as , pests, diseases, as well as nonoptimal management. Thus, YPG can also be seen as an input data bias correction parameter normalized from 0 to 1, which will be also applied for projections (Fig. 7).

It is found that the parameter set P2 is less problematic than parameter set P1. Very heterogeneous YGPs in the two districts Tay Giang and Phuoc Son are an indication that the model is not calibrated optimally. P2 shows a reduced spatial variability in the YPG compared to P1. This indicates an overall better performance of P2 for the VGTB basin. The simulated annual yields can be seen in Fig. 8, exemplarily shown for district Bac Tra My. P1 heavily overestimates the observed yields, while P2 underestimates. It can be seen that P2 performs better in reproducing the trend of the yield time series.

The spatial groundnut yields are illustrated in Fig. 9. Mean groundnut yield (2004–2011) is found to be highly variable across the VGTB basin using both parameterizations. Thereby, P2 shows less variability between neighboring grid

**Table 4** Summary of irrigation technology and schedule for **scenario 1**

| Outflow ($m^3/s$) | Block | Area 1 (Secondary canal 2–3) | | | | Area 2 (Secondary canal 4–5) | | | | Area 3 (Secondary canal 6–7) | | | |
|---|---|---|---|---|---|---|---|---|---|---|---|---|---|
| | | 1 | 2 | 3 | 4 | 1 | 2 | 3 | 4 | 1 | 2 | 3 | 4 |
| 0.3 | Start | 7 | 3 | 2 | 9 | – | – | 8 | 4 | 5 | 6 | – | 12 |
| | Tech | CF | CF | CF | CF | – | – | CF | CF | CF | CF | – | AWD |
| 0.4 | Start | 9 | 6 | 11 | 5 | 13 | 7 | | 15 | 14 | 1 | – | 10 |
| | Tech | CF | CF | CF | CF | CF | CF | | AWD | CF | CF | – | CF |
| 0.5 | Start | 8 | 13 | 1 | 11 | 7 | 14 | 10 | 15 | 12 | 9 | – | – |
| | Tech | AWD | AWD | CF | AWD | AWD | AWD | AWD | AWD | AWD | AWD | – | – |
| 0.6 | Start | 11 | 8 | – | 13 | – | 1 | 8 | 1 | 12 | 15 | 14 | 15 |
| | Tech | AWD | CF | – | AWD | – | CF | CF | CF | AWD | AWD | AWD | AWD |
| 0.9 | Start | – | 15 | 14 | 14 | 1 | 9 | 1 | 15 | 13 | – | 11 | 1 |
| | Tech | – | AWD | AWD | AWD | AWD | AWD | CF | AWD | AWD | – | CF | AWD |

**Table 5** Summary of irrigated area and irrigation times for **scenario 2**

| Outflow ($m^3/s$) | Total irrigated blocks | Number of blocks applied | | Average irrigation times | | Water productivity ($kg/m^3$) | |
|---|---|---|---|---|---|---|---|
| | | AWD | CF | AWD | CF | AWD | CF |
| 0.3 | 8 | 1 | 7 | 19 | 30 | 0.52 | 0.36 |
| 0.4 | 8 | 5 | 3 | 20 | 28 | 0.51 | 0.36 |
| 0.5 | 8 | 7 | 1 | 20 | 33 | 0.50 | 0.36 |
| 0.6 | 8 | 6 | 2 | 20 | 35 | 0.51 | 0.36 |
| 0.9 | 8 | 7 | 1 | 21 | 33 | 0.51 | 0.35 |

cells, and is therefore expected to be more reliable. The simulated mean planting dates range from November, 1st in the South to November, 8th (not shown).

Based on the performed climate simulation in the LUCCi project using the regional climate model WRF, crop yield projections are performed in order to identify region, which will be preferential for groundnut copping (Fig. 10).

Based on both parameterizations, more favorable conditions are found for the districts Tay Giang and Phuoc Son.

## Summary and Conclusions for Local Stakeholders

The study summarizes different scientific approaches to assess the hydrological and agricultural impacts of climate change and climate variability in the VGTB river basin in Central Vietnam. The scientific foundation is laid for the region, but specific tools are still under development such as the AGRO-ECO. Therefore, additional stakeholder meetings will be necessary beyond the course of the LUCCi project. The following conclusions are drawn:

**Process-based and spatially distributed hydrological modeling (*JAMS/J2000* & WaSiM):**

- The distributed hydrological model J2000 is well suited to represent and evaluate scenarios that describe future climate- and land use changes. Due to the utilized HRU concept the model is able to describe the physical properties, like land use, in a detailed, but computationally efficient manner.
- WaSiM, in concert with Extreme Value Theory provides useful approach to derive suitable climate change adaptation measures such as the design of river and dyke embankment, reservoirs, and other flood control measures.
- Both changes, the land use and the climate, suggest an increase of runoff in future. Nevertheless, at the same time the tendency for extreme droughts will increase as well. The majority of considered future climate projections suggest heavy increases of peak flows, however, high uncertainties exist due to different regional climate projections and emission scenarios.

**Table 6** Summary of irrigation technology and schedule for **scenario 2**

| Outflow ($m^3/s$) | Block | Area 1 (Secondary canal 2–3) | | | | Area 2 (Secondary canal 4–5) | | | | Area 3 (Secondary canal 6–7) | | | |
|---|---|---|---|---|---|---|---|---|---|---|---|---|---|
| | | 1 | 2 | 3 | 4 | 1 | 2 | 3 | 4 | 1 | 2 | 3 | 4 |
| 0.3 | Start | – | – | 2 | 10 | 4 | – | 7 | 12 | 13 | – | 15 | 1 |
| | Tech | – | – | CF | CF | CF | – | CF | CF | CF | – | AWD | CF |
| 0.4 | Start | 14 | 9 | 13 | – | – | – | – | 12 | 10 | 15 | 11 | 1 |
| | Tech | AWD | AWD | AWD | – | – | – | – | AWD | CF | AWD | CF | CF |
| 0.5 | Start | – | 14 | 15 | 1 | – | 9 | 12 | 2 | – | – | 7 | 10 |
| | Tech | – | AWD | AWD | AWD | – | AWD | AWD | CF | – | – | AWD | AWD |
| 0.6 | Start | 8 | 7 | 15 | 14 | 1 | 13 | – | – | – | 1 | – | 6 |
| | Tech | AWD | AWD | AWD | AWD | CF | AWD | – | – | – | CF | – | AWD |
| 0.9 | Start | 1 | 1 | 14 | 14 | 1 | – | – | – | 15 | 13 | – | 15 |
| | Tech | AWD | AWD | AWD | AWD | CF | – | – | – | AWD | AWD | – | AWD |

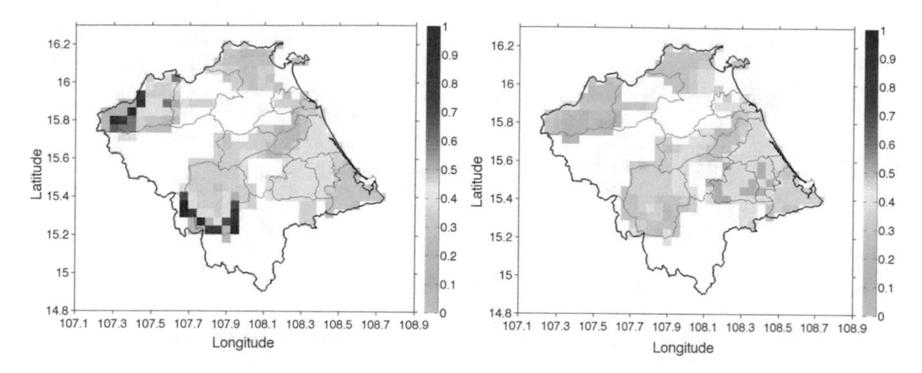


**Fig. 7** Yield Gap Parameter (YGP) [0, …,1] for groundnut indicating the deviations between simulated and observed yields; YGP based on the parameter set P1 (*left*) and P2 (*right*). High (*low*) values show high (*low*) deviation. The YGP is calibrated locally (grid specific)

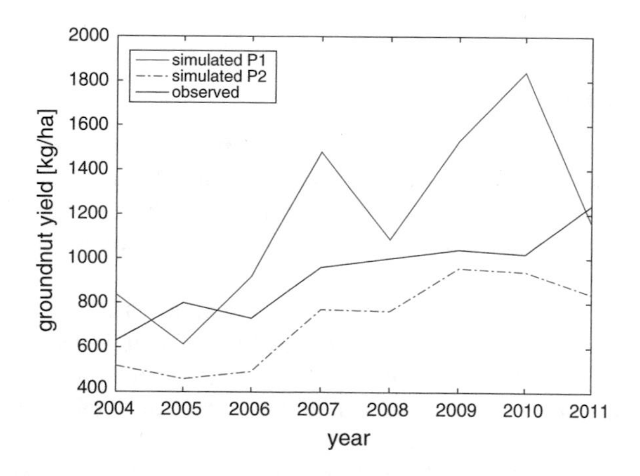


**Fig. 8** Simulated groundnut yields Bac Tra My based on the parameter set P1 and P2

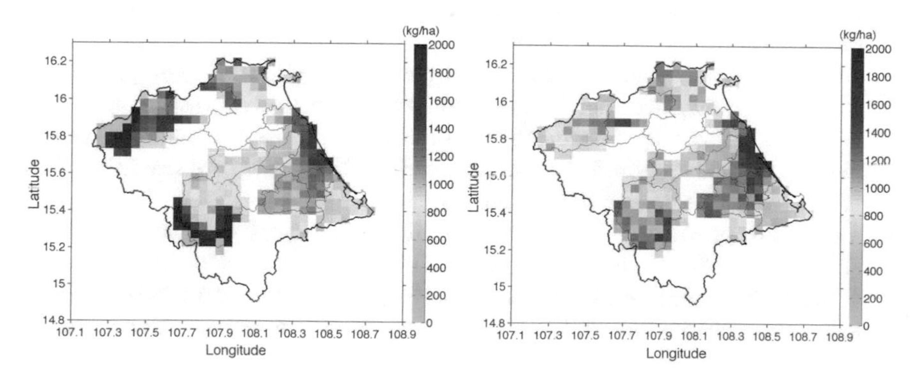


**Fig. 9** Simulated mean groundnut yields (2004–2011) based on the parameter set P1 (*left*) and P2 (*right*)

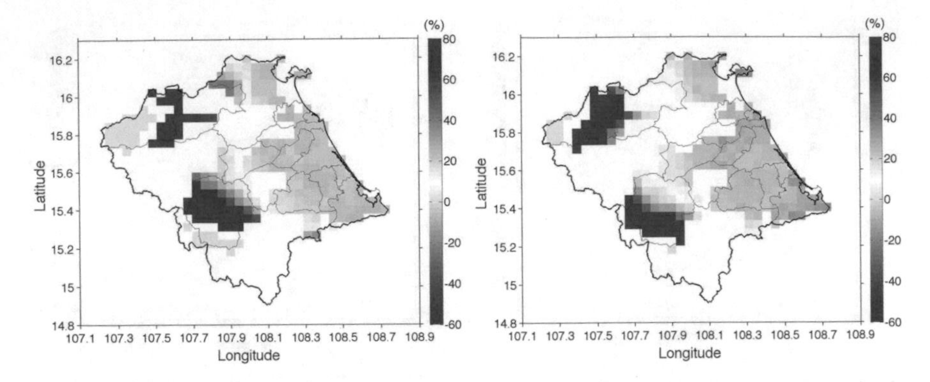


**Fig. 10** Projected groundnut yield changes for 2001–2030 compared to the baseline period 1971–2000 (future *minus* baseline) using the ECHAM5 data and the A1B emission scenario based on parameter set P1 (*left*) and P2 (*right*)

**Agro-economic modeling**:

- AGRO-ECO, a multivariate mathematical optimization tool to increase the gross margins from rice production by optimizing the area for rice cropping, irrigation technology, and irrigation schedule is still under development. The model is being discussed and further adapted with local stakeholders in the irrigation scheme.
- Under water limitations during the summer–autumn season, the results suggest the usage of AWD rather than CF, if the outlet discharge is increased. This may help to buffer water shortages and to reduce methane emissions at the same time. However, additional investments is required to transform the canal system into concrete material to fully benefit from the advantages of AWD.
- AGRO-ECO can be applied in concert with long-term climate projections and/or seasonal climate forecasts. Before its application and transfer to similar irrigation schemes, the new modeling system must be validated and tested thoroughly.
- GLAM has been applied for the VGTB basin with regional climate projections without any postprocessing, i.e., bias correction/statistical downscaling. GLAM is process-based with a high number of empirically based model parameters, thus the model is well suited for regions where observations are scarce and thus process-based field scale models are not applicable.
- This study provides first steps toward further analyses of how to improve attainable yields for the VGTB basin, i.e., the application of the *Optimized Planting Date algorithm* (Laux et al. 2010; Waongo et al. 2013, 2015) will be tested for groundnut across the VGTB basin during future research activities.

## References


Dang T, Laux P, Kunstmann H (2016a). Future high and low flow estimations for Central Vietnam: a complex hydrometeorological modeling chain approach (submitted to Hydrological Science Journal)

Dang T, Pedroso R, Laux P, Kunstmann H (2016b) Development of an integrated Hydrological-Agronomic-Economic modeling system for a typical rice irrigation scheme in Centrel Vietnam (submitted to Agriculture Water Management)

Fink M, Fischer C, Führer N, Firoz AMB, Viet TQ, Laux P, Flügel W-A (2013) Distributive hydrological modeling of a monsoon dominated river system in central Vietnam. In: Piantadosi J, Anderssen RS, Boland J (eds) MODSIM2013, 20th international congress on modelling and simulation. Modelling and simulation society of Australia and New Zealand, December 2013, p 1826–1832, ISBN: 978-0-9872143-3-1

Flügel WA (1996) Hydrological Response Units (HRU) as modelling entities for hydrological river basin simulation and their methodological potential for modelling complex environmental process systems. Erde 127:42–62

Führer N (2016) Geospatial evaluation of fluvial flooding in central Vietnam for spatial planning, Doctoral dissertation. RUHR University Bochum, Germany. 147

Haberland U (2010) From hydrological modelling to decision support. Adv Geosci 27:11–19

Kralisch S, Krause P (2006) JAMS—a framework for natural resource model development and application. In: Voinov A, Jakeman A, Rizzoli AE (eds) Proceedings of the iEMSs Third Biannual Meeting, Burlington, USA

Kralisch S, Krause P, Fink M, Fischer C, Flügel W-A (2007) Component based environmental modelling using the JAMS framework. In: Kulasiri D,Oxley L (eds) Proceedings of the MODSIM 2007 international congress on modelling and simulation christchurch, New Zealand

Lampayan RM, Rejesus RM, Singleton GR, Bouman BA (2015) Adoption and economics of alternate wetting and drying water management for irrigated lowland rice. Field Crops Res 170:95–108

Laux P, Phan VT, Lorenz Ch, Thuc T, Ribbe L, Kunstmann H (2012) Setting up regional climate simulations for Southeast Asia. In: Nagel WE, Kröner DB, Resch MM (eds) High performance computing in science and engineering '12, Conference proceedings, Springer, pp 391–406

Laux P, Thuc T, Kunstmann H (2013) High resolution climate change information for the Lower Mekong River Basin of Southeast Asia. In: Nagel WE, Kröner DB, Resch MM (eds) High performance computing in science and engineering '13, Conference proceedings, Springer, pp 543–551

Laux P, Jäckel G, Munang R-T, Kunstmann H (2010) Impact of climate change on agricultural productivity under rainfed conditions in Cameroon—a method to improve attainable crop yields by planting date adaptations. Agric For Meteorol 150:1258–1271

Laux P, Nguyen PNB, Cullmann J, Van TP, Kunstmann H (2016) How many RCM ensemble members provide confidence in the impact of land-use land cover change? Int J Climatol. doi:10.1002/joc.4836

MONRE (2009) Climate change, sea level rise scenarios for Vietnam. Ministry of natural resources and environment. Hanoi, Vietnam, p 34

Ngo DT, Kieu C, Thatcher M, Nguyen LD, Phan VT (2014) Climate projections for Vietnam based on regional climate models. Clim Res 60(3):199–213

Pandey A, Mai VT, Vu DQ, Bui TPL, Mai TLA, Jensen LS, Neergaard A (2014) Organic matter and water management strategies to reduce methane and nitrous oxide emissions from rice paddies in Vietnam. Agr Ecosyst Environ 196:137–146

Pfennig B, Kipka H, Wolf M, Fink M, Krause P, Flügel W-A (2009) Influence of different HRU linkage approaches on spatial variability of predicted runoff and associated processes. In: Yilmaz KK et al (2009) New approaches to hydrological prediction in datasparse regions. IAHS Publication 333, Proceedings of 8th IAHS Scientific Assembly & 37th IAHS Congress, Hyderabad, India, pp 37–43

Quang Nam Statistical Office (2014) Quang Nam statistical yearbooks. Statistical Publishing House, Tam Ky. http://www.gso.gov.vn/default_en.aspx?tabid=509&itemid=2716

Siopongco J, Wasmann R, Sander BO (2013) Alternate wetting and drying in Philippine rice production: feasibility study for a Clean Development Mechanism, International Rice Research Institute, Los Baños (Philippines), IRRI Technical Bulletin no. 17, p 14

Souvignet M, Laux P, Freer J, Cloke H, Thinh DQ, Thuc T, Cullmann J, Nauditt A, Flügel W-A, Kunstmann H, Ribbe L (2014) Recent climatic trends and linkages to river discharge in Central Vietnam. Hydrol Process 28(4):1587–1601

Salmi T, Maatta A, Anttila P, Ruoho-Airola T, Amnell T (2002) Detecting trends of annual values of atmospheric pollutants by the Mann-Kendall test and Sen's slope estimates. Finnish Meteorological Institute, Helsinki

Salvini G, Avitabile V (2013) Modeling of land change processes for 2010–2020 in the Vu Gia Thu Bon river basin, central Vietnam, Technical ReportLUCCi WT 7.4 Department of Earth Observation Friedrich Schiller University Jena, Germany

Schulla J (2012) Model description WaSiM. Technical report, p 305. Available at http://www.wasim.ch/en/products/wasim_description.htm. Accessed Jan 2014

Trinh QV (2014) Estimating the impact of climate change induced salinity intrusion on agriculture in estuaries—the case of Vu Gia Thu Bon, Vietnam. Doctoral dissertation. RUHR University Bochum, Germany, p 175

UNDP (2008) Human development report 2007/2008. Climate change and human development in Vietnam. Greet Ruysschaert, Peter Chaudhry, p 18

Waongo M, Laux P, Challinor AJ, Traoré SB, Kunstmann H (2013) Optimize crop planting date in West Africa: a crop model and fuzzy rule based approach for agricultural decision support. J Appl Meteorol Climatol 53:598–613

Waongo M, Laux P, Kunstmann H (2015) Adaptation to climate change: the impacts of optimized planting dates on attainable maize yields under rainfed conditions in Burkina Faso. Agric For Meteorol 205:23–39

# Impacts of Land-Use/Land-Cover Change and Climate Change on the Regional Climate in the Central Vietnam

**Patrick Laux, Phuong Ngoc Bich Nguyen, Johannes Cullmann and Harald Kunstmann**

**Abstract** Nowadays, it is widely accepted that both elevated greenhouse gas (GHG) concentration as well as land-use/land-cover change (LULCC) can influence the regional climate dynamics. It is a matter of fact that changes in the land use/land cover are often ignored in long-term regional climate projections. Even worse, often an outdated (RCM default) LU map is applied for modeling. In the framework of the LUCCi project, we applied the Weather Research and Forecasting Model WRF in combination with an updated LULC map to study (1) the impacts of an improved and updated Land-Use Land-Cover map on the regional climate in the VuGia-ThuBon basin in Central Vietnam; (2) the impacts separately of both the changed LU map and climate change (CC) on the regional climate; and (3) the sensitivity of land-use conversions in WRF simulations. It is found that the impacts of the outdated LU map exceed those of climate change, at least for the period 2001–2030. In addition, the deforestation scenario does not provide statistically significant signals of the most crucial surface climate variables, whereas the urbanization scenario provides evidence for a temperature signal (temperature increase) over the converted area, but no clear signal for precipitation is found.

---

P. Laux (✉) · P.N.B. Nguyen · H. Kunstmann
Institute of Meteorology and Climate Research, Karlsruhe Institute of Technology, Kreuzeckbahnstrasse 19, Garmisch-Partenkirchen 82467, Germany
e-mail: patrick.laux@kit.edu

P.N.B. Nguyen
IHP/HWRP Secretariat, Federal Institute of Hydrology, Am Mainzer Tor 1, Koblenz 56068, Germany

J. Cullmann
World Meteorological Organization, 7 bis, avenue de la Paix, Geneva 2 CH 1211, Switzerland



A. Nauditt and L. Ribbe (eds.), *Land Use and Climate Change Interactions in Central Vietnam*, Water Resources Development and Management,
DOI 10.1007/978-981-10-2624-9_9

## Introduction

Mankind has modified more than 50 % of the global land surface, thereby directly altering the exchange of energy, momentum, and water between the atmosphere and the land surface. Albeit it is well known that land-use land cover change (subsequently referred to as LULCC) plays a crucial role for the climate system, far more attention is still paid to the impact of climate change (CC) due to the changing atmospheric GHG composition. The LULCC impacts may vary depending on the region.

For example, urbanization typically leads to increased air temperature at surface as well as changes in the spatial patterns and intensities of precipitation, but their magnitudes depend on day time, season, geographical location, climate regime, circulation feedback, and surrounding land cover (e.g., Giannaros et al. 2013; Grimmond et al. 2011; Niyogi et al. 2011; Stewart and Oke 2012; Vargo et al. 2013). Deforestation may lead to decreased surface air temperatures in temperate regions, but increased temperatures in tropical regions (e.g., Costa and Pires 2010; Davin and de Noblet-Ducoudré 2010).

According to Mahmood et al. (2014), LULCC impacts are comparatively larger on local and regional than on global scales. Regional climate models (RCMs) comprise both terrestrial and atmospheric compartments and are therefore considered to be suitable tools to study land–atmosphere interactions. Consequently, RCMs can be used to analyze the impact of LULCC and CC on the regional climate system. Besides the advantages of using the RCMs, their results are affected by a number of uncertainties, which center on the applied land-use/land-cover (LULC) data, boundary conditions, and RCM parameterization schemes. Default land-use maps used in RCMs are often out-of-date.

Due to the high computational demands for regional climate modeling, only few studies exist which try to separate the impact of LULC from the impact of CC. For instance, Moore et al. (2010) compared the effects of projected future Greenhouse Gases and future LULCC on the spatial variability of crop yields in East Africa using a RCM. Their results provide evidence for significant impacts of both CC and LULCC, which can be of the same order of magnitude. To the knowledge of the authors no similar study exists for Southeast Asian domain so far.

This work tries to fill this gap. It has been conducted in the framework of the project Land Use and Climate Change Interactions in Central Vietnam (LUCCi). During the first phase of the project, long-term regional climate projections (based on the WRF-default LULC map) have been conducted using the regional climate model WRF to derive high resolution climate information for this poorly gaged region (Souvignet et al. 2014). The provision of new LULC maps during the course of the project, based on supervised classification of remote sensing data and ground truthing activities (Schultz and Avitabile 2012), offered new possibilities to analyze the impacts of the updated LULC information. In the WRF modeling system, the default LULC map dates back to 1992, and it is questionable if it can be used for climate projections for a region, which is characterized by tremendous land-use changes such as deforestation and urbanization.

## Specific Objectives

Based on modeling studies using the regional climate simulation model WRF, the following specific research questions will be addressed:

1. What are the impacts of an improved LULC representation on the representation of the most crucial climate variables?
2. What is the most dominant forcing for regional climate in the region, LULC or CC?
3. What is the sensitivity of LULCC in the regional climate model WRF for Central Vietnam?

## Data and Methods

In order to address the specific research questions, the following experimental design (Fig. 1) consisting of 3 different scientific approaches has been applied. WRF simulations are performed based on two different boundary forcing data, i.e., the ERA-interim reanalysis run and the ECHAM5 model (A1B emission scenario). The same domain as well as physical parameter setup has been applied for the 3 approaches (Laux et al. 2012, 2013).

<table>
<tr><th>Boundary forcing<br>LULC forcing</th><th>ERA-interim</th><th colspan="3">ECHAM5 (A1B)</th><th>ERA-interim</th></tr>
<tr><td></td><td>2010-2014</td><td>2001-2010</td><td>2011-2020</td><td>2021-2030</td><td>2010</td></tr>
<tr><td rowspan="2">WRF LULC-default</td><td rowspan="2">Control (5)</td><td colspan="3">Control (1)</td><td rowspan="2"></td></tr>
<tr><td>CASE 1</td><td>CASE2</td><td>CASE 3</td></tr>
<tr><td rowspan="7">WRF LULC-LUCCi</td><td rowspan="7">CASE-2010 (5)</td><td colspan="3" rowspan="2">CASE-2001 (1)</td><td>Control (15)</td></tr>
<tr><td>Urban-20 (15)</td></tr>
<tr><td rowspan="5">CASE 4</td><td rowspan="5">CASE 5</td><td rowspan="5">CASE 6</td><td>Urban-14 (15)</td></tr>
<tr><td>Urban-09 (15)</td></tr>
<tr><td>Deforest-20 (15)</td></tr>
<tr><td>Deforest-14 (15)</td></tr>
<tr><td>Deforest-09 (15)</td></tr>
</table>

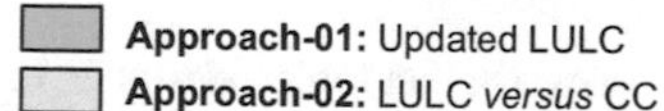

**Approach-01:** Updated LULC
**Approach-02:** LULC *versus* CC
**Approach-03:** Urbanization and deforestation

**Fig. 1** Experimental design of regional climate model WRF simulations to study the interactions of land use and climate change on regional climate in the Vu Gia-Thu Bon river basin of Central Vietnam. The colors show the applied approaches corresponding to the specific objectives outlined before. The number in brackets is the number of performed WRF simulations using perturbed initial conditions (PICs) of the forcing boundary data (Nguyen 2016)

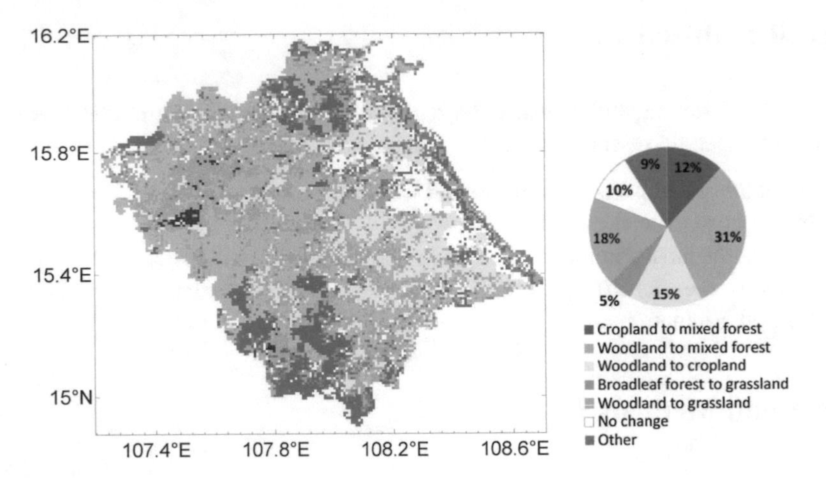


**Fig. 2** Changes in the land-use representation from WRF LULC-default to WRF LULC-LUCCi (Ph.D. Dissertation Thesis of Nguyen Phuong, under revisions)

To study the **impacts of the improved and updated Land Use Land Cover (Approach-01)** on the regional climate across the VGTB basin of Central Vietnam, WRF simulations driven with ERA-Interim have been applied for the period 2010–2014, both using the WRF LULC-default as well as the updated LULC map produced within the framework of the LUCCi project (subsequently referred to as WRF LULC-LUCCi map). Significant differences between both maps are found (Fig. 2). To account for RCM inherent uncertainties, 5 WRF simulations are performed for each land-use maps, respectively. The results are based on the average statistics of the 5 runs.

**Separating the impacts of LULC and CC on climate variables (Approach-02)**
The experimental design allows separating the impacts of LULC from CC (Fig. 1). The LULC signal for different periods can be extracted by the calculation of CASE 4 *minus* CASE 1, CASE 5 *minus* CASE 2, and CASE 6 *minus* CASE 3, whereas the CC-induced signals can be calculated as CASE 2 *minus* CASE 1, CASE 3 *minus* CASE 1, CASE 5 *minus* CASE 4, and CASE 6 *minus* CASE 4. Combined LULC and CC impacts correspond to the CASE 5 *minus* CASE 1 and CASE 6 *minus* CASE 1. Please note that the calculated differences are based on single WRF simulations only, PICs have not been possible due to the high computational demands.

**Potential impacts of urbanization and deforestation on regional climate in Central Vietnam (Approach-03)**
Focus of Approach-03 is the sensitivity as well as uncertainties of different land-cover scenarios on regional climate in the VGTB basin. For this reason, a new methodology has been developed to separate the land-use change-induced signal in the WRF simulations from the noise caused by PICs. Based on the simulations for 2010, the signal-to-noise ratio (SNR) in concert with statistical tests is applied to identify

regions with pronounced LULC signals. The impact of the PICs ensemble size is analyzed to derive robust recommendations for LULCC-induced climate modeling studies. In this case, a relative large number of PICs, i.e., 15 members, have been applied to systematically study the modeled impacts of LULCC with different magnitudes, expressed as different radii around a station, i.e., 20, 14, and 9 km.

## Results

In the following the results of the aforementioned approaches are presented.

### *Impacts of Improved and Updated Land Use Land Cover (Approach-01)*

The WRF simulations using the WRF LULC-LUCCi map shows remarkable differences in the heat fluxes and the albedo across the domain compared to the WRF-default LULC map: While latent heat is significantly reduced for the highland regions in the Western parts, the sensible heat flux is reduced in the Eastern parts of the basin. Surface air temperature is increased up to 1 °C across the VGTB basin. Annual precipitation is found to be increased of about 500 mm for the whole basin, with regional differences ranging from precipitation increases up to 1500 mm in the Northern basin to decreased of about 500 mm in the Western parts of the basin. The differences are also analyzed with respect to the seasonality and land-use type. For surface air temperature, the changes are highest for July and August, also showing the highest spread between the different LU classes (Fig. 3). For precipitation, mostly precipitation increases are simulated over the year except for November/December.

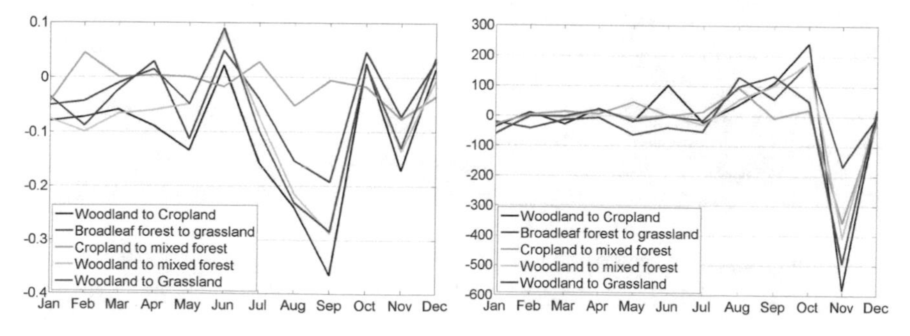


**Fig. 3** Simulated response of the improved WRF LULC-LUCCi on surface air temperature [°C] (*left*) and precipitation [mm] (*right*) averaged over the VGTB basin for the period 2010–2014. The differences are shown compared to WRF LULC-default simulations (WRF LULC-LUCCi minus WRF LULC-default) based on the mean of the PICs simulations (Nguyen 2016)

During this period, up to 500 mm of precipitation decreases are simulated, also having a high spread between the land-use classes (not shown).

## *Separating the Impacts of LULC and CC on Climate Variables (Approach-02)*

Figure 4 illustrates the results of the performed transient WRF simulations for 2001–2030 for 2 m air temperature and precipitation. It can be seen that the effects of the updated LULC map partly increase the effects caused by climate change. Due to the higher magnitudes of the LULC signals, the combined signal resembles much stronger the LULC signal compared to the CC signal. One can observe strong regional differences for temperature and precipitation. For temperature, the LULC effects are much stronger for the costal regions (Eastern part of the basin) than for

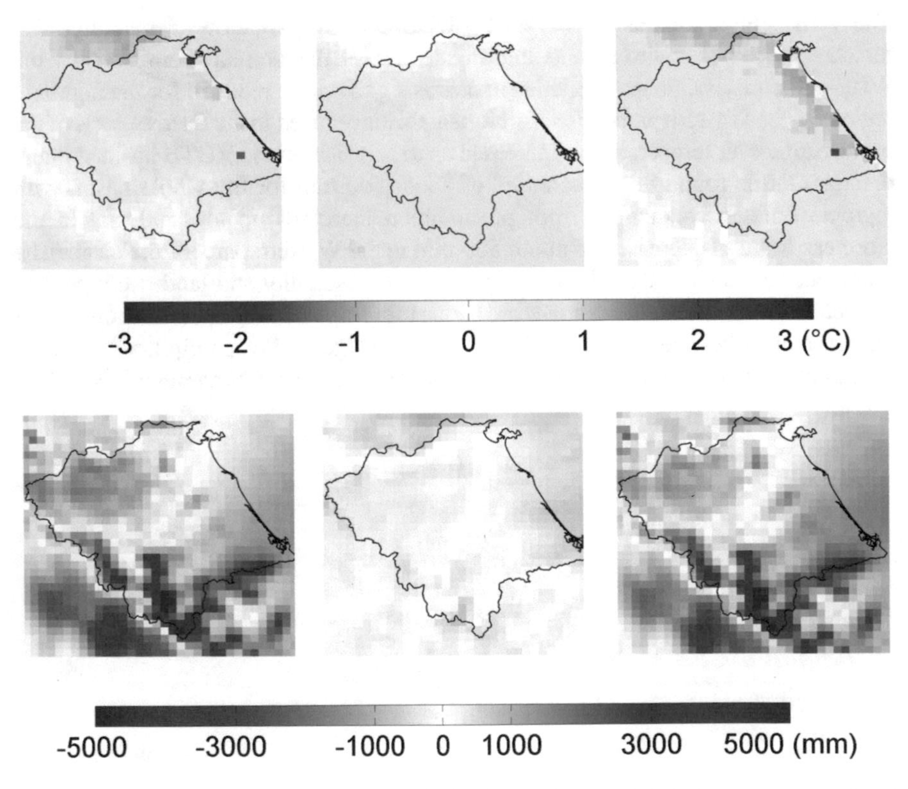


**Fig. 4** Signals, caused by: LULC update only (CASE 6 minus CASE 3) for temperature (**a**) and precipitation (**d**), CC only (CASE 6 minus CASE 4) for temperature (**b**) and precipitation (**e**), as well as combined LULC update and CC signals (CASE 6 minus CASE 1) for temperature (**c**) and precipitation (**f**) (Nguyen 2016)

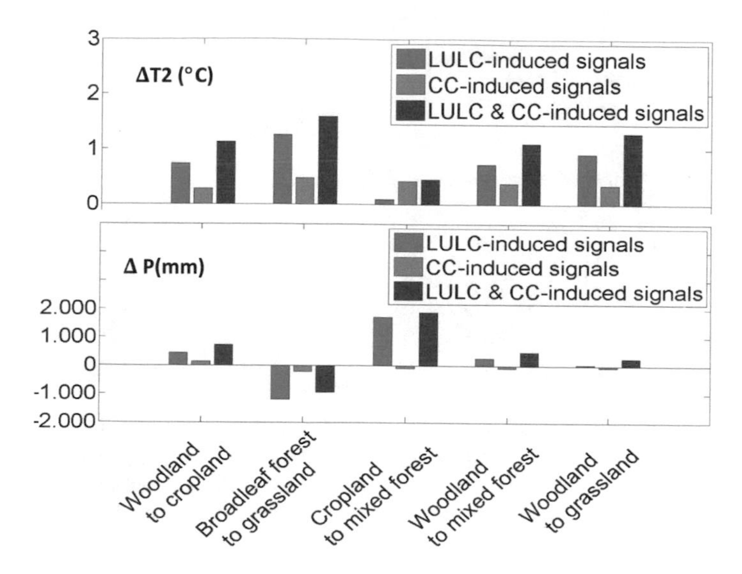


**Fig. 5** Signals for temperature (*top*) and precipitation (*bottom*), averaged for different LU replacements in the WRF LULC-LUCCi map compared to the outdated WRF LULC-default map (Nguyen 2016)

the highland (Western and Southernmost regions). In the highlands, the CC-induced impact is found to be slightly higher than those of the updated LULCC map. This is related to the relatively small differences in the parameterization values between the LU classes within the Western and the southernmost regions. For the simulations based on the updated LU map, strong decreases are found for the Northwestern part of the basin (up to 1500 mm/year), whereas the Southern parts exhibits locally increases of more than 4000 mm. Compared to those signals, the expected impacts caused by CC are small and range between +500 and −500 mm.

Figure 5 confirms the relatively larger effects of the changed LULC compared to CC. For temperature only positive signals (temperature increases) are observed, whereas precipitation also shows strong negative signals for the regions in which broadleaf forest is replaced by grassland.

## *Potential Impacts of Urbanization and Deforestation on Regional Climate in Central Vietnam (Approach-03)*

By using ensemble simulations consisting of 15 members, we identified pronounced signals for the urbanization scenario (urbanization around DaNang station), leading to increased latent heat fluxes in the surrounding of DaNang. Due to energy conservation, the sensible heat fluxes are reduced. The overall impact of the urbanization is a mean increase of temperature of about 2 °C (Fig. 6). This increase

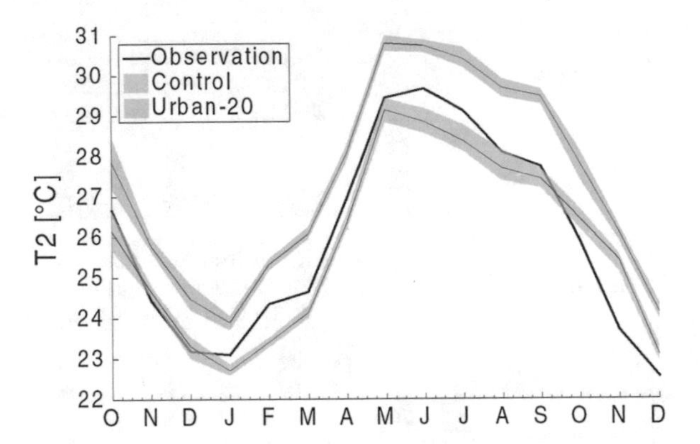


**Fig. 6** Temperature signal from control and sensitivity simulations for the urbanization scenario (20 km radius around DaNang) at DaNang station based on WRF ensemble simulations for 2010 (Ph.D. Dissertation Thesis of Nguyen Phuong, under revisions). Since the envelopes of the control and sensitivity experiment do not overlap, a clear temperature signal can be concluded (Nguyen 2016)

is found to be relatively constant throughout the year with slightly increased values (about 2.5 °C) for the summer months (JJA). No clear signals are found for precipitation.

The deforestation scenario did not lead to significant changes in the partitioning of the heat fluxes. As a consequence no clear signals are found for temperature and precipitation in the study region.

Based on signal-to-noise analysis and bootstrap significance tests, recommendations can be given about how many RCM ensemble simulations are necessary to robustly demonstrate the existence or non-existence of a signal. For more information, the reader is referred to Laux et al. (2016).

## Summary and Conclusions

Based on the results presented above, it is concluded that regional climate simulations should not be performed with the default land-use descriptions of the RCMs. In such regions with heavily altered land surface conditions mostly caused by intensifications of agriculture and urbanizations we recommend to use updated land-use information, i.e. more recent land-use maps to model recent climate conditions. The effects caused by outdated land-use information can exceed the effects caused by climate change. Long-term climate projections using default LU information may potentially lead to erroneous conclusions about climate change. Since the RCM internal variability (model noise) may be larger than the signals caused by LULCC and/or CC, it is also concluded that ensemble climate simulations are required to derive robust signals.

## References


Costa MH, Pires GF (2010) Effects of Amazon and Central Brazil deforestation scenarios on the duration of the dry season in the arc of deforestation. Int J Climatol 30:1970–1979

Davin EL, de Noblet-Ducoudré N (2010) Climatic impact of global-scale deforestation: radiative versus nonradiative processes. J Clim 23:97–112

Giannaros TM, Melas D, Daglis IA, Keramitsoglou I, Kourtidis K (2013) Numerical study of the urban heat island over Athens (Greece) with the WRF model. Atmos Environ 73:103–111

Grimmond C, Blackett M, Best M, Baik J-J, Belcher S, Beringer J, Bohnen- stengel S, Calmet I, Chen F, Coutts A et al (2011) Initial results from Phase 2 of the international urban energy balance model comparison. Int J Climatol 31:244–272

Laux P, Phan VT, Lorenz C, Thuc T, Ribbe L, Kunstmann H (2012) Setting up regional climate simulations for southeast Asia. In: Nagel WE, Kröner DB, Resch MM (eds) High Performance Computing in Science and Engineering '12, Conference Proceedings. Springer, pp 391–406

Laux P, Thuc T, Kunstmann H (2013) High resolution climate change information for the lower mekong river basin of southeast Asia. In: Nagel WE, Kröner DB, Resch MM (eds) High Performance Computing in Science and Engineering '13, Conference Proceedings. Springer, pp 543–551

Laux P, Nguyen PNB, Cullmann J, Van TP, Kunstmann H (2016) How many ensembles provide confidence in the impact of land cover change. Int J Climatol (online first), doi:10.1002/joc.4836

Mahmood R, Pielke RA Sr, Hubbard KG, Niyogi D, Dirmeyer P, McAlpine C, Carleton A, Hale R, Gameda S, Beltran-Przekurat A, Baker B, McNider R, Legates DR, Shepherd M, Du J, Blanken P, Frauenfeld OW, Nair US, Fall S (2014) Land cover changes and their biogeophysical effects on climate. Int J Climatol 34:929–953. doi:10.1002/joc.3736

Moore N, Torbick N, Lofgren B, Wang J, Pijanowski B, Andresen J, Kim D-Y, Olson J (2010) Adapting MODIS-derived LAI and fractional cover into the RAMS in East Africa. Int J Climatol 30:1954–1969

Nguyen PNG (2016) Biogeophysical impacts of land-use/land-cover on regional climate. A study for the Vu Gia—Thu Bon basin in Central Vietnam. Ph.D. dissertation thesis (to be submitted to Augsburg University)

Niyogi D, Pyle P, Lei M, Arya SP, Kishtawal CM, Shepherd M, Chen F, Wolfe B (2011) Urban modification of thunderstorms: an observational storm climatology and model case study for the Indianapolis urban region. J Appl Meteorol Climatol 50:1129–1144

Schultz M, Avitabile V (2012) VGTB Land Cover 2010 v2—Updated. Project Report. Vu Gia Thu Bon Information System, 2012. http://leutra.geogr.uni-jena.de/vgtbRBIS/metadata/start.php

Souvignet M et al (2014) Recent climatic trends and linkages to river discharge in Central Vietnam. Hydrol Process 28:1587–1601. doi:10.1002/hyp.9693

Stewart ID, Oke TR (2012) Local climate zones for urban temperature studies. Bull Am Meteorol Soc 93:1879–1900

Vargo J, Habeeb D, Stone B (2013) The importance of land cover change across urban–rural typologies for climate modeling. J Environ Manage 114:243–252

# Integrated River Basin Management in the Vu Gia Thu Bon Basin

**Lars Ribbe, Viet Quoc Trinh, A.B.M. Firoz, Anh Thu Nguyen, Uyen Nguyen and Alexandra Nauditt**

**Abstract** The Vu Gia Thu Bon River basin sustains important functions and services, which provide water, food, and energy resources for the people living in the basin as well as for people beyond the basin boundaries. Furthermore, the river basin system provides regulating services like flow attenuation, and various cultural and recreational services. During the last decades, significant infrastructure, like reservoirs, channels, water treatment, and distribution networks, was built to increase the benefits generated from the basin in a reliable and secure way. In order to maintain these functions many institutions are responsible to manage certain sectorial aspects of natural resources, often depending on each other to perform their tasks adequately. During the last decades an intensification of resources uses occurred and the trade offs between different sectors like hydropower, agriculture and urban water has increased—in particular during years or seasons which are characterized by climate extremes like droughts or floods. This situation calls for a more coordinated management approach leading to the collaboration of different sectors throughout the river basin. Providing the right information for decision support is a crucial element of river basin management. In this line, Vu Gia Thu Bon River Basin Information Center (VGTB-RBIC) was established by the research consortium making a reliable scientific database available for all different stakeholders as a first but crucial step toward a coordinated approach to river basin management.

## Introduction

The integrated management of water and related resources at the basin level is widely considered as the right approach to cope with challenges like floods, droughts, pollution or water resources allocation as these issues are typically characterized by a high level of complexity and multiple interactions and feedbacks of the environmental and social systems involved (Bruns et al. 2002).

L. Ribbe (✉) · V.Q. Trinh · A.B.M. Firoz · A.T. Nguyen · U. Nguyen · A. Nauditt
Institute for Technology and Resources Management in the Tropics and Subtropics, Technische Hochschule Köln - University of Applied Sciences, Köln, Germany
e-mail: lars.ribbe@th-koeln.de



A. Nauditt and L. Ribbe (eds.), *Land Use and Climate Change Interactions in Central Vietnam*, Water Resources Development and Management,
DOI 10.1007/978-981-10-2624-9_10

Integrated Water Resources Management (IWRM) should foster the connection of the water sector to other sectors and support the overall development process. In this line, IWRM is mentioned as one of the targets of the Sustainable Development Goals (SDGs; target 6.5[1]). Achieving Water Security at river basin level can be considered as the crosscutting theme that also has direct linkages to several other development goals. Considering the central role of water development for other sectors, there is also a call for a more intersectoral vision of resources management at river basin level—beyond or rather in addition to the IWRM approach—giving more independent consideration to the other key sectors involved. This approach, known as the "water-energy and food security nexus" (WEF Nexus), is increasingly considered for application in the context of river basin management (UNECE 2015).

Prior to being formulated as a target of the Sustainable Development Goals, there was already ample reference to IWRM in many water policies worldwide (for examples see the EU Water Framework Directive 2000; Brazil National Water Resources Policy 1997).

While the concept of IWRM or WEF-Nexus is convincing, its implementation is posing significant challenges to national and local water governance. In particular the question of who has the coordination role and how to achieve coherence between different actors and sectors is difficult to address (Bateman and Racier 2012). The need to consider various natural and human environment subsystems poses a challenge to implementing IWRM. Environmental elements comprise ground and surface water interactions, water quality and water quantity linkages, natural variability and trends of the climate system, upstream and downstream relationships, among others. Regarding the human system interactions of institutional, socioeconomic and financial functions in relation to strategic and operational decision-making need to be addressed and are typically related to diverse sectors like irrigation and agriculture, water supply and sanitation, flood and drought management, hydropower and navigation etc.

At river basin level IWRM should start with an assessment of the water related services which are provided by river basin management like agricultural production, energy production, domestic and industrial supplies, fisheries, recreation, wildlife and freshwater protection, and forest conservation. The next section analyzes these functions for the Vu Gia Thu Bon river basin.

## Functions and Services the Vu Gia Thu Bon Basin Provides

The various catchments of the Vu Gia Thu Bon river basin provide significant services regarding the provision of water, energy, and food resources as well as recreational and cultural activities. The following paragraphs provide some key facts about the services based on natural resources.

[1]https://sustainabledevelopment.un.org/sdg6.

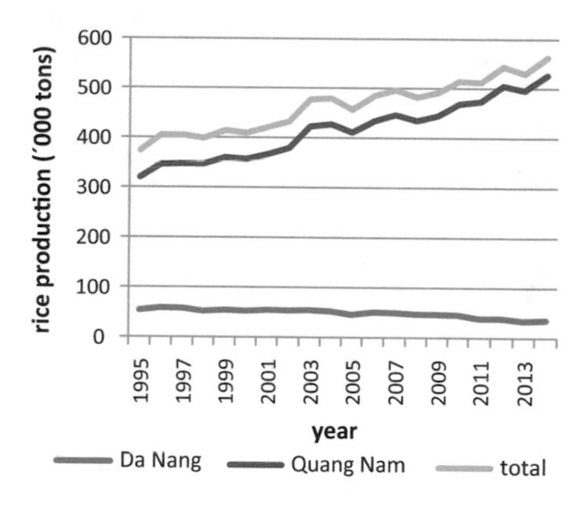


**Fig. 1** Total production of cereals in Da Nang and Quang Nam

## *Food*

In particular, the lower part of the basin is endowed with fertile soils, which together with the usually abundant water resources yield high harvests for a diversity of crops, with rice being by far the dominant crop. Rice represents the most important staple food in the whole of Vietnam. Per capita consumption is 214 kg year on average[2] and accounts for 60 % of caloric supplies of the population. Considering this national average the total rice consumption of the 2.5 million inhabitants in the VGTB basin is an estimated 535,000 tons, which matches quite well the annual rice production (compare Fig. 1) in the basin. In fact, the region is self-sufficient regarding rice production but does not contribute to rice exports of the country. While the population of VGTB represents 2.8 % of the Vietnamese population, the total rice production in VGTB represents around 1.3 % of the total rice production in Vietnam (45 million tons) (GSO 2014).

In past decades higher rice yields resulted in a higher rice production even though the area planted with rice diminished slightly. This gave rise to the increase of other crops like cassava, maize, and vegetables (compare Chapter "Rice-Based Cropping Systems in the Delta of the Vu Gia Thu Bon River Basin in Central Vietnam" of this book).

## *Energy*

High amounts of rainfall, particularly in the mountainous part of the VGTB basin, gives rise to a large hydropower potential. The total installed capacity (2013) is

[2]http://www.uark.edu/ua/ricersch/pdfs/per_capita_rice_consumption_of_selected_countries.pdf.

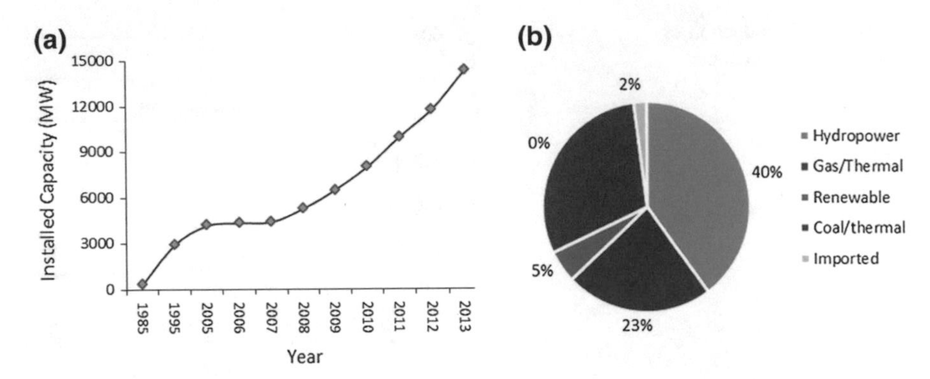


**Fig. 2** Hydropower development and energy production pattern in Vietnam. **a** Temporal development of hydropower plants in Vietnam, cumulative installed capacity of hydropower plants in MW from 1985 to 2013. **b** Energy production pattern in Vietnam in year 2013. *Data source* MOIT (2015)

1026 MW with an additional 4698 MW planned to be built until 2020. The Vu GiaThu Bon (VGTB) river basin is ranked fourth in Vietnam for potential hydropower generation capacity after the Da, Dong Nai, and Se San river systems. The total generation of hydropower was 3985 TWh in 2013 (MOIT 2014). Compared to an average consumption of 1415 KWh per capita at the country level, the total electricity produced by hydropower in the VGTB outweighs the estimated consumption (total 3.5 TWh).

Hydropower is traditionally an important sector of Vietnam's energy system. Approximately 40 % of the total electricity (2014: 128.4 TWh; ADB 2015) is generated by hydropower (Fig. 2b). According to MOIT (2014), the total national hydropower potential that can be exploited from a technical and economic perspective amounts to 26 GW, with a potential hydropower production of 100 TWh (MOIT 2014). Currently a capacity of 14240 MW is installed, i.e., around 55 % of the potential is already used (Fig. 2).

## *Water*

Considered as a long term annual average, water is a very abundant resource in the VGTB basin. With an average discharge of $12 \times 10^9$ m$^3$ the Vu Gia Thu Bon river basin is the fifth most water rich river basin in the country. However, run off generation is characterized by high inter- and intra-annual variability. The driest recorded year so far created a discharge of $5 \times 10^9$ m$^3$ (1982) and 70 % of discharge usually occurs during the five months of rainy season (Oct–Feb). Furthermore, low flow during the dry months causes intrusion of saltwater to the rivers and channels, limiting the availability of freshwater even more (compare chapter "Hydrological and Agricultural Impacts of Climate Change in the Vu Gia-Thu Bon River Basin in Central Vietnam" in this book).

**Table 1** Designed and actual irrigation capacity in the Vu Gia Thu Bon Basin

| Irrigation structure | Amount | Designed capacity (ha) | Actual capacity (ha) | Actual/designed |
|---|---|---|---|---|
| Reservoir | 95 | 41,632 | 23,622 | 0.57 |
| Small dam | 561 | 7951 | 5421 | 0.68 |
| Pumping station | 216 | 18,147 | 12,854 | 0.71 |
| Sum | | 66,329 | 41,029 | 0.62 |

*Sources* Quangnam (2012) and Danang (2012)

Water supply to agriculture constitutes the largest water allocation in the basin. According to Quangnam (2012) and Danang (2012) water supplied for irrigated agriculture in the basin in the year 2012 amounted to 244.7 $Mm^3$ for a total irrigated area of 41,029 ha. By 2010, Quangnam and Danang had altogether 95 irrigation reservoirs with a total capacity of $497.4 \times 10^6$ $m^3$. The irrigated area of these reservoirs was 23,622 ha in that year, equal to 57 % of designed capacity. The basin has 561 small weirs, mainly located in the upland areas, providing water for 5420 ha. In the lowlands, 216 pumping stations irrigate 12,854 ha (71 % of designed capacity). Altogether irrigation works provide water for 39,707 ha or 63 % of rice cultivated areas in the province (Table 1).

The total consumptive water demand in the basin amounts to well over $1100 \times 10^6$ $m^3$ per year while the estimated demand needed to control salt water intrusion is estimated by Viet (2014) as more than three times this amount (compare Table 2).

## *Tourism*

The number of tourists visiting the region is constantly rising, particularly in and around Danang city—a growing and prospering city at the side of the sea. Quang Nam Province is well-known for its many and diverse festivals during a year. The

**Table 2** Water demand of different sectors (average estimate based on 2012 data)

| Water use activity | Yearly discharge volume required ($10^6$ $m^3 \times year^{-1}$) |
|---|---|
| Agriculture | 892 |
| Domestic uses | 145 |
| Industry | 79 |
| Preventing saltwater intrusion and maintaining ecological flows | 3690 |
| Total | 4805 |

**Table 3** Number of tourists in Danang and Quang Nam

| Year | Danang (1000 visitors) | Quang Nam (1000 visitors) | Sum (1000 visitors) |
|---|---|---|---|
| 2009 | 1131 | 1764 | 2895 |
| 2010 | 1499 | 2097 | 3596 |
| 2011 | 2229 | 2160 | 4389 |
| 2012 | 2571 | 2548 | 5119 |
| 2013 | 2939 | 3208 | 6147 |
| 2014 | 3800 | 3700 | 7500 |
| 2015 | 4600 | 3900 | 8500 |

*Source* www.vietnamtourism.gov.vn

major urban center attracting tourism in Quang Nam is Hoi An city while many tourists also visit the region for leisure at the beach and visiting the attractive landscape around the urban centers (Table 3).

## Water Resources Development in VGTB

Population grew rapidly during the past three decades and the demand for urban water supplies, for agricultural and for energy production increased dramatically spurring a development of related water infrastructure. Figure 3 summarizes the developments for the lower VGTB basin based on digitized maps since 1925 which enabled the reproduction of information on infrastructure development in the region (based on Viet 2014). By 1925, the Vu Gia drained to the Thu Bon through the Quang Hue River. To supply water for Da Nang city, a canal was dredged to connect the Vu Gia and Tuy Loan rivers. Since then, the Vu Gia has been flowing to the Cua Han mouth in Da Nang as seen today. Afterward, in order to regulate river flows, the An Trach barrage system was constructed on the Vu Gia and its tributaries. In 2006, the Duy Thanh barrage was built on the Ba Ren River to prevent saltwater intrusion. After 1975, a number of pumping stations were constructed in the region for irrigation purposes while more recently 13 pumping stations have been removed due to the expansion of urban areas into former irrigation areas. The water works in the VGTB lowland are mostly constructed for irrigation purposes, however, the Cau Do and Hoi An pumping stations were constructed to supply drinking water to the cities Da Nang and Hoi An.

Agricultural production plays an important role in the economic structure of VGTB basin. In 2013, half of the population in the region economically relied on agricultural production. Although there is an increasingly diversified crop pattern, paddy rice is still the dominant crop as it accounts for approximately 70 % of the irrigated area. Demand for irrigation water therefore is high and seasonal. Water sources are under increasing stress due to human interventions and climate

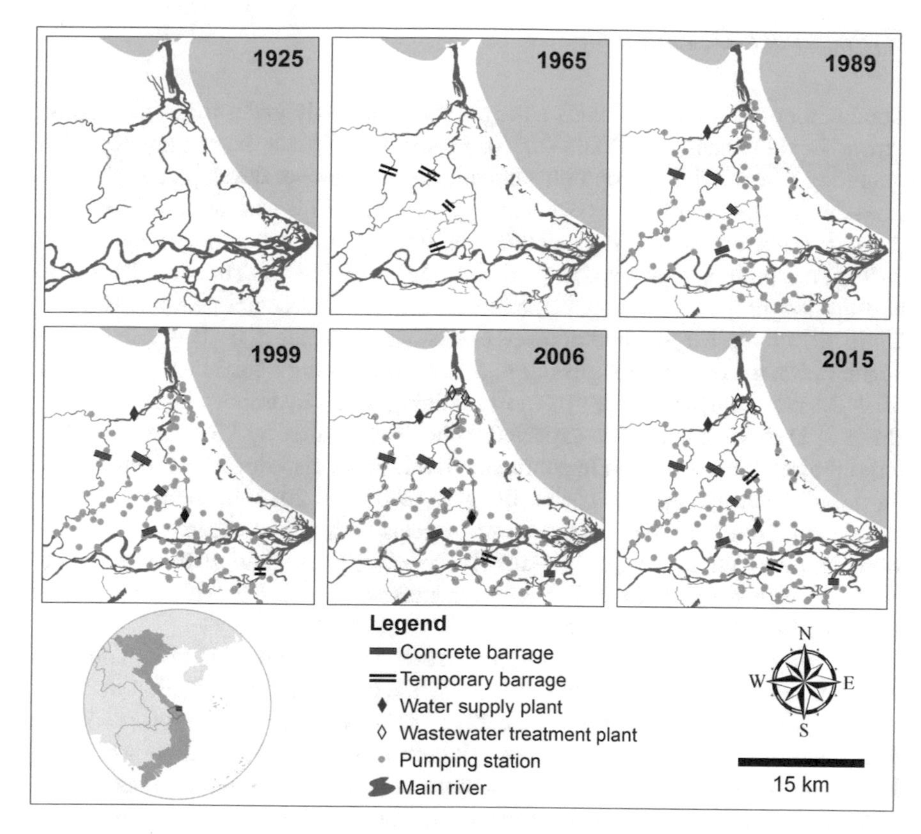


**Fig. 3** Changes in river courses and water works in the estuary during 1925–2015. An cities. *Source* Based on Viet (2014)

change and the agricultural production, particularly rice production, is becoming more vulnerable. Water stress in drought periods is defined as the major constraint to agricultural production in the region. Low river discharge exacerbates saltwater intrusion into rivers, which threatens normal activities of irrigation pumping stations. Meanwhile, existing irrigation structures have not yet met production requirements. Up to 2010, the basin has constructed 820 irrigation works including 72 reservoirs, 546 weirs, and 202 pumping stations.

However, the effective capacity of the irrigation system is only 41,029 ha of agricultural land, compared to a designed capacity of 66,329 ha (Table 1). Except for the Phu Ninh reservoir, the regulatory capacity of the system is very low, keeping the dependence on direct pumping from the river the main irrigation water source. Therefore, improving irrigation infrastructure, increasing water use efficiency and reusing return flow are considered as priority strategies in the VGTB basin (Van 2013; Viet 2014).

## *Hydropower Development*

The hydropower sector in VGTB emerged rather recently under the Sixth National Power Development Plan (2006–2010). From 2003, Quang Nam province began preparing a provincial hydropower plan as part of its power development plan. The cascade hydropower development planning for the VGTB was approved by MOIT through Decision No. 875/QĐ-KHĐT identifying 10 large hydropower projects (>30 MW) with a total capacity of 1147 MW (Table-1). The first large-scale hydropower project, A Vuong, came into operation in December 2008 (210 MW). Until 2015 a further six hydropower plants were constructed (Dak Mi 4, Song Tranh 2, Song Con, and Song Bung 4, 5, and 6) reaching a total installed capacity of 940 MW (compare Table 4). The remaining three hydropower projects Song Bung 2, Dak Mi 2 and 3 are expected to be in operation by 2020. Next to these large projects, the small and medium size rivers were also developed with a focus on run-off-river hydropower plants. Between 2008 and 2013, 36 small (<10 MW) and medium (>10–30 MW) hydropower projects with a total capacity of 436 MW were approved under the Quang Nam Hydropower plan. Until 2014 nine projects with a total capacity of 130 MW have been finished. Table 4 provides an overview of the hydropower plants which are currently operated.

Currently, due to highly variable and seasonal climatic conditions combined with low storage capacities of the reservoirs, the hydropower production fluctuates significantly. In the unusual dry year 2013, for example, total energy production from hydropower was extremely low. Figure 4 classifies the hydropower plants according to type, installed capacity, storage volume, and operation mode.

While most large-scale hydropowers are now in operation, there are several medium and small HPP under construction or planning. Storage plants with a capacity of larger than 30 MW represents a large proportion (72 %) of all

**Table 4** Operated large hydropower plants

| Reservoir | First year of operation | Reservoir total storage (Mm$^3$) | Installed capacity (MW) | Annual average energy potential GWh) |
|---|---|---|---|---|
| A Vuong | 2008 | 343.6 | 210 | 825 |
| Song Con 2 | 2009 | 1.2 | 46 | 168 |
| Song Tranh 2 | 2011 | 733.4 | 162 | 621 |
| Dak Mi 4 Upper | 2011 | 310.0 | 141 | 582 |
| Dak Mi 4 Lower | 2011 | 2.6 | 39 | 161 |
| Song Bung 5 | 2014 | 20.3 | 57 | 220 |
| Song Bung 6 | 2014 | 3.3 | 29 | 151 |
| Song Bung 4 | 2015 | 510.8 | 156 | 618 |
| Song Bung 2 | 2016 | 230.0 | 100 | 426 |

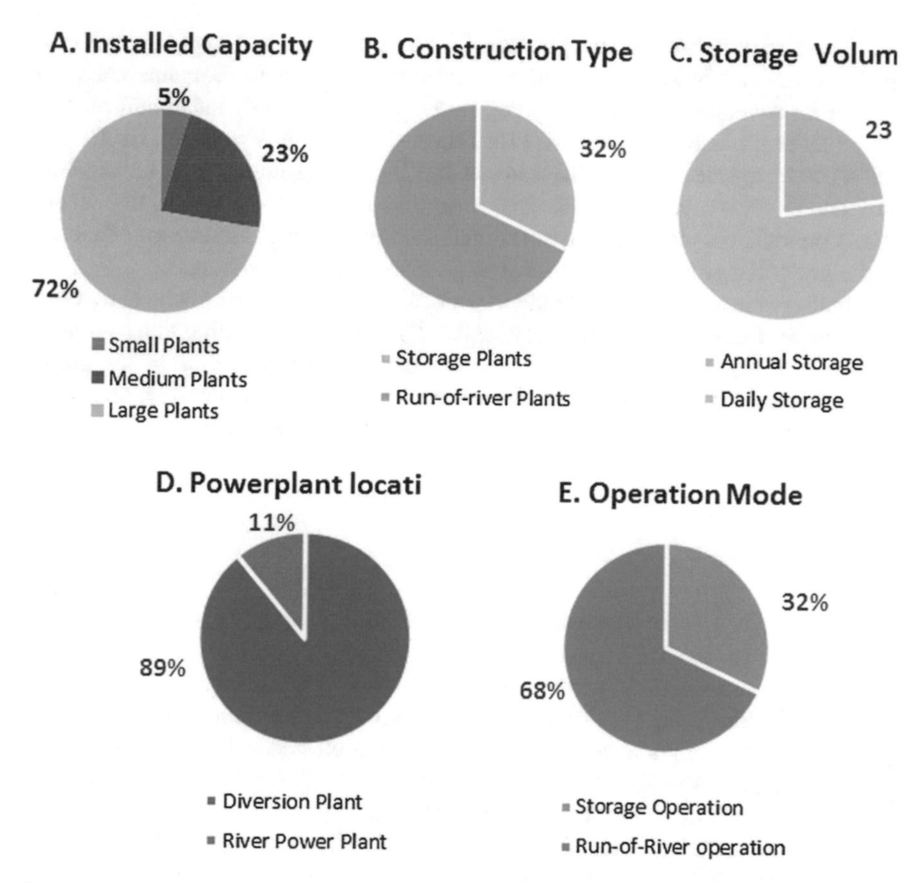


**Fig. 4** Classification of hydropower plants in VGTB. Statistical analysis of hydropower plants in VGTB with the existing and planned hydropower which will be feeding into the grid according to the criteria **a** installed capacity, **b** construction type, **c** storage volume, **d** location of the power house and operation mode; illustrated is the respective share dependent on the total number of hydropower ($n = 44$). All the data for this analysis was collected from Quang Nam department of Trade and Industry (DTID) 2013

hydropower projects. Whereas some large reservoirs are still under construction, there is a discussion on investments in large hydropower projects which would allow to add functions of water storage for the purpose of flood and drought risk management and even increase irrigation security.

While the hydropower sector has been developing extremely fast since 2008, criticism regarding the design of the plants, operational management of the hydropower and impact on downstream water availability was eventually brought up in the media.

The controversy regarding hydropower development was further aggrevated by the dam construction failure of the Song Tranh 2 hydropower project. Several cracks were observed in the main dam leading to the immediate shut down of the dam just 6 months after its inauguration. The authority, under the pressure of news media, civil society and the scientific community, started to reorganize and reevaluate its development plan of hydropower. As a result, nine small and medium hydropower projects (e.g. Dak Sa, Dak Pring, Cha Val, A Vuong 3, Dak Di 1, 2, and 4) were rejected by the Chairman of the People's committee of Quang Nam province on December 22, 2013. The cause was thought to be due to the adverse effects on local people, land, environmental disturbance, low investment efficiency, or the projects not implemented according to schedule commitments.

Another concern arose due to the impact of the dam Dak Mi 4 whose inflow is from the Vu Gia river but the outlet is to the Thu Bon River leading to a significant reduction of flow to the Vu Gia river mid- and downstream. One of the consequences is the further intrusion of salt water to the Vu Gia branch with one of the effects being the salinity increase at Cau Do water treatment plant.

## *Domestic Water Supply*

Da Nang Water Supply Company (DAWACO) and Quang Nam Water Drainage and Supply Joint Stock Company are responsible for water supply in Da Nang city and Quang Nam Province, serving the urban areas in the basin. In the rural area there is no access to this service and people mainly exploit ground water from wells or surface water from streams and rivers to meet their water demands.

According to CVIWR (2014), 92.9 % of the people in the center of Da Nang had access to a continuous supply of clean water as opposed to only 38.8 % of the people in the suburban area of Hoa Vang district. Thus, a major challenge is to achieve water security for people in particular in the rural areas.

Table 5 lists the main water supply infrastructure and its supply capacities.

### Drainage Infrastructure

Da Nang has a combined drainage system covering most of the urban area. There are several recently established residential areas which have a separated drainage system collecting wastewater to be conducted to the Wastewater Treatment Plant (WWTP) (Table 6). Most other households are equipped with septic tanks.

Quang Nam also has largely a combined drainage system while the capital Tam Ky is equipped with a separated collection system and a modern WWTP. Other areas do not have a functioning drainage system. Wastewater from the suburban and rural areas are to a large degree discharged directly into rivers, streams, or channels.

**Table 5** Water treatment plants in 2012

| Province | Water treatment plant (WTP) | Capacity ($10^3$ $m^3$/day) | Water sources |
|---|---|---|---|
| Danang | Cau Do (old) | 50.0 | Vu Gia River |
| | Cau Do (new) | 120.0 | Vu Gia River |
| | San Bay | 30.0 | Vu Gia River |
| | Son Tra | 5.0 | Son Tra streams |
| Quangnam | Dien Nam-Dien Ngoc | 5.0 | Ground water |
| | Hoi An | 21.0 | Vinh Dien River |
| | Phu Ninh (Phase 1) | 100.0 | Phu Ninh Reservoir |
| | Tam Hiep | 15.0 | Trau River |
| | Others (less than 5000 $m^3$/day) | 12.0 | |

**Table 6** Wastewater treatment plants in 2014

| Province | Wastewater treatment plant (WWTP) | Capacity ($10^3$ $m^3$/day) | Technical description |
|---|---|---|---|
| Danang | Phu Loc | 36.0 | Anaerobic Pond |
| | Son Tra | 10.0 | Anaerobic Pond |
| | Ngu Hanh Son | 10.0 | Anaerobic Pond |
| | Hoa Cuong | 36.0 | Anaerobic Pond |
| | Hoa Xuan | 20.0 | Sequence Batch Reactor (SBR) |
| Quangnam | Dien Nam- Dien Ngoc | 5.0 | Anaerobic Pond |
| | Bac Chu Lai | 1.9 | Anaerobic Pond |
| | Co Khi Chu Lai Truong Hai | 3.8 | Anaerobic Pond |
| | Tam Ky | 15.0 | Anaerobic Pond |

## History of River Basin Management in Vietnam and Significance for the VGTB Basin

In 1998, the Law on Water Resources was enacted including the first reference to water management at river basin level. Article 5 (Sect. 1) regulated that “The protection, exploitation, and use of water resources shall follow river basin planning approved by one agency irrespective of the administrative borders crossing the basin (National Assembly of Viet Nam 1998). The law determined that MARD is responsible to form such basin agencies and detailed the organization and operation of River Basin Planning and Management Boards (RBMPB, Sect. 4, Article 64). On this basis, MARD established RBPMBs for key rivers namely Mekong, Dong Nai, Red-Thai Binh and Vu GiaThu Bon, and Dong Nai. RBPMBs functioned as public service delivery units led by a representative of MARD at Vice Minister level for collaborating and harmonizing benefits of related stakeholders in water resources management within each river basin. However, in 2002, the newly established Ministry of Natural Resources and Environment (MoNRE) was assigned by the government to act as a focal point of water resources at national scale. MoNRE is accountable for regulating and directing the execution of water resources protection as well as holding the role of "standing member of the National Water Resources Council". As implementation of this law MoNRE established the River Basin Environment Protection Committee (RBEPC) for Cau, Dong Nai and Nhue-Day rivers.

Since the Law on Water Resources was adopted, two types of river basin organizations were established while neither the RBPMBs nor the RBEPC followed a truly integrated approach to water resources management. The former focused too much on irrigation and flood issues while the latter puts an emphasis on water quality. Furthermore, in both cases neither the administrative implementation of stakeholder dialog nor a financial mechanism—two essential elements of successful RBOs- were operational.

In December, 2008, after 10 years of implementation of the Law on Water Resources, the GoV enacted the Decree no. 120/2008/NĐ-CP regarding River Basin Management as the most significant milestone in transforming traditional water resources management into IWRM following a river basin approach. Based on this Decree, a new form of RBOs can be established, the so-called River Basin Committee (RBC) which holds the function of "supervising and coordinating activities carried out under the responsibility of different ministries, sectors, and localities in implementing river basin planning and furthermore "proposing policies, recommending measures for water protection, water resources exploitation, use, and development" (GoV 2008). In line with this Decree, MoNRE was officially assigned to keep the state management role of river basin management throughout the country and is authorized to establish RBC in river basins based on the requirement of water resources management for each river basin. The Decree No. 120/2008/ND-CP specifies that a River Basin Committee (RBC) has the following functions:

- Monitor and coordinate activities of ministries and sectors related to the implementation of a river basin plan;
- Propose and issue policies and recommendations on measures for water environment protection, water resources exploitation, utilization and development and the prevention and mitigation of damage caused by water in the river basin.

Kellogg Brown and Root Pty Ltd (2008) cited in Burry and Oliver (2009) observed that the roles and responsibilities of MARD and MoNRE and their respective departments, in view of the requirements of an Integrated Water Resources Management and the reality of river basin management in Vietnam in general but in the VGTB in particular have not been clarified. For the last decade, the unclear roles and responsibilities between MoNRE and MARD has been the prime argument in water resources management and in fact both Ministries claim "state management functions" for water, showing their overlap of functions (Molle and Chu 2007). This has hindered progress of the countries water resources management.

New Law on Water Resources in 2012 (LWR 2012)[3] specifies that MoNRE has the coordinating function regarding the basic survey, strategy, and master plan at national level and for interprovincial water resources (LWR 2012, chapter "Biophysical and Socio-economic Features of the LUCCi—Project Region: The Vu Gia Thu Bon River Basin"). They are also responsible to elaborate lists of interprovincial and intraprovincial basins (LWR 2012, Article 7). However, as MARD holds key responsibility for the essential issues at the river basin level like irrigation, floods and rural water management, there will be no approximation to IWRM in the basin unless both ministries cooperate closely.

## *Institutional Roles and Basin Governance*

The RBOs are often initially set up to address one or two critical problems in the river basin, e.g., navigation management, water pollution, or disaster prevention (GWP and INBO 2009). The roles of RBOs have changed over time as they become more integrated in their approach to managing water (GWP and INBO 2009). To date, many RBOs have been established throughout the world with a history of over 100 years. In Germany, the Emscher River Association was founded in 1904 and the Ruhr reservoir association in 1913 (private foundation dates even back to 1899) (Ruhrverband 2016). Spain has a long history in developing formal Governmental authorities on the river basin scale and it adopted the system of *Confederaciones Hdrográficas* in 1926 (Bhat and Bloomquist 2004). In the United States, the powerful Tennessee Valley Authority (TVA) was created in 1933 (Barrow 1998).

For the VGTB River Basin so far no functional river basin organization exists. Thus the different institutions with a role in water resources management are working rather independently without a joint coordination. Table 7 lists the different institutions depicting its role for water management and the existing linkages with other organizations.

Many institutions are involved in decisions with impact on the consumption, allocation, or protection of water resources. Even though their interactions are many and in many cases mutual dependence exists, information exchange between these institutions is not systematic and not centrally coordinated. As can be seen in the above table, the political system provides two institutions with a coordination role at the national and at the provincial level: The National Water Resources Council and the Provincial People Committee. As the VGTB is strategic for two provinces, the provincial level is not really adequate to takeover a coordination role. The NWRC on the other hand is politically and geographically rather far away from the local requirements in the VGTB basin. Thus the setup of any kind of

[3]Law on Water Resources 2012: http://www.faolex.fao.org/docs/pdf/vie117928.pdf.

**Table 7** Key stakeholders, their role, mandate and interaction for the VGTB basin

| Stakeholder | Overall role | Mandate in the river basin | Interaction with stakeholders |
|---|---|---|---|
| State management group | | | |
| GoV[1] | The highest political level in the country | Approves and provides strategies and plans for socio-economic gdevelopment, natural resources and environment protection for the nation, which gives strong potential impact to the development of the river basin | Relates to different water related sectors through NWRC and through ministries |
| NWRC[2] | Consults the GoV on policies, strategies, plans, and conflict resolution in major river basins | Has high influence on strategies and plans to the river basin | The NWRC has close relationship with MoNRE (its permanent member), and other relevant ministries |
| MoNRE[3] | Ministry with state management function for water resources management | Important stakeholder that has coordinating function at the river basin level | MoNRE interacts with the Government, relevant ministries, PPCs and DoNREs to manage the river basin |
| MARD[4] | Undertakes irrigation, flood protection and rural domestic/industrial water supply | Although MARD has lost the water resources coordination function to MoNRE it still is a crucial player for river basin management | MARD works with the Government, relevant ministries, PPCs, and DARDs in the river basin |
| PPC[5] | The highest political level in each province | Approves and provides strategies and plans for socioeconomic development, natural resources protection for their provinces, has strong influence on the development of the river basin | PPC through its Departments provides and implements strategies and plans for activities of the river basin in their province area |
| DoNRE[6] | State management function on natural resources and environment at province level | Direct influence to the development of the river basin | Administered by its PPC and under technical control by MoNRE |
| DARD[7] | State management function for agriculture, aquaculture, forestry, salt management, flood protection, and rural development at province level | Traditionally with important role in water management in a province with crucial influence on key water management functions in a river basin | administered by the PPC and under technical control by MARD |

(continued)

**Table 7** (continued)

| Stakeholder | Overall role | Mandate in the river basin | Interaction with stakeholders |
|---|---|---|---|
| DoIT[8] | State management function for planning and implementing hydropower projects | Strong influence in the basin as they plan and submit hydropower projects for PPC approval | Administered by its PPC and under technical control by MoIT (Ministry of Industry and Trade) |
| VG-TB RBPMB[9] | to provide the basin plan, coordinate stakeholders, and conflict resolution | A consultative body which does not have any state management function | The RPMB works under MARD and PPCs, and has close relationship with DARDs |
| CMCDIWR[10] | Basin Agency Established by Quang Nam PPC | A consultative body which does not have a clear governmental mandate | |
| Water exploitation and use group | | | |
| EVN[11] | A state-owned utility and the largest electricity company in Vietnam | Owning and operating most large hydropower projects in the river basin and being responsible for developing the national power development plan | EVN interacts with several stakeholders in the river basin though information exchange is limited |
| IMCs and WUGs[12] | IMCs manage major irrigation works and main channel systems. WUGs manage second level water works and channels at their lowest level | as irrigation accounts for more than 80 % of the total water use, this group is highly relevant for river basin management | IMCs and WUGs works closely with DARDs, PPCs, NGOs, other relevant Departments and Ministries |

[1]Government of Vietnam
[2]National Water Resources Council
[3]Ministry of Natural Resources and Environment
[4]Ministry of Agriculture and Rural Development
[5]Provincial Pegople Committee
[6]Department of Natural Resources and Environment
[7]Department of Agriculture and Rural Development
[8]Department of Industry and Trade
[9]RBPMB is the River Basin Planning Management Board under the management of MARD
[10]CMCDIWR is the "Committee for the Management, Control and Development of Integrated Water Resources for the Vu Gia—Thu Bon Basin" was established under the chairmanship of the Quang Nam PPC, with the support of ADB
[11]Electricity Group of Vietnam
[12]Irrigation Management Company and Water User Group

coordinating institution as a River Basin Organization would for sure be beneficial. Hooper (2006) distinguishes nine types of RBOs.

In order to achieve coordination at river basin level different institutional approaches are possible.

(a) Exchange of existing institutions towards a common goal (River basin coordinating committee or council)
(b) A more formally constituted body which takes decisions on the objectives and strategy of basin development (river basin commission)
(c) An institution which takes fundamental decisions and is responsible for the planning and implementation of major development projects in the basin (like hydropower or irrigation development).

Considering the fact that no functioning river basin organization is established so far the option (a) would probably be the most suitable approach for the VGTB basin as it does not require major legal changes. For any solution towards an RBO for the VGTB basin there should be clear definitions of different functions like decision making, implementation, conflict resolution, participation/stakeholder processes, financing, monitoring, information sharing, and capacity development.

## *Decision Making and Information Management Towards IWRM*

A fundamental basis for a development towards achieving IWRM at the River Basin scale is the provision of relevant information. The decision-making process should be linked to adequate monitoring and information generation as the quality of decisions depends on the availability of timely and crucial information. Several institutions may require the same or similar information like water availability or variability, water quality, water uses, and demands. On the other hand, some institutions may depend on the result of other institutions. As for the Vu Gia Thu Bon before now no river basin organization or system of information exchange was active. The LUCCi research consortium created the River Basin Information Center (RBIC) linked with the project database at its heart (RBIS, see chapter "Vu Gia Thu Bon River Basin Information System (VGTB RBIS)—Managing Data for Assessing Land Use and Climate Change Interactions in Central Vietnam"). RBIC provides open access to certain information elements while other information is available for registered users only. The Center provides a couple of predefined maps, figures, and tables, other information products can be synthesized based on the available data on demand. Figure 5 presents the general concept and flow of data of RBIC. The Center is hosted at the Vietnam Academy for Water Resources in Da Nang.

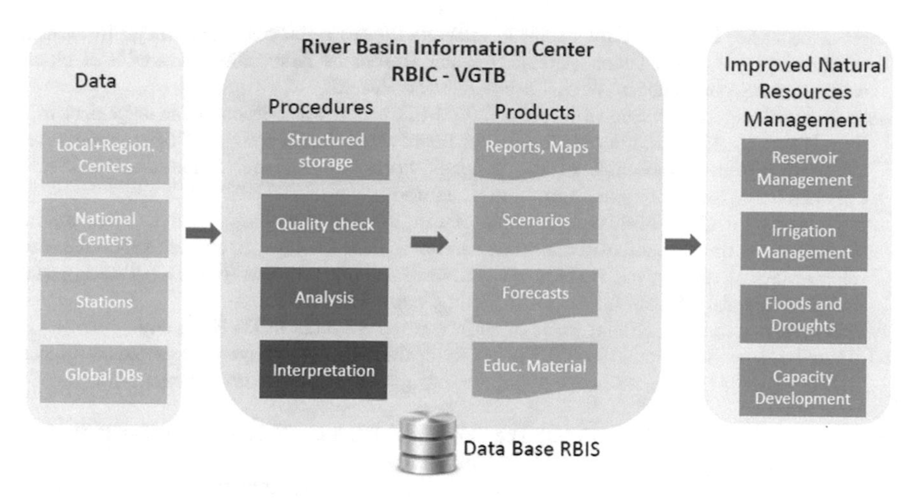


**Fig. 5** Data elements and information flow in the River Basin Information Center

## *Conclusion and Outlook*

The VGTB river basin sustains a large population regarding water, food, and energy supply. Additionally, it provides resources which are exported to other regions. Floods and droughts are challenging the provision of services and endanger water, food, and energy security. In order to maintain the functions of the basin in an adequate way an improved collaboration between many institutions, which are related to provide these essential services, is needed. As mentioned above the goal should be to establish a more formal collaboration structure like a river basin committee or authority. As information sharing is a crucial step for integration, the establishment of a River Basin Information Center (RBIC) through the LUCCi project could serve as a starting point in this direction. The RBIC includes an openly accessible data base with data on the natural and human environment which are regarded as an essential element in the decision making process towards a better integration of stakeholders in the region.

## References

ADB (2015) Vietnam energy sector assessment, strategy, and road map. ISBN 978-92-9257-312-6 (Print), 978-92-9257-313-3 (e-ISBN). Available at http://www.adb.org/sites/default/files/institutional-document/178616/vie-energy-road-map.pdf

Barrow CJ (1998) River basin development planning and management: a critical review. World Develop 26(1):171–186

Bhat A, Blomquist W (2004) Policy, politics, and water management in the Guadalquivir River Basin, Spain. Water Resour Res 40

Brazil National Water Resources Policy (1997). Available at http://www.gwp.org/Global/ToolBox/About/IWRM/America/Brasil%20National%20Water%20Resources%20Policy.pdf

Bateman B, Racier R (2012) Case studies in integrated water resources management: from local stewardship to national vision. American Water Resources Association. Available at http://www.awra.org/committees/AWRA-Case-Studies-IWRM.pdf

Bruns B, Bandaragoda DJ, Samad M (eds) (2002) Integrated water-resources management in a river-basin context: institutional strategies for improving the productivity of agricultural water management. In: Proceedings of the regional workshop, Malang, Indonesia Jan 15–19. International Water Management Institute, Colombo

Burry B, Oliver P (2009) Situation analysis of the VG-TB river basin, central Vietnam. In: Education and training, international water centre. Written for keys for success with integrated water resource management. 5th network of River Basin Organization (NARBO) Training. Hoi An, Viet Nam 18–25 Feb, International Water Centre, Brisbane

CVIWR (2014) Report on current status of surface water resource in Da Nang city

EU Water Framework Directive (2000) The EU water framework directive—integrated river basin management for Europe. Available at http://ec.europa.eu/environment/water/water-framework/index_en.html

GoV (Government of Vietnam) (2008) The Government Decree No.120/2008/ND-CP on river basin management. Government of Viet Nam, Hanoi

GSO (General Statistics Office of Vietnam) (2014) Statistics on agriculture, forestry and fisheries by province in 2014. Available at http://www.gso.gov.vn/default.aspx?tabid=717

GWP and INBO (Global Water Partnership and International Network of Basin Organizations) (2009) A handbook for integrated water resources management in basins. Global Water Partnership and International Network of Basin Organizations, Sweden

Hooper BP (2006) Key Performance indicators of river basin organizations. Visiting scholar program, US Army Corps of Engineers. http://www.iwr.usace.army.mil/Portals/70/docs/iwrreports/2006-VSP-01.pdf

IMC Danang (2012) Inventory of irrigation works in Danang Province (working report)

IMC Quangnam (2012) Inventory of irrigation works in Quangnam Province (working report)

Kellogg Brown and Root Pty Ltd. (2008) TA 4903-VIE water sector review project, draft final report. The Office of the National Water Resources Council, on behalf of the Government of Vietnam and the Asian Development Bank. South Australia

LWR (2012) Law on Water Resources, Government of Vietnam. URL: http://faolex.fao.org/docs/pdf/vie117928.pdf

MOIT (2014) Policies on sustainable hydropower development in Vietnam. Hydropower Department, General Department of Energy, Ministry of Investment and Trade, Vietnam

Molle M, Chu TH (2007) Implementing integrated river basin management: lessons from the Red River basin, Vietnam. Working paper, IRD, M-Power. Integrated Water Management Institute, Bangkok

National Assembly of Viet Nam (1998) Law on water resources 08/1998/QH10. National Assembly, Hanoi. Available at http://thuvienphapluat.vn/van-ban/Tai-nguyen-Moi-truong/Luat-Tai-nguyen-nuoc-1998-08-1998-QH10-41679.aspx

Ruhrverband (2016) How the Ruhrverband developed. http://www.ruhrverband.de/en/ueber-uns/chronik/. Visited 16 July 2016

UNECE (2015) Reconciling resource uses in transboundary basins: assessment of the water-food-energy-ecosystems nexus (Newyork and Geneva; United Nations, 2015). Available at http://www.unece.org/fileadmin/DAM/env/water/publications/WAT_Nexus/ece_mp.wat_46_eng.pdf

Van TTH (2013) Irrigation efficiency for paddy field in considering reuse potential of return flow in the lowland area of the Vu Gia Thu Bon Basin, Central Vietnam. University of Applied Sciences, TH Köln

Viet TQ (2014) Estimating the impact of climate change induced saltwater intrusion on agriculture in estuaries—the case of Vu Gia Thu Bon. RUHR University Bochum Publisher, Vietnam

# Land Use Adaption to Climate Change in the Vu Gia–Thu Bon Lowlands: Dry Season and Rainy Season

**Harro Stolpe, Nils Führer and Viet Quoc Trinh**

## Introduction

Coastal areas in Vietnam are facing impacts of climate change, e.g. sea level rise (SLR), increasing frequency of severe storms as well as impacts of land use change, e.g. urbanization, tourism etc. The Vu Gia Thu Bon (VGTB) lowlands are a typical example of such area.

The following part describes the water management in the dry season and the rainy season in the VGTB lowlands. It bases on Trinh (2014) and FÜHRER (2015); it summarizes the main results.

The VGTB river system is formed by the two main rivers Vu Gia and Thu Bon. The VGTB basin is the fifth largest river basin in Vietnam with a catchment area of around 10,350 km$^2$.

The VGTB lowlands form a dynamic coastal system in terms of natural characteristics as well as socio-economic development. Large parts of the coastal lowlands are densely populated and characterized by an intensive use of land. Located in the savannah tropics, the VGTB lowlands feature distinctive dry and rainy seasons with accordingly human activities and hazards. Therefore, the management of water and land in the lowlands differs strongly between the seasons.

The VGTB lowlands cover 410 km$^2$. The region has tropical monsoon climate with high temperatures. The rainfall in this area reaches 2400 mm/year, while the actual evaporation rate is 1050 mm/year. The rainfall occurs primarily in the 4 months of the rainy season, which spans from September to December, and causes

H. Stolpe (✉) · N. Führer · V.Q. Trinh
EE+E Environmental Engineering+Ecology, Ruhr University of Bochum, Bochum, Germany
e-mail: harro.stolpe@rub.de



A. Nauditt and L. Ribbe (eds.), *Land Use and Climate Change Interactions in Central Vietnam*, Water Resources Development and Management,
DOI 10.1007/978-981-10-2624-9_11

flooding. Little rainfall occurs during the long dry season, which spans from February to August, and often causes water shortages.

The Vu Gia and Thu Bon are the primary rivers that provide fresh water for the region. The Thu Bon drains into the East Sea at Cua Dai river mouth. The Vu Gia diverges into several estuarial tributaries after entering the lowlands before converging again and draining into the East Sea at the river mouths of Cua Han. Numerous weirs are constructed on the estuarial tributaries to regulate the flow for irrigation and domestic use.

The hydrological system of the VGTB lowlands is completely different in dry season and rainy season and is therefore analysed separately.

## Dry Season

The hydrological system in the dry season, the derived modelling, the definition of significant hydrological and land use scenarios and risk assessment are described in terms of saltwater intrusions. The key results are summarized.

### *Dry Season Hydrological System*

The VGTB lowland is an intensive agricultural area; half of its surface area is used to cultivate rice and other annual crops. Rice is the dominant crop and is cultivated on 14,300 ha, while the annual crops are cultivated on 6100 ha. Almost all of the irrigated areas in the region obtain water directly from the estuarial rivers.

The impact of saltwater intrusion is reflected in both water supply for domestic use and irrigation. The saltwater intrusion in the VGTB is partly a natural phenomenon due to changing hydrological conditions and partly a man-made problem due to water allocation measures.

According to the recent report on the status quo of the ten largest river basins in Vietnam, saltwater intrusion is one of the largest concerns in the estuaries of these river basins (MARD 2005). Saltwater intrusion has caused adverse impacts on irrigation and urban water supply in recent years in the VGTB lowlands (Lai et al. 2011).

Saltwater intrusion in the lowlands is expected to change dramatically in the future due to the changes in river flow, sea level rise, downstream water use, river morphology as well as the construction of large reservoirs in the mountainous areas of the basin.

The seasonality is divided into rainy season from September to December and dry season from February to August. There is a secondary rainfall peak in May and June, which usually occurs in the north-western parts of the basin and may cause heavy rains, so-called Tieu Man floods. The seasonality of rain influences the behaviour of saltwater intrusion in the lowlands strongly.

It is expected that climate change-induced saltwater intrusion will cause increasing problems for the irrigation in the investigated area, which is shown in Fig. 1. (Trinh 2014).

The Vu Gia system, which consists of the Vu Gia River and its tributaries, is characterized by the weirs in its main streams. These weirs block the transition between freshwater and saltwater, and they have become the boundaries of saltwater intrusion. Downstream saltwater intrusion responds rapidly to the changes in weir operation. The salinity boundaries of 1, 2 and 5‰ can reach to 16.0, 15.5 and 14.5 km from Cua Han river mouth.

The VGTB lowlands often experience saltwater intrusion during dry periods. Saltwater intrusion is an interactive result of many combined factors, such as of hydrological regimes, the operation of reservoirs in the upstream mountainous area, river morphology and tidal activity. The dense network of irrigation canals and drainage systems allows the salinity to spread to many lowland parts. Salt spreads into the lowlands through three waterways: Vu Gia, Thu Bon and Vinh Dien. The Vu Gia and Thu Bon bring saltwater directly from the East Sea.

The salinity boundaries vary over time as the inflow from upstream and tides fluctuate. The 1‰ of salt level is an important indicator when predicting the use of

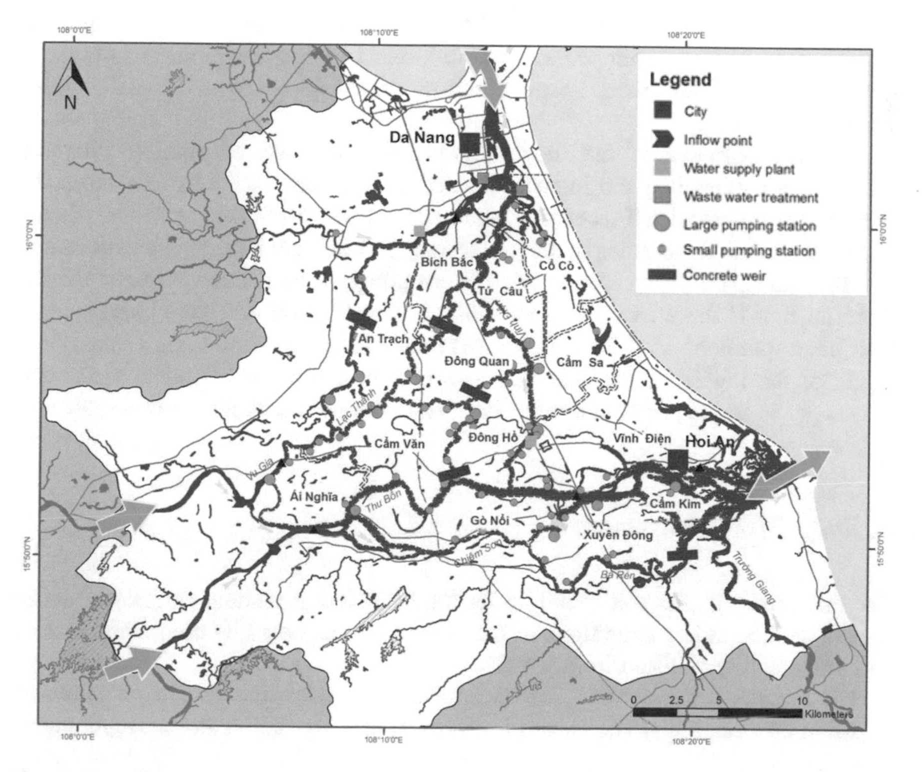


**Fig. 1** Investigation area and dry season hydrological system Trinh 2014

water resources from the rivers during irrigation. When the salt concentration exceeds 1‰, the river water becomes unsuitable for irrigation.

## Suface Water Model

For the dry season, MIKE 11, a 1D full hydrodynamic model was applied to simulate the surface water flow and saltwater intrusion for the current state as well as for the scenarios of potential future development. Figure 2 shows the model, which is characterized by the following variables: (1) inflow from upstream, (2) water allocation in the river network, (3) water pumping for irrigation, (4) industrial and drinking water supply, (5) water flow by weir management and (6) water level of the sea.

The model integrates a hydrodynamic module to simulate the flow and an advection–dispersion module to simulate the salinity. The modelled system is defined with attributes as follows: 20 river branches and tributaries, 129 river or channel cross sections, 3 downstream water level boundaries, 4 upstream discharge boundaries, 29 points of lateral sources, 7 weirs, 15 large pump stations, 2 water works, and 2 wastewater treatment plants. The model was calibrated by measured data like water levels, water flows and water salinity at seven observation points.

Five scenarios are developed as listed in Table 1 representing relevant situations in the VGTB lowlands due to upstream inflow, water use, sea level rise and possible adaption measures.

Essential objective of the modelling is the discussion of possible adaption measures for mitigating the impact of salinity intrusion. Two adaption measures scenarios were calculated (scenario 3 and scenario 5). Both scenarios include the following adaption measures in the lowlands (1) modification of the weir operation, (2) construction of Quang Hue Closure interrupting the connection between Vu Gia and Thu Bon in the west of the lowlands, (3) construction of Thanh Quyt syphon weir interrupting the Vinh Dien River, (4) rehabilitation of Binh Long Canal in the South of the lowlands, (5) enhanced use of return flow of irrigation water and (6) supply of the required minimum flow from the mountainous upstream area.

## Salinity Risk Assessment

The risk caused by the water salinity for the irrigation dependent agriculture under the different boundary conditions of the scenarios is assessed by combining hazard and vulnerability as listed in Table 2.

The duration of exceeding salt is applied to classify the hazard of saltwater intrusion into the rivers. The hazard levels of a pumping station and the agricultural

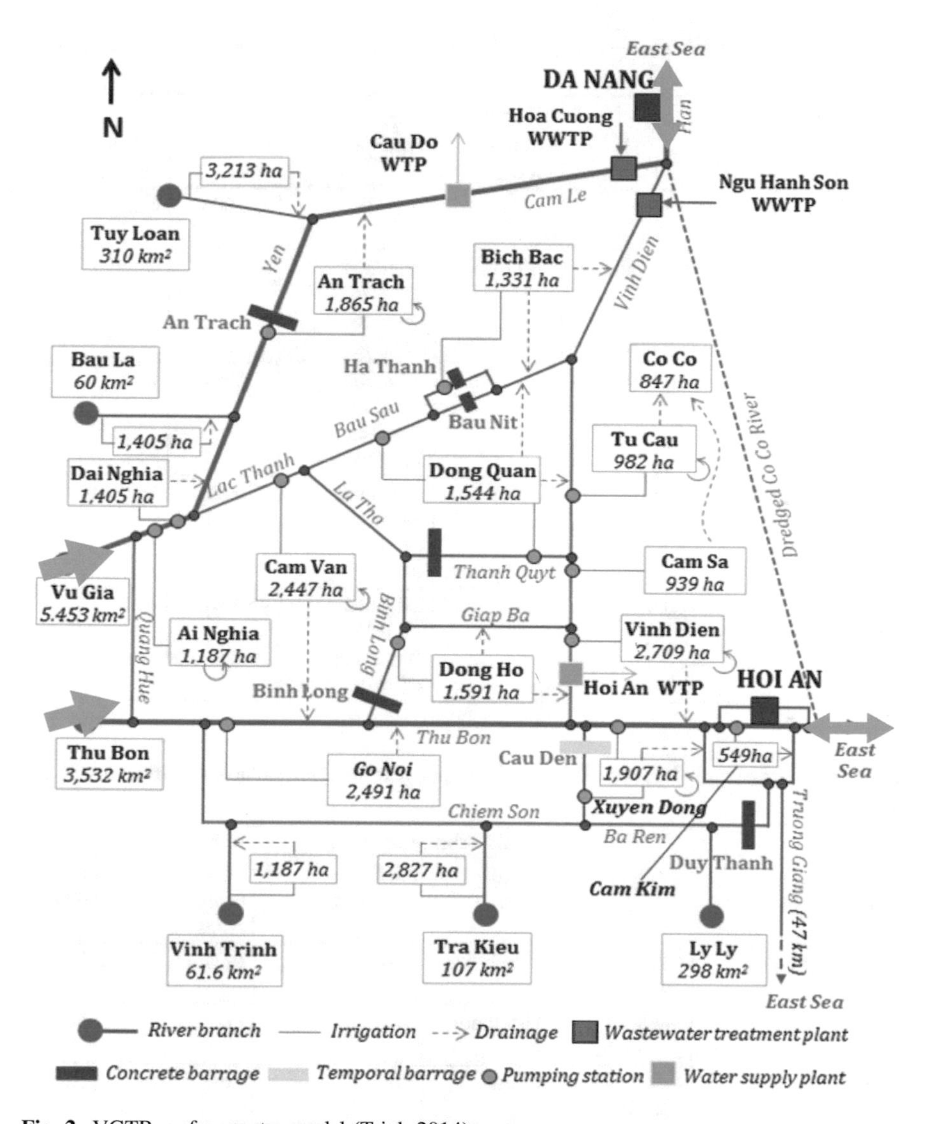


**Fig. 2** VGTB surface water model (Trinh 2014)

land irrigated by this pumping station will receive corresponding values like the river segment at the inlet location of the pump station.

The vulnerability is assessed by classifying the land use based on its irrigation demand, the higher the demand the higher the vulnerability. The existing land use (2009) is applied to define the vulnerability for the current state scenario and the planned land use in 2030 is applied for the future scenarios.

**Table 1** Dry season scenarios—changes in variables refer to the baseline scenario no. 1, Trinh (2014)

| No | Scenario | Variable | | | |
|---|---|---|---|---|---|
| | | Upstream inflow | Water use | Sea level | Adaption measures |
| 1 | Current state (baseline scenario) | as 2005 | as 2005 | as 2005 | without |
| 2 | Midterm without adaption measures | +40 m$^3$/s | Irrigation: −1 m$^3$/s Domestic: +1 m$^3$/s | +15 cm | without |
| 3 | Midterm with adaption measures | +40 m$^3$/s | Irrigation: −1 m$^3$/s Domestic: +1 m$^3$/s | +15 cm | with |
| 4 | Long term without adaption measures | +40 m$^3$/s | Irrigation: −1 m$^3$/s Domestic: +3 m$^3$/s | +50 cm | without |
| 5 | Long term with adaption measures | +40 m$^3$/s | Irrigation: −1 m$^3$/s Domestic: +3 m$^3$/s | +50 cm | with |

**Table 2** Risk definition for irrigation (d/a = days per year water not suitable for irrigation), Trinh (2014)

| | | Vulnerability of crops | | | |
|---|---|---|---|---|---|
| | | High rice | Medium annual crops | Low perennial crops | No other |
| HAZARDS >1 g/L | High >90 d/a | High | Medium | Low | No |
| | Medium 30–90 d/a | Medium | Medium | Low | No |
| | Low <30 d/a | Low | Low | Low | No |
| | No 0 d/a | No | No | No | No |

Examples of the scenario modelling results are given in Fig. 3 (current state) and Fig. 4 (long term without adaptation measures). The figures show a significant increase of areas with risk for irrigation due to sea level rise without measures. The areas of risk can be reduced by implementing adaption measures.

Figure 5 is a summary of the results of the scenario analyses. It shows in particular that the risk can be reduced for irrigation through appropriate measures (scenario 3, scenario 5).

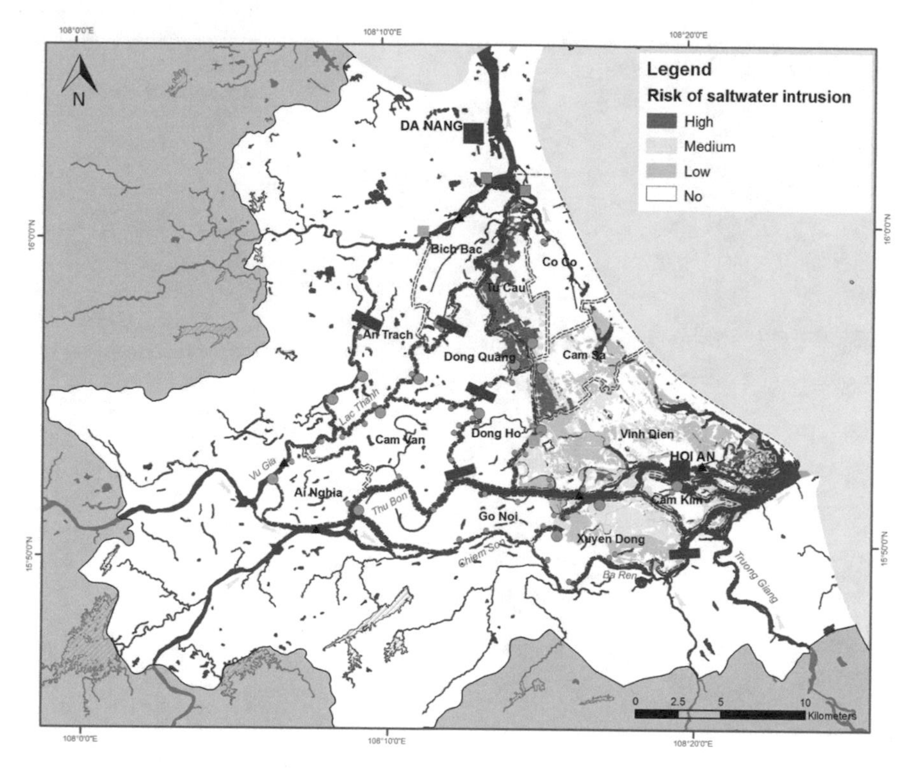


**Fig. 3** Salinity risk for scenario 1 (current state), Trinh (2014)

## *Key Results*

The key results of the analysis of the dry season hydrological system are as follows:

- GIS/numerical model based risk assessment: The dry season water system is adequately represented through scenarios, measures can be evaluated soundly.
- Mountainous area, minimum flow: A defined minimum flow from the mountainous areas is the main precondition for a proper management of the lowlands water system.
- Water allocation upstream of the weirs: The defined minimum flow is to be used in praxis as request to water management, design and dimension of the weirs.
- Water flow over the weirs: If the defined minimum flow is met, the necessary water flow over the weirs can be defined to meet the water quality in the irrigation areas downstream of the weirs.

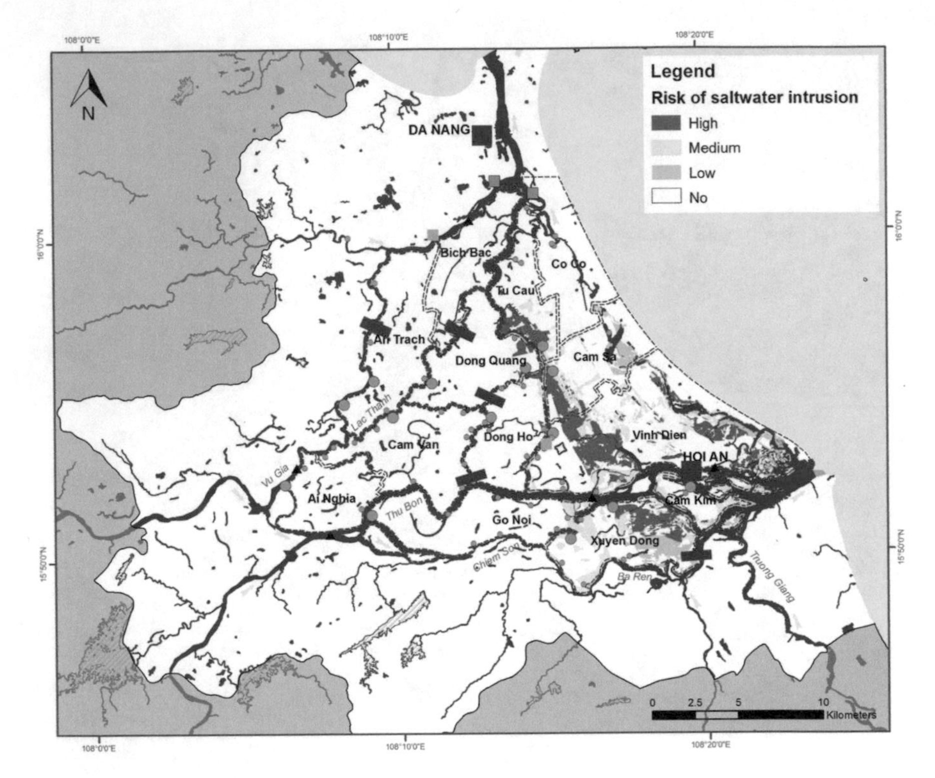


**Fig. 4** Salinity risk for scenario 4 (long term without adaption measures), Trinh (2014)

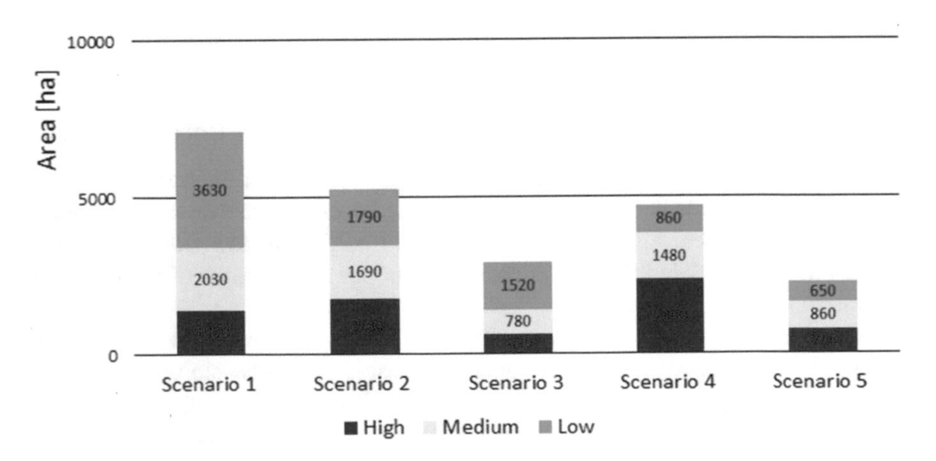


**Fig. 5** Scenario Results: Salinity intrusion risks (Trinh 2014)

- Defined minimum flow not occurring: If the defined minimum flow is not met, the salinity possibly, progressively with the SLR, cannot be pushed back for covering the irrigation water demand downstream the weirs. This causes an increasing risk for the agriculture.
- Measures: Possible measures besides the implementation of minimum flow are e.g. weir management, additional weirs etc.

## Rainy Season

The hydrological system in the rainy season, the derived modelling, the definition of significant hydrological scenarios and risk assessment are described in terms of flooding. The key results are summarized.

### *Rainy Season Hydrological System*

The rainy season in the VGTB basin is characterized by extreme basin wide rainfall events, which locally may exceed 600 mm/d. Main causes are orographic rainfalls at the Trung Son range, singular weather phenomena, typhoons, and accompanying weather conditions. The rainy season lasts from September to December, reaching its peak in late October to early November.

The hydrological system of the VGTB lowlands in rainy season is characterized by short-term and large-scale fluvial floods. Minor fluvial floods may occur more than one time per season while extreme fluvial floods (over warning level 3) occur around every 20–30 years. The flood event stated as most disastrous occurred in 1964, followed by events in 2007, 1999 and 2009 consecutively. Impacted by typhoons and tropical storms, the lowlands are not just prone to fluvial floods but also to coastal and pluvial floods.

Major inflows, as in dry season, are the Vu Gia and Thu Bon rivers, originating from the mountainous upstream area. Maximum monitored inflow by the Thu Bon river is around 10,600 $m^3$/s. The maximum inflow by the Vu Gia river is just monitored for around 47% of the upstream sub-basins and reaches around 7230 $m^3$/s. Based on the corresponding return period, another 12,200 $m^3$/s have to be added for the ungauged 53% of upstream basin area.

While damages to the agriculture are low, floods in the VGTB lowlands are causing significant damages to settlement areas. The total damages by hazardous events in November 2007 and November 1999 including the flood events mentioned above in the provinces Quang Nam and Da Nang reached a total of about 230 Billion VND (10.4 Million USD) and 400 Billion VND (18 Million USD), respectively.

In the context of climate change, it is expected that the flood risk in Central Vietnam will increase due to the intensification of extreme weather conditions.

However, flood management in the lowlands, as in whole Central Vietnam, mainly focuses on the preparation, actions directly in advance of, while and directly after a flood event, the response and recovery. The land elevation by infilling for settlement areas is the only mitigation and prevention measure currently taken.

The VGTB lowlands are the part of the basin intensively used and developed. The further development of settlements, agriculture and infrastructure causes hydraulic changes in the city of Da Nang and further reduces floodplains. As the development and in special the extension of urban areas is ongoing, the mitigative and preventive aspects of flood management coordinated and integrated with urban planning are in great demand to ensure that flood risk will not increase and if possible is decreasedis characterized by a wide open system, providing large floodplains (Fig. 5). This system is only interrupted by scattered settlements and line elements like the National Highway No. 1A and the railway in north–south direction. The major floodway is the Thu Bon river in the south of the lowlands. Vu Gia river is flowing to the north east to Da Nang City, it is the secondary flood way. Several streams, partly man-made, partly natural connect the two main floodways (Fig. 6).

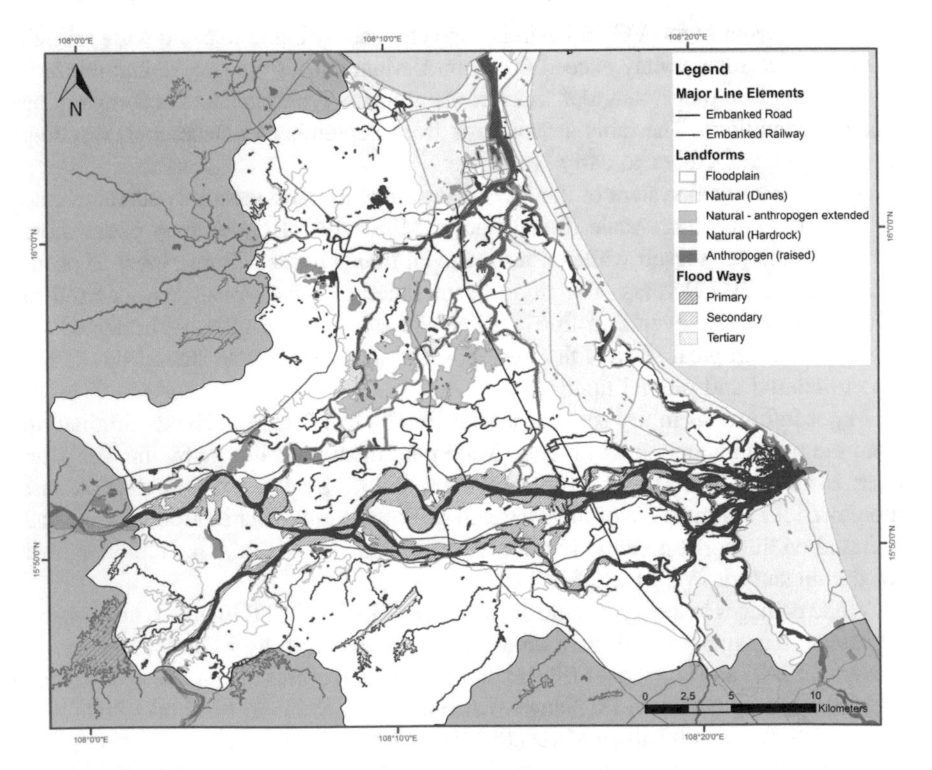


**Fig. 6** The hydrological system during the rainy season for the baseline year 2009 (flood year). *Orange* and *yellow* colours indicate elevated areas (FÜHRER 2015)

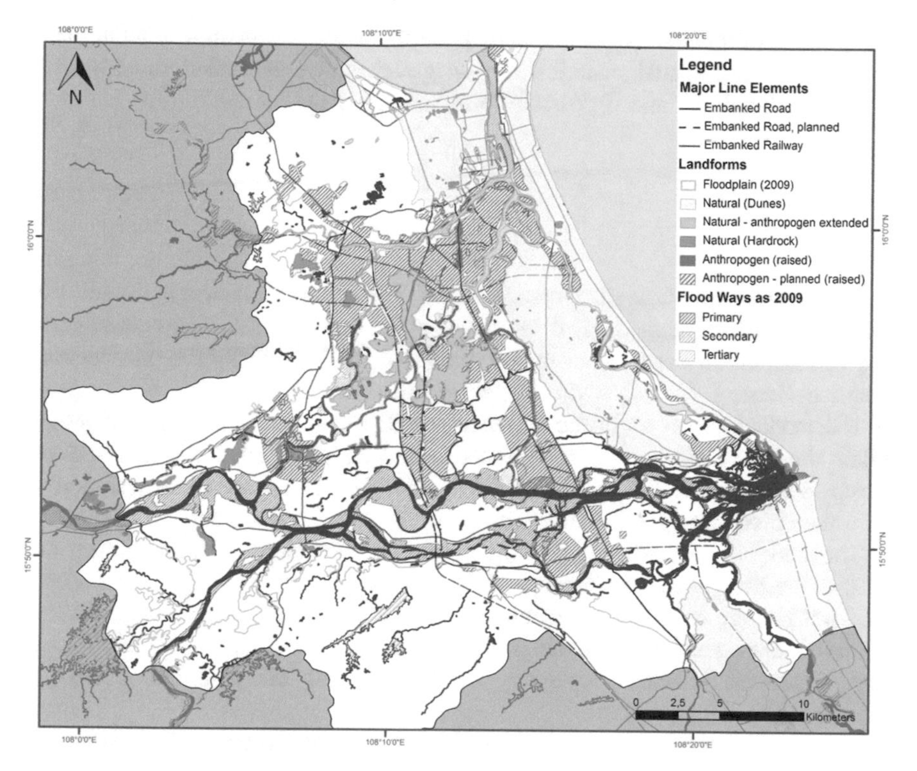


**Fig. 7** Flood system 2030 (based on spatial planning, *orange*, *yellow* and *red*: elevated areas), FÜHRER (2015)

Flood protection is not or just local and in minor context existing. Major protection is on-site and based on an over centuries adapted way of living with floods as natural hazard. Settlements are located on naturally elevated land like dunes near to the coast (Fig. 7). Rice is just planted during dry season, and perennial crops are usually planted outside flood prone areas.

This system will be changed by constructions according to the Master Plans of the Departments of Construction of Quang Nam and Da Nang (Fig. 7). The plans foresee major changes focusing on an enlargement of urban and industrial areas. This intensive expansion takes place within the floodplains. Flood protection is based on land elevation and embankment for transport infrastructure.

New urban areas and industrial areas, if planned in flood prone areas, have to be protected by elevating the land above the peak water level of a once-in-a-hundred-years flood event (HQ100, probability 0.01). The area of the floodplains will reduce, floodways will be changed. A shift and increase of flood water levels and flood duration in flood prone areas will be potentially induced. As the current flood management is not integrated and coordinated with spatial and urban planning potential negative impacts are not considered up to now.

The risk evaluation presented here focuses on the evaluation whether flood management and spatial planning are integrated and coordinated to prevent an increase of flood risk and if possible to reduce flood risk.

## *Flood Model*

The rainy season hydrological system in the VGTB lowlands is represented by a numerical model. The model is determined on the basis of field surveys, calibration and verification for the flood events of 1999 and 2007, basically the most extreme historical floods for which hydrological information exists.

The model was set up in the software MIKE Flood 2011. The area of interest within the VGTB lowlands is represented as linked 1D-2D model. The model was set up at the scale 1:50,000. The DEM as base for the 2D approach was derived from the official topographic maps in a scale between 1:2,000 and 1:10,000.

The model is used to investigate the impacts of the official spatial planning until 2030 to the flood situation. Impacts of land use and climate change primary due to sea level rise are considered directly. Potential mitigation measures were derived. Reservoir management upstream, land use change upstream and also the climate change induced increasing intensity of rainfall were analysed and taken into account, too.

The focus is on a medium term, investigating the impacts up to 2030, to provide measures and planning suggestions, which then also prevent an increase of risk and potentially reduce risk in the medium to long term view. The investigated scenarios are given in Table 3.

Three potential hydraulic measures within the lowlands were identified and simulated: (1) Diversion of the flood wave upstream of Quang Hue from Vu Gia to Thu Bon river, lowering flow through Vu Gia river, (2) Removal of aggradation in a temporary tributary of Thu Bon river, increasing flow capacity and volume, (3) Unregulated outflow to the East Sea in the dune area, increasing outflow.

**Table 3** Rainy Season Scenarios—changes in variables refer to the baseline scenario (FÜHRER 2015)

| No. | Scenario | Land use/spatial planning | Sea level | Measures |
|---|---|---|---|---|
| 1 | Current State (baseline scenario) | as 2009 | as 2009 | without |
| 2 | Medium term, planned w/o measures | 2030 | 0 cm | without |
| 3 | Medium term, planned with measures | 2030 | 0 cm | with |
| 4 | Long term, planned w/o measures | 2030 | +75 cm | without |
| 5 | Long term, planned with measures | 2030 | +75 cm | with |

## *Land Use Based Flood Risk Evaluation*

The basic approach of the risk analysis was adapted in 2014 based on discussions with authorities in Da Nang and Quang Nam. The risk analysis was based on the protection goals defined in the Vietnam Building Code on Regional and Urban Planning and Rural Residential Planning (Ministry of Construction of Vietnam 2008) and the Technical Standard for the Design of Sea Dykes (Ministry of Agriculture and Rural Development, Vietnam 2012).

The protection goals are defined depending on land use and population. The protection goals were defined as shown in Table 4 and spatially calculated and represented as maps (e.g. Fig. 8 for scenario 1: current state—2009).

Figure 9 shows a comprehensive evaluation of the risk areas for the scenarios as introduced in Table 3. The area for which the protection goal fails is increasing sharply in case the planning for 2030 is implemented. The area for which the protection goal fails will only increase by 150 ha in case of a sea level rise as expected by the Ministry of Natural Resources And Environment of Vietnam for 2100 (+75 cm, baseline 1980–1999) (MONRE 2009). This increase in area refers to the temporary inundation by fluvial floods, low areas may constantly be inundated by increasing sea level. From the assessment it can be concluded that the impact due to sea level rise to the area at risk is low compared to the impacts due to the spatial planning.

All flood mitigation measures tested here were simplified, intentionally over-designed in a first step and partly simulated just one-dimensionally to receive a clear statement for the impact. The impact of all measures was assessed low, based on the simulation result. Only the removal of aggradation is expected to positively influence floods, if further water bodies would be included. However, the measure is assessed as economically not feasible due to the need for regular maintenance. Hydraulic measures on a regional scale within the lowlands are not applicable. The reservoir management upstream of the lowlands has to be considered in the context of hydraulic measures on regional scale. On-site protection for certain areas by, e.g. dykes or mobile flood walls are local scale measures and not considered here.

**Table 4** Flood risk—protection goal (FÜHRER 2015)

| Class | Land use type | Protection goal | Flood risk—protection goal failed |
|---|---|---|---|
| I | Annual agriculture | no | |
| II | Perennial agriculture | HQ 1 | |
| III | Rural settlement | HQ 20 | |
| IV | Urban settlement | HQ 100 | |
| V | Industrial area | HQ 100 | |

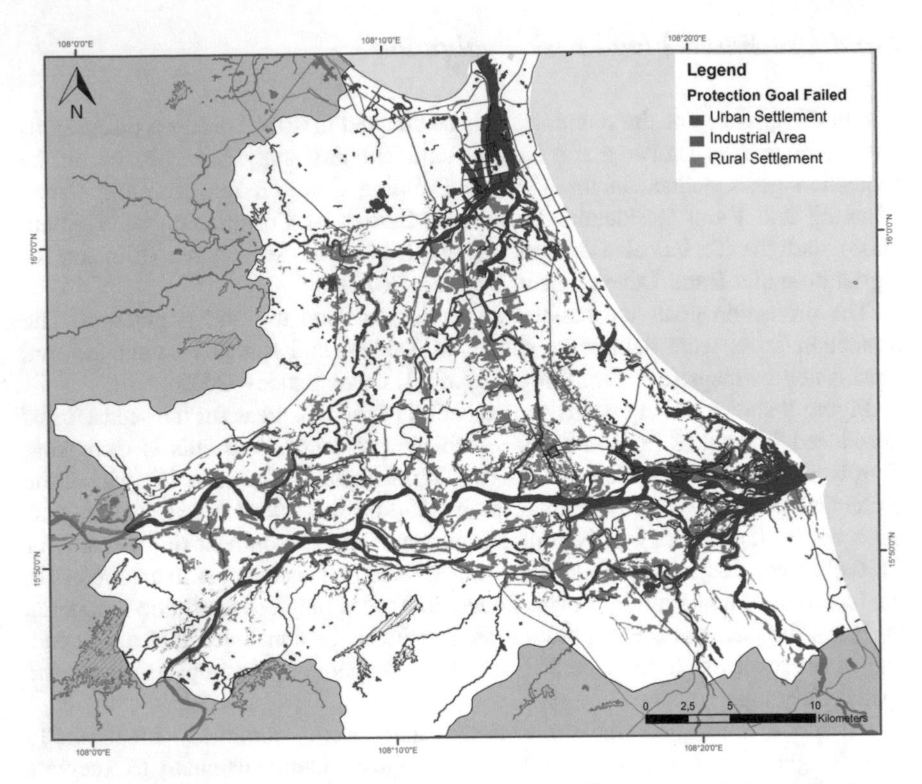


**Fig. 8** Scenario results: Flood risk areas, protection goal failed for scenario 1 (current state—2009), FÜHRER (2015)

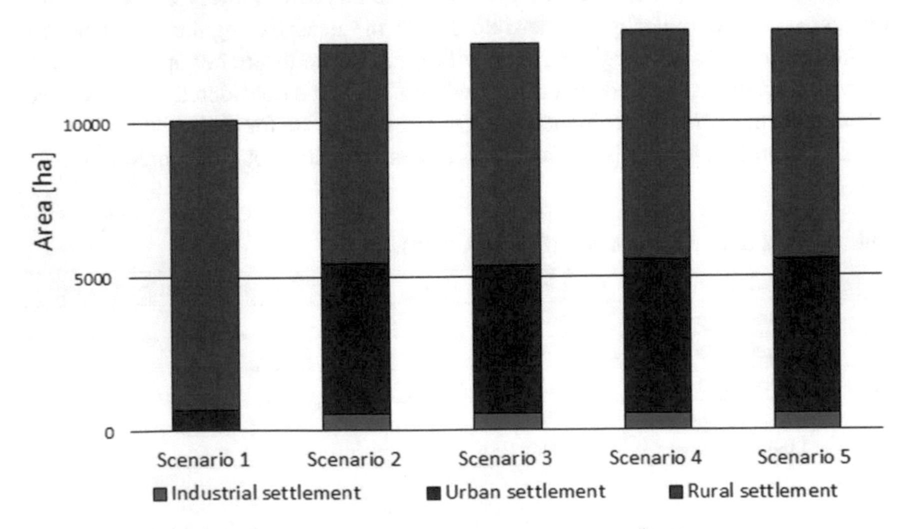


**Fig. 9** Scenario results: Flood risk areas, protection goal failed (FÜHRER 2015)

## *House Based Risk Assessment*

Measures to reduce flood risk and to ensure a theoretical fulfilment of protection goals should be respected in planning a priori as preventive and mitigative part of flood management. FÜHRER (2015) developed a method to respect the protection goals (failed/passed) as introduced above on the level of single houses in order to ensure an appropriate planning. Two further criteria were introduced in addition to the protection goal in order to ensure the handover to the local planners.

The first new criterion is the location suitability. It is determined based on the nearest distance to higher land (landform, Fig. 10), the nearest distance to major infrastructures and the type of settlement. The location suitability represents positive and negative characteristics of settlements in the context of flooding. Furthermore, it provides a potential link to further spatial evaluation of additional sectors like environment.

The second new criterion defines risk hot spots based on the density of houses especially vulnerable to flooding. The risk hot spots express the relative demand for action. This criterion provides the possibility to spatially prioritize measures and strategies.

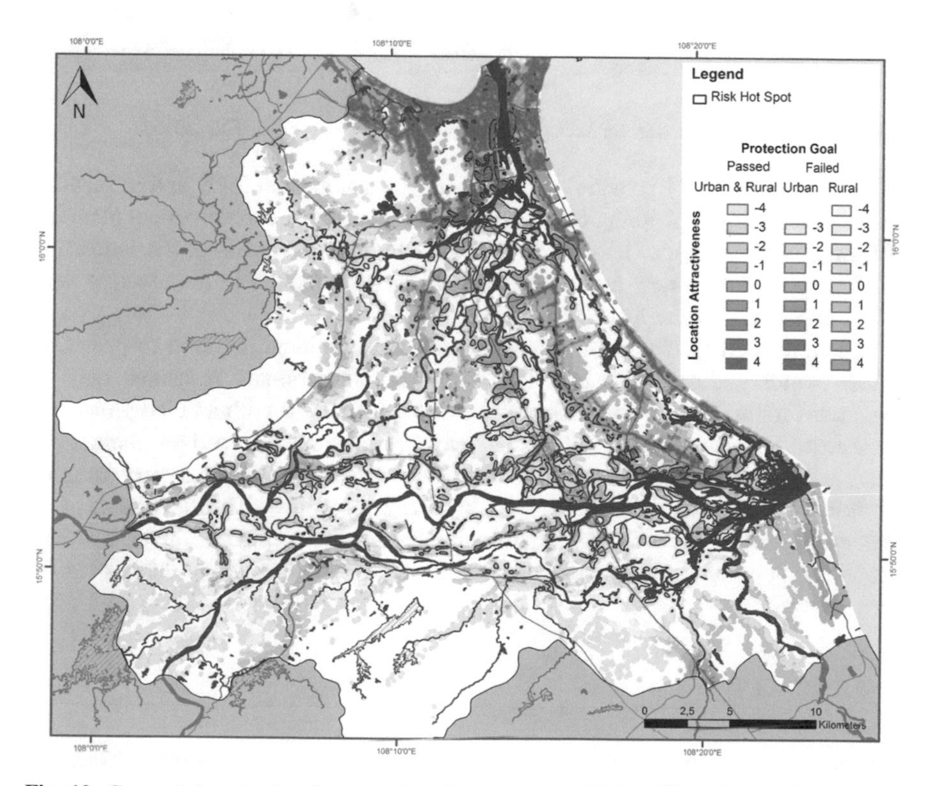


**Fig. 10** Geospatial evaluation for scenario 1 (current state—2009), FÜHRER (2015)

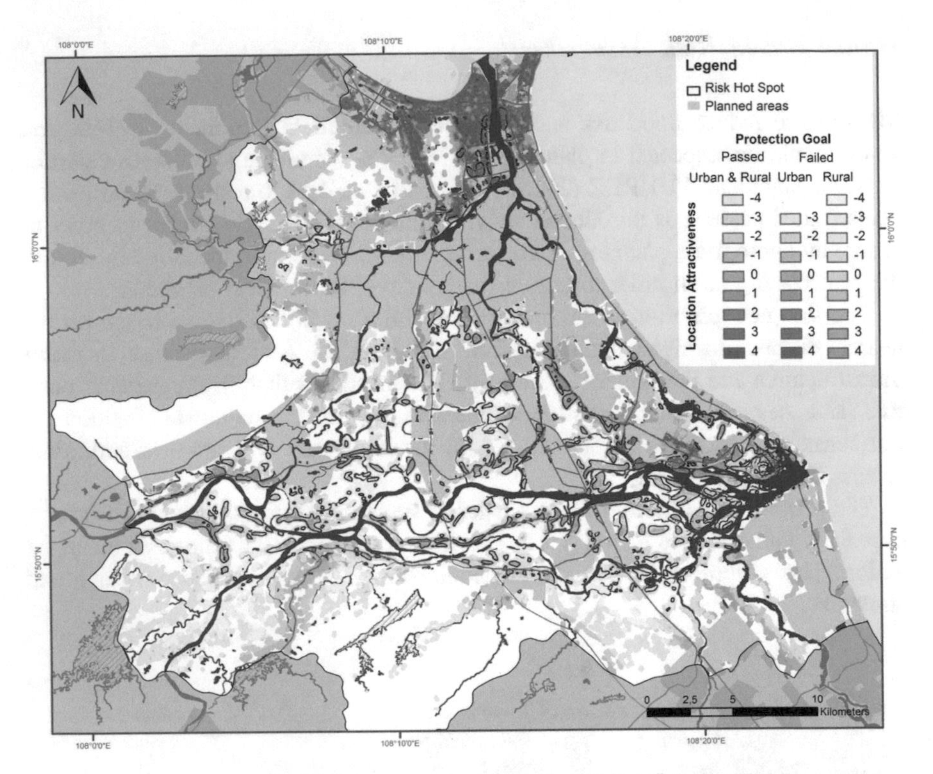


**Fig. 11** Geospatial evaluation for scenario 2 (as planned—2030), FÜHRER (2015)

The Figs. 10 and 11 provide the results for the status as in 2009 and as planned by the general urban planning of Da Nang and the regional constructional planning of Quang Nam (scenario 1 and 2 in Table 3). Figure 11 visualizes the increasing risk hot spots due to new urban areas (grey) and the change in the location suitability expressed by the location attractiveness.

Locations suitable for further development may be identified and respected in spatial planning. Local measures like dykes or other on-site measures may be considered for areas with high risk, risk hot spots, and areas with a failed protection goal. Areas less suitable for settlement may be recognized by urban planners as areas, where people will potentially move away from. Vice versa very suitable locations will potentially develop into local urban centres. The development of settlements and the use of land may be optimized in the context of fluvial flooding through this evaluation. An increase of flood risk can be prevented or the risk can be even reduced.

## *Key Results*

The key results of the analysis of the rainy season hydrological system considering the current state of research are as follows:

- GIS/numerical model based risk assessment: The rainy season hydrological system is adequately represented, scenarios/measures can be evaluated.
- Land use change and SLR: The flood risk will be locally reduced by the planned land use change ("new" areas, elevated land) and in other areas locally intensified ("old" areas). The flood risk will be additionally locally intensified by the climate change induced SLR.
- Land use change, climate change in the mountainous area: The ongoing land use change and climate change in the upstream mountainous area and the ongoing reservoir management will also intensify the flooding.
- Measures: The simulated hydraulic measures will be not efficient enough.
- Flood adapted land use: It is recommended to focus on a "flood adapted land use". This means keeping the regularly flooded areas as free as possible from sensitive land uses (instead: mainly agriculture) and concentrating sensitive land uses in the N-S and E-W development axes and in the urban development areas of Da Nang and Hoi An.

## Outlook

### *Dry Season*

The experiences and knowledge gained regarding the institutional set-up, the identification of stakeholders responsible for land and water resources management on national, provincial and district level will allow a systematic transfer of the results to other river basins and provinces.

The developed methods and tools which can be transferred comprise assessment, planning and management tools, support for improved monitoring and data storage for hydrometeorology, land resources and water quality, conservation strategies as well as capacity building measures.

The developed tools are appropriate for similar environments with low data availability and similar challenges, for example: River Basin Information System, River Basin Status Report Template, modelling and DSS, handbook for sustainable land and water resources management strategy. These products can be used by leading institutions at national level like MARD and MONRE and the PPCs of the involved provinces.

### *Rainy Season*

Flood management and spatial planning in Central Vietnam are facing increasing challenges due to the urban extension into the floodplains. Flooding and flood risk are shifting and potentially increasing. Additionally negative impacts by climate change may trigger those challenges.

The flood risk evaluation as presented above provides the potential to integrate and coordinate flood management and spatial planning. Land use can be optimized by evaluating scenarios including measures. An increase of flood risk can be prevented and the risk even can be potentially reduced.

Using the VGTB lowlands as exemplary area, the evaluation was developed mostly on information existing for all provinces in Central Vietnam and can be transferred to lowlands throughout this region.

## References

Führer N (2015) Geospatial evaluation of fluvial flooding in central vietnam for spatial planning. Dissertation, Ruhr University of Bochum

Lai NV et al (2011) Studying and evaluating the relationship between sea level rise and saltwater intrusion in the coastal zone in Quang Nam Province. Project P1-08 VIE. Institute of Geography (VAST), Vietnam Academy of Science and Technology (VAST). Available at: http://ecoenvi.org/

MARD (Ministry of Agriculture and Rural Development) (2005) The 10 largest river basins in Vietnam. Scientific report. pp 330. Hanoi

Ministry of Construction of Vietnam (2008) Decision on promulgating the Vietnam building code on regional and urban planning and rural residential Planning

Ministry of Agriculture and Rural Development of Vietnam (2012) Technical standard for the design of sea dykes, technical, 1613/QĐ-BNN-KHCN, Hanoi

MONRE (Ministry of Natural Resources and Environment) (2009). Climate change, sea level rise scenarios for Vietnam, Hanoi. Available at: http://vgbc.org.vn/vi/nc/169-climate-change-sea-level-rise-scenarios-for-vietnam-2009

Trinh QV (2014) Estimating the impact of climate change induced saltwater intrusion on agriculture in estuaries—the case of Vu Gia Thu Bon, Vietnam. Dissertation, Ruhr-University of Bochum

# Distributed Assessment of Sediment Dynamics in Central Vietnam

**Manfred Fink, Christian Fischer, Patrick Laux, Hannes Tünschel and Markus Meinhardt**

**Abstract** Central Vietnam is located within the Southeast Asia monsoon. It is affected by extreme climatic phenomena, like Typhoon storm events. This combined with the deeply weathered bedrock, typical for subtropical regions, resulting in higher vulnerability for erosion and therefore resulting high sediment rates. Additionally, local human activities and the effects of global climate change amplify this higher vulnerability. The presented analysis contains the assessment of sheet erosion by means of the J2000-S eco-hydrological model, where the effect of climate and land use scenarios were analyzed. The results show that land use change has a higher effect then the climate chance on sheet erosion in the analyzed future period. Additionally, the landslide activity in the area was assessed, using a landslide inventory and bivariate statistical analysis. A landslide susceptibility map was the result of this assessment.

## Introduction

Central Vietnam is located within the Southeast Asia monsoon. This region is suspected to be strongly impacted by climate change (Miola and Simonet 2014); it may be especially susceptible to climate extremes, which are presumed to intensify in frequency and magnitude (IPPC 2012). The combination of the changing climate and the deeply weathered bedrock, typical for tropical regions, can lead to a higher vulnerability for increased sediment dynamics. Additionally, this higher expected vulnerability is amplified by human activities in the region as the building of infrastructure, settlements and agricultural expansion in the entire study area

M. Fink (✉) · C. Fischer · H. Tünschel · M. Meinhardt
Department of Geography, Friedrich-Schiller University Jena, Jena, Germany
e-mail: manfred.fink@uni-jena.de

P. Laux
Regional Climate and Hydrology, Karlsruhe Institute of Technology, Karlsruhe, Germany
e-mail: patrick.laux@kit.edu



A. Nauditt and L. Ribbe (eds.), *Land Use and Climate Change Interactions in Central Vietnam*, Water Resources Development and Management,
DOI 10.1007/978-981-10-2624-9_12

continues to increase. The sediment (mass movement) problems in the area can be classified into three different process categories:

- Areal water erosion (sheet, rill and interrill erosion) related to raindrop energy and surface runoff.
- Mass movement (landslides and rock falls) related to the instability of parts of the hill slope.
- River bank and riverbed erosion related to the energy of the flowing water.

We focus on the assessment of sheet erosion and landslide activity in the area. Climate and land use scenarios were analyzed for sheet erosion.

For the analysis of sheet erosion, numerous methods have been developed and used all over the world. Methods used for the erosion assessment include a variety of different physically based dynamic models such as WEPP Water Erosion Prediction Project (WEPP) model (Nearing et al. 1989; Pieri et al. 2007) and EUROpean Soil Erosion Model (EUROSEM; Morgan et al. 1998). These models are mainly used in small to medium sized basins that have a good collection data. Many applications describe single erosion events.

The empirical Universal Soil Loss Equation (USLE) is the most commonly used assessment technique (Wischmeier and Smith 1978). This static method uses the most important physiographic factors to estimate the sediment delivery on agricultural fields. It is based on numerous field measurements all over the world and provides a static erosion potential for the analyzed area.

The SWAT (Arnold and Fohrer 2005) and AnnAGNPS (Bingner et al. 2015) models are a combination of dynamic physically based models and the empirical static USLE methods. They utilize the empirical factors of the USLE and replace the static averaged rain factor with a surface runoff component derived from a hydrological model. This combination allows for a dynamic estimation of the erosion process that is dependent of the actual weather conditions.

A variety of methods have been developed worldwide for landslide susceptibility assessment. The bivariate and multivariate methods belong to the quantitative statistical approaches (Carrara 1983; Bui et al. 2011). In addition, there are training and membership techniques based on quantitative methods, such as applying neural networks or fuzzy sets (Kanungo et al. 2006; Tien Bui et al. 2012a, b, c; Park et al. 2013). Another group is deterministic or process-based models, which apply laws of physics to calculate the slopes stability (Montgomery and Dietrich 1994; Claessens et al. 2007). These approaches often use precipitation data, because it is the main landslide triggering factor. Additionally they require geo- and soil physical parameters, which are sometimes not available. Less input data is necessary for an approach, which is relying mostly on precipitation thresholds, above which landslides are initiated (Crosta 1998; Tien Bui et al. 2013). These approaches assess landslide hazard, if a temporal component is present in the model (Meinhardt et al. 2015).

## Sediment Generation Due to Sheet Erosion in VGTB

### *Current Status*

Sheet erosion is one of the most important sediment generating processes. Several time series and geo data were used to estimate the erosion rates. Within the catchment, 21 precipitation stations and 3 climate stations with data from 1979 to 2014 were available. The climate stations were located in the lowlands and only a few precipitation stations were located at higher altitudes. For the study area's river system, 10 gauging stations were available, but only two of them measure discharge values continuously. The other eight stations were located in the lowlands: as the rivers there are tidal influenced, they measure water level only (cf. chapter "Biophysical and Socio-economic Features of the LUCCi—Project Region: The Vu Gia Thu Bon River Basin").

For the description of soils, a map at the scale of 1:100,000 from the National Institute of Agriculture Planning and Protection from the year 2005. In addition, 150 soil profile descriptions in the catchment were available to derive soil-model parameters for the various soil classes described in the map. The geology is also described by a map in a scale of 1:100,000 from the Department of Geology and Minerals of Vietnam, Hanoi from the year 1997. For land use, a land cover classification of Landsat images for the year 2010 was used. The digital elevation model (DEM) was derived from contour lines and points from a digital map (scale 1:50,000) using the topography-to-raster algorithm of ArcGIS. The resulting DEM had a resolution (cell size) of 25 m (Duong et al. 2014).

The spatial representation of the model is based on the HRU (Hydrological Response Units) approach (Flügel 1996). This approach is enhanced by individual polygons (cf. Fig. 2, detail window) with a multidimensional routing scheme (Pfennig et al. 2009). The HRUs are the result of overlay analysis of the DEM derivatives, Topographical Wetness Index, Mass Balance Index and Annual Solar Radiation Index and the individual soil classes. The land use and geology maps were just used to assign these properties to the resulting polygons using a maximum membership function. This had two reasons: firstly land use classification is too fine grained and noisy for an overlay analysis. Their use would result in too many HRU polygons for the modelling. Secondly, the geology features are quite similar to the soil map, such that the variability is already represented in the soil map. The HRU delineation process resulted in 477,888 HRUs and 24,192 reach segments for the entire VGTB basin (Fink et al. 2013) (c.f. Fig. 1).

J2000 can be classified as a distributed, process-oriented hydrological model for hydrological simulations of meso-scale and macro-scale catchments. It is implemented in the Jena Adaptable Modelling System (JAMS) framework (Kralisch and Krause 2006; Kralisch et al. 2007), which is a software framework for component-based development and application of environmental models. The model describes the hydrological processes as encapsulated or independent process modules. These modules describe for example input data regionalization and

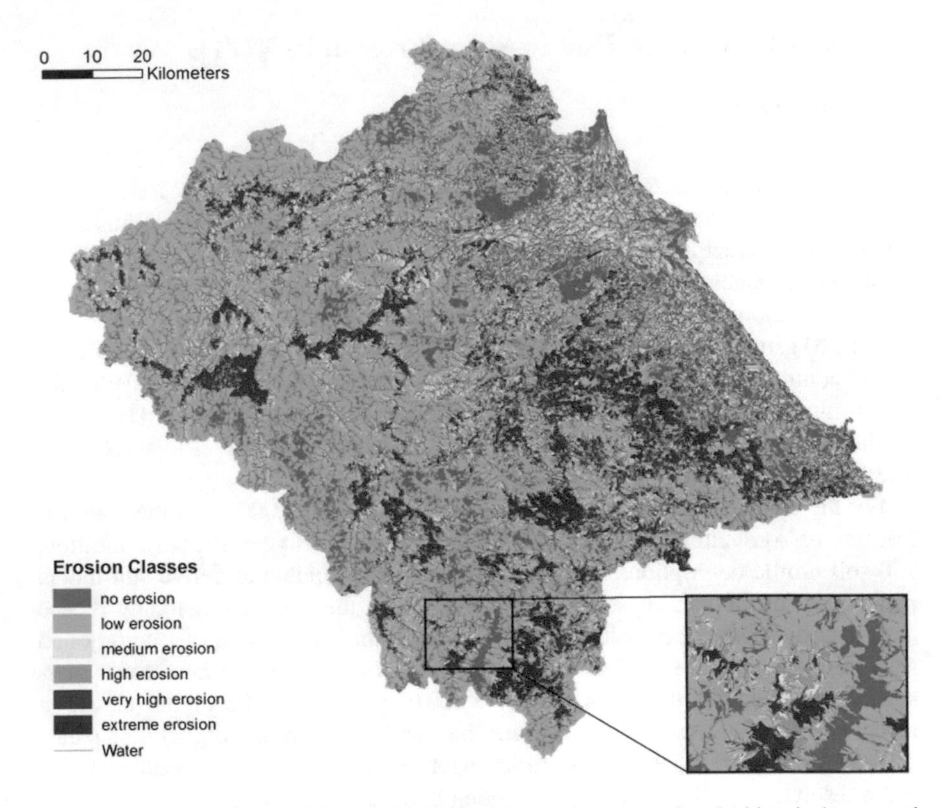


**Fig. 1** Areal results (477,888 HRUs) for 1997 of the erosion rates using the historical measured climate data

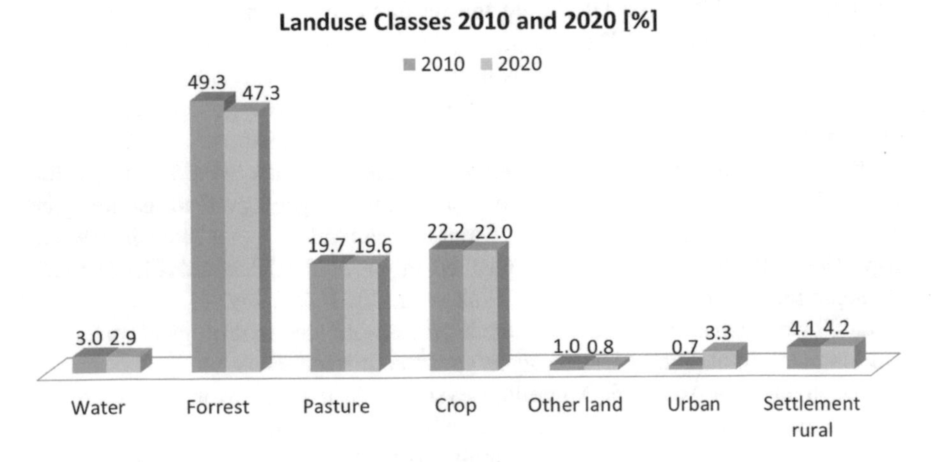


**Fig. 2** Proportions of land use classes in the baseline (2010) and the future (2020) scenario

correction, calculation of potential and actual evapotranspiration, canopy interception, soil moisture, groundwater and irrigation processes (Fink et al. 2013). With the hydrological modelling system J2000 and the integrated module for sediment yield/soil loss prediction, the different linkage approaches were implemented. The calculation of sediment yield (SY in tonnes) used the Modified Universal Soil Loss Equation (MUSLE) (Williams 1975) with a dynamic factor. The runoff factor was calculated in J2000. The MUSLE follows the structure of the Universal Soil Loss Equation (USLE), with the exception that rainfall factor R is replaced with the runoff factor. The equation calculates sediment yield/soil loss from a rainfall event for each HRU. The structure of used MUSLE is:

$$\mathrm{SY} = R * K * L * S * C * P * \mathrm{ROKF}$$

where $R = 11.8\ (Q * \mathrm{qp} * A_{\mathrm{HRU}})^{0.56}$ with the factors: $Q$ = volume of runoff in mm, qp = peak flow rate in $m^3/s$, $K$ = soil erodibility factor, $L$ = slope-length factor, $S$ = slope gradient factor, $C$ = cover and management factor, $P$ = support practice factor, ROKF = coarse fragment content factor, $A_{\mathrm{HRU}}$ = area of the HRU in $m^2$. Different factors of the MUSLE have been delineated using the above-mentioned geo data combined typical values from the literature. The distributed results from the modelling exercise are shown in Fig. 1 for each of the 477,888 HRUs.

The shown classes use the gross output of every HRU including the water a sediment flux from the above located HRUs. Therefore the highest values of sediment output are located at the foot slopes of the agricultural used areas in the uplands. Further reasons for very high rates, besides the mentioned cascading effect, are high rainfall, respectively, surface runoff rates and steep hill slopes in the uplands. Low rates are associated with permanent vegetated land cover classes like forest and grassland. Compared to typical results from the temperate zone, most of the values are very high. The very high precipitation rate (approx. 3500 mm/a) in the area is the major reason for this. Additionally the literature values for the soil and crop factors are mainly delineated with results of experiments conducted in the temperate climate zone. This can lead to misleading results, especially for croplands and grasslands (Labrière et al. 2015) where there is a tendency to overestimate the erosion rates in deeply weathered tropical soils. Thus the erosion results should be used in a relative manner like shown in Fig. 1. As expected, croplands with steep slopes are most prone to erosion.

## Future Development

For the erosion modelling the scenarios for land use and climate were analyzed. The land use scenarios consists of a baseline and a near future scenario. The baseline scenario is based on a classification of remote sensing images from 2010 (cf. Sect. 4.2.1). The future scenario consists of a combination of a land cover map of the VGTB river basin in the year 2020 at 30 m spatial resolution, as a result of the projection of the past land cover change dynamics occurred in the period 2001–2010 to the reference year 2020 (cf. Sect. 4.2.6), and the master plan of the Department of Natural Resources DONRE for 2030. Since the projection (Sect. 4.2.5) was only created for the mid and highlands. The master plan is used to represent the lowlands. A summary of the changing proportions is shown in Fig. 1. The proportion of urban settlements is increasing and the forest is decreasing. The growth of the settlement basically takes place in the lowlands, which results in decreasing cropland. This loss of cropland is compensated in the upland where forest is converted to cropland.

The land use scenario was combined with climate change scenarios. These are provided by WRF (Weather Research and Forecasting) (Skamarock and Klemp 2008) downscaling models in spatial resolution of approx. 5 km by 5 km based on ECHAM5 (European Centre for medium-range weather forecasts and HAMburg) (Roeckner et al. 2003) A1B and B1 scenarios (cf. chapter "Integrated River Basin Management in the Vu Gia Thu Bon Basin").

The scenario analyses have been conducted using four different model configurations:

- Baseline land use; climate downscaling scenario A1b; period (1971–2033)
- Baseline land use; climate downscaling scenario B1; period (1971–2050)
- Future land use; climate downscaling scenario A1b; period (1971–2033)
- Future land use; climate downscaling scenario B1; period (1971–2050).

In Table 1 the quantiles of the daily sediment load from the entire study area for the different scenarios partitioned into two periods from 1972 to 2002 and from 2002 to 2032 are shown. The values indicate that most of them are higher for the near future period. Only the maximum and 99% values are exceptions with different tendencies. Because the erosion is an event related process, the lower quantiles are always zero. The statistical more stable quantiles between 50 and 95%, the future

**Table 1** Quantiles of sediment loads

| Sediment [kt/d] | | | | | | | | |
|---|---|---|---|---|---|---|---|---|
| | Land use 2010 (baseline) | | | | Land use 2020 (scenario) | | | |
| | 1972–2002 | 2002–2032 | 1972–2002 | 2002–2032 | 1972–2002 | 2002–2032 | 1972–2002 | 2002–2032 |
| | B1 | B1 | A1B | A1B | B1 | B1 | A1B | A1B |
| min | 0.00 | 0.00 | 0.00 | 0.00 | 0.00 | 0.00 | 0.00 | 0.00 |
| 1% | 0.00 | 0.00 | 0.00 | 0.00 | 0.00 | 0.00 | 0.00 | 0.00 |
| 5% | 0.00 | 0.00 | 0.00 | 0.00 | 0.00 | 0.00 | 0.00 | 0.00 |
| 10% | 0.00 | 0.00 | 0.00 | 0.00 | 0.00 | 0.00 | 0.00 | 0.00 |
| 50% | 0.50 | 0.57 | 0.50 | 0.66 | 0.70 | 0.72 | 0.70 | 1.63 |
| 90% | 361.03 | 495.36 | 335.30 | 570.91 | 469.25 | 635.00 | 471.68 | 1833.00 |
| 95% | 2920.09 | 3151.88 | 2826.05 | 3865.46 | 3691.75 | 3963.90 | 3644.61 | 5743.98 |
| 99% | 18,689 | 19,052 | 18,696 | 22,656 | 23,703 | 23,938 | 23,596 | 22,402 |
| max | 202,600 | 105,922 | 202,944 | 258,909 | 257,165 | 134,273 | 257,483 | 229,686 |

**Table 2** Quantiles of runoff

| Runoff [m³/s] | | | | | | | | |
|---|---|---|---|---|---|---|---|---|
| | Land use 2010 (base line) | | | | Land use 2020 (scenario) | | | |
| | 1972–2002 | 2002–2032 | 1972–2002 | 2002–2032 | 1972–2002 | 2002–2032 | 1972–2002 | 2002–2032 |
| | B1 | B1 | A1B | A1B | B1 | B1 | A1B | A1B |
| min | 80.2 | 62.6 | 81.5 | 72.3 | 76.6 | 62.6 | 80.8 | 87.9 |
| 1% | 94.3 | 90.4 | 95.6 | 90.4 | 93.4 | 89.5 | 94.8 | 101.5 |
| 5% | 105.4 | 110.6 | 106.8 | 108.8 | 104.4 | 109.5 | 105.6 | 116.5 |
| 10% | 114.9 | 121.6 | 116.5 | 122.2 | 113.5 | 120.8 | 115.2 | 126.5 |
| 50% | 370.8 | 376.9 | 374.3 | 371.6 | 372.5 | 378.5 | 377.3 | 424.9 |
| 90% | 1655.9 | 1739.5 | 1654.2 | 1907.8 | 1658.2 | 1742.6 | 1660.8 | 1957.4 |
| 95% | 2460.6 | 2588.7 | 2422.0 | 2907.6 | 2461.0 | 2603.1 | 2432.6 | 2718.7 |
| 99% | 6389.4 | 6275.0 | 6404.5 | 7047.8 | 6385.8 | 6302.3 | 6438.6 | 5329.6 |
| max | 28,307.2 | 21,376.8 | 28,431.5 | 36,203.4 | 28,354.5 | 21,407.3 | 28,478.5 | 28,546.0 |

scenarios are always higher, especially the combination of the land use scenario from 2020 and the climate from 2002–2032 show very high values. The land use scenario 2020 feature in most cases higher values (in average 45% higher the in the scenario for 2010).

In Table 2 the result for the simulated runoff for the above-mentioned scenario settings is displayed. The differences between the future and the present climate and land use scenarios are not as systematic as in the case of the sediments.

To look closer at the climate scenarios a Mann–Kendall trend test (Salmi et al. 2002) of yearly values have been performed (cf. Table 3). For the A1b scenario, the test indicates positive trends for the sediment load and precipitation at the 5% significance level whereas the temperature features a very high significant (0.1% level) positive trend. The less intensive scenario B1 shows the same tendencies then the trends of the A1b scenario, but only the temperature feature a significant result. The calculated scenarios gives valuable indications that climate as well as land use change are factors that increase erosion, respectively, sediment load. Regarding Table 1 it is obvious that the influence of the land use change is dominating this increase. The major reason for this is that the settlement development in the lowlands replaces cropland. This will be compensated by an increase of cropland in the mid and upland where suitable locations are already occupied by existing agricultural use. Therefore the new cropland will be in less favourable locations that are in many cases located on higher slopes, which leads to progressive erosion rates. The assessment of climate scenarios shows that rainfall intensities (cf. Table 1) will increase and therefore also lead to higher erosion rates (cf. Table 1). This increase is less substantial as in the case of land use change. That means that measures against increasing erosion and sediment load should focus on the prevention of the vegetation cover of the erosion prone areas.

**Table 3** Trends of sediment load, runoff, precipitation and temperature obtained with the Mann Kendall test

| | A1B 1971–2033 | | B1 1971–2050 | |
|---|---|---|---|---|
| | Significance | Trend slope | Significance | Trend slope |
| Sediments [t] | * | 6654 | | 2625 |
| Runoff [$m^3/s$] | * | 2.68 | | 1.08 |
| Precipitation [mm] | * | 0.0193 | | 0.006594 |
| Mean temperature [°C] | *** | 0.0115 | *** | 0.00963 |

*** trend at $\alpha = 0.001$ level of significance, ** trend at $\alpha = 0.01$ level of significance, * trend at $\alpha = 0.05$ levelof significance

## *Landslide Susceptibility*

Another important source of sediments are landslides. For the assessment of the landslide susceptibility, an extended version of the statistical index method (SI) has been used. The SI approach was first applied by Yin and Yan (1988) but named Information Value method. This method is based on the distribution of mapped landslides in the classes of physical parameters influencing the occurrence of landslides (Meinhardt et al. 2015). The method is a calculation of the landslide density for the mapped areas were the resulting weights of the factors using the natural logarithm. The SI method was improved using a parameter reduction method based on the validation criteria of reduced parameter sets (Meinhardt et al. 2015).

The results from the extended SI method show that the best results were achieved using the following parameters; aspect, valley depth, profile curvature, distance to roads, distance to waterbodies, land use, soil, lithology and precipitation increase. Within the resulting weights, the distance to roads is by far indicated as the most important parameter, followed by slope, lithology and precipitation increase (Meinhardt et al. 2015). Precipitation increase is defined as the spatial interpolated long year average precipitation (1979–2006) compared with the precipitation in the recent years (2007–2011) based on the measured data (Meinhardt et al. 2015).

Concerning the resulting susceptibility map (Fig. 3), it is obvious that the highest values are located in the mountainous zones of the study area. Furthermore some parameters, e.g. distance to roads and rivers are clearly visible in the map because they receive a high relevance from the calculated factors.

For validation, the area under the curve (AUC) is utilized. To calculate the AUC, the resulting susceptibility values of the whole study area need to be sorted in descending order on the X-axis and plotted against the cumulative landslide occurrence on the Y-axis (Lee and Pradhan 2007; Tien Bui et al. 2012b). If the total area is 100, it indicates perfect prediction accuracy. Figure 4 shows for example AUCs around 92. Furthermore, it is necessary to differentiate between prediction and success rates. In case of the latter, the landslides from method training are used for validation. On the other hand the landslides that were kept for validation are applied to calculate the prediction rate (Meinhardt et al. 2015).

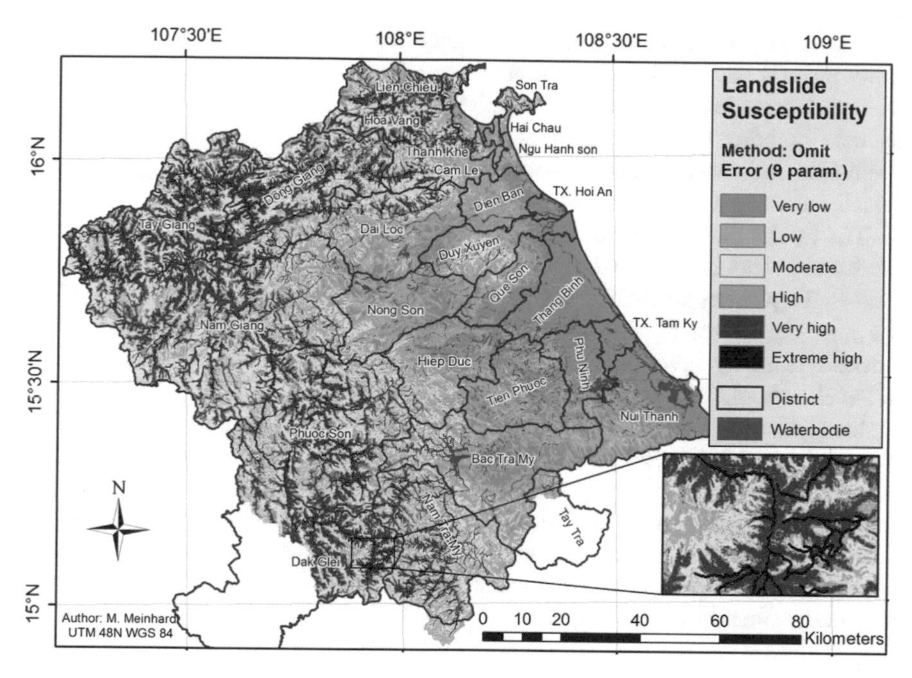


**Fig. 3** Landslide susceptibility map based on the extended SI method applied with the 9 input parameters (Meinhardt et al. 2015)

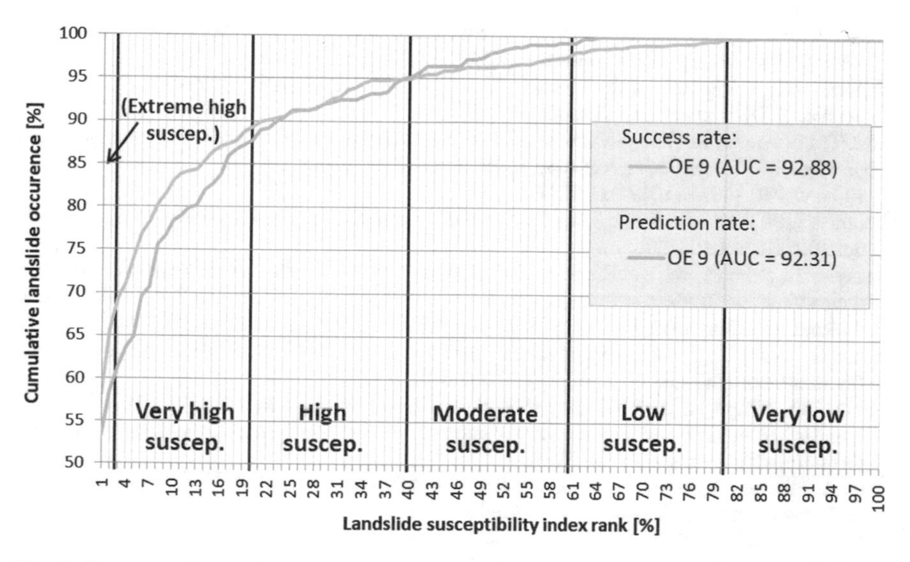


**Fig. 4** Success and prediction rate of the extended SI method (Meinhardt et al. 2015, modified)

The most practicable approach to validate and compare the different susceptibility maps is the calculation of the AUC. In addition to the AUC-value, the curves in Fig. 4 provide some more information about the quality of the results of the success (training dataset) and the prediction (validation dataset) rate. The shown prediction and success rates are close to 100, the ideal result, and are quiet similar which indicates a good and stable model performance.

The results gathered by the landslide assessment show where the likelihood of landslide events is high, and gives also information of the most important control factors. To prevent potential damages, relevant slopes should be secured, especially the ones cut by roads. Moreover the maps should be already used during infrastructure planning, avoiding zones of high susceptibility values where possible. What else should be avoided in this area is deforestation because this leads to much higher landslide susceptibility, as it is confirmed by the slide density recorded on crop and grassland. Besides, the effects of the vegetation removal, provisional road cuts are a factor contributing to landslides. Regarding the ongoing extension of hydropower reservoirs in the region, the achieved results from the conducted sediment dynamic assessments show increasing trends in sediment loads. Therefore we can conclude that the prevention of erosion and landslides is an important topic to prevent siltation of reservoirs.

## References

Arnold JG, Fohrer N (2005) SWAT2000: current capabilities and research opportunities in applied watershed modelling. Hydrol Process 19(3):563–572. doi:10.1002/hyp.5611

Bingner RL, Theurer FD, Yuan Y (2015) AnnAGNPS Technical Processes: Technical_Documentation. Version 5.4

Bui DT, Lofman O, Revhaug I, Dick O (2011) Landslide susceptibility analysis in the Hoa Binh province of Vietnam using statistical index and logistic regression. Nat Hazards 59(3):1413–1444. doi:10.1007/s11069-011-9844-2

Carrara A (1983) Multivariate models for landslide hazard evaluation. Math Geol 15(3):403–426. doi:10.1007/BF01031290

Claessens L, Schoorl JM, Veldkamp A (2007) Modelling the location of shallow landslides and their effects on landscape dynamics in large watersheds: an application for Northern New Zealand. Geomorphology 87(1–2):16–27. doi:10.1016/j.geomorph.2006.06.039

Crosta G (1998) Regionalization of rainfall thresholds: an aid to landslide hazard evaluation. Environ Geol 35(2–3):131–145. doi:10.1007/s002540050300

Duong DQ, Stolpe H, Jolk C, Greassidis S, Führer N, Zindler B, Fink Vo, M DP (2014) Standardize geodata management in Vietnam—an urgent need. International Symposium on Geoinformatics for Spatial, Da Nang

Fink M, Fischer C, Führer N, Firoz, AMB, Viet TQ, Laux P, Flügel W (2013) Distributive hydrological modeling of a monsoon dominated river system in central Vietnam. International Congress on Modelling and Simulation (20th). pp 1826–1832

Flügel W (1996) Hydrological response units (HRU's) as modelling entities for hydrological river basin simulation and their methodological potential for modelling complex environmental process systems—results from the Sieg catchment. Die Erde 127:43–62

IPPC (2012) Managing the risks of extreme events and disasters to advance climate change adaptation: a special report of Working Groups I and II of the Intergovernmental Panel on Climate Change. Cambridge, New York

Kanungo DP, Arora MK, Sarkar S, Gupta RP (2006) A comparative study of conventional, ANN black box, fuzzy and combined neural and fuzzy weighting procedures for landslide susceptibility zonation in Darjeeling Himalayas. Eng Geol 85(3–4):347–366. doi:10.1016/j.enggeo.2006.03.004

Kralisch S, Krause P (2006) JAMS—A framework for natural resource model development and application. In: Proceedings of the iEMSs Third Biannual Meeting.

Kralisch S, Krause P, Fink M, Fischer C, Flügel W-A (2007) Component based environmental modelling using the JAMS framework. In: Proceedings of the MODSIM 2007 International Congress on Modelling and Simulation, 812–818

Labrière N, Locatelli B, Laumonier Y, Freycon V, Bernoux M (2015) Soil erosion in the humid tropics: a systematic quantitative review. Agric Ecosyst Environ 203:127–139. doi:10.1016/j.agee.2015.01.027

Lee S, Pradhan B (2007) Landslide hazard mapping at Selangor, Malaysia using frequency ratio and logistic regression models. Landslides 4(1):33–41. doi:10.1007/s10346-006-0047-y

Meinhardt M, Fink M, Tünschel H (2015) Landslide susceptibility analysis in central Vietnam based on an incomplete landslide inventory: comparison of a new method to calculate weighting factors by means of bivariate statistics. Geomorphology 234:80–97. doi:10.1016/j.geomorph.2014.12.042

Miola A, Simonet C (2014) Concepts and metrics for climate change risk and development: towards an index for climate resilient development. EUR, Scientific and technical research series, vol 26587. Publications Office, Luxembourg

Montgomery DR, Dietrich WE (1994) A physically based model for the topographic control on shallow landsliding. Water Resour Res 30(4):1153–1171. doi:10.1029/93WR02979

Morgan RPC, Quinton JN, Smith RE, Govers G, Poesen JWA, Auerswald K, Chisci G, Torri D, Styczen ME (1998) The European soil erosion model (EUROSEM): a dynamic approach for predicting sediment transport from fields and small catchments. Earth Surf Proc Land 23:527–544

Nearing MA, Foster GR, Lane LJ, Finkner SC (1989) A process-based soil erosion model for USDA-Water Erosion Prediction Project technology. Trans ASAE 32(5):1587–1593. doi:10.13031/2013.31195

Park S, Choi C, Kim B, Kim J (2013) Landslide susceptibility mapping using frequency ratio, analytic hierarchy process, logistic regression, and artificial neural network methods at the Inje area, Korea. Environ Earth Sci 68(5):1443–1464. doi:10.1007/s12665-012-1842-5

Pfennig B, Kipka H, Wolf M, Fink M, Krause P, Flügel W (2009) Development of an extended routing scheme in reference to consideration of multi-dimensional flow relations between hydrological model entities. In: Proceedings of the 18th World IMACS Congress and MODSIM09 International Congress on Modelling and Simulation, 1972–1978

Pieri L, Bittelli M, Wu JQ, Dun S, Flanagan DC, Pisa PR, Ventura F, Salvatorelli F (2007) Using the Water Erosion Prediction Project (WEPP) model to simulate field-observed runoff and erosion in the Apennines mountain range, Italy. J Hydrol 336(1–2):84–97. doi:10.1016/j.jhydrol.2006.12.014

Roeckner E, Bäuml G, Bonaventura L, Brokopf R, Esch M, Giorgetta M, Hagemann S, Kirchner I, Kornblueh L, Manzini E, Rhodin A, Schlese U, Schulzweida U, Tompkins A (2003) The atmospheric general circulation model ECHAM5—Part I: Model description. Max-Planck-Institut für Meteorologie, Hamburg (Report No. 349)

Salmi T, Maatta A, Anttila P, Ruoho-Airola T, Amnell T (2002) Detecting trends of annual values of atmospheric pollutants by the Mann–Kendall test and Sen's slope estimates

Skamarock WC, Klemp JB (2008) A time-split nonhydrostatic atmospheric model for weather research and forecasting applications. J Comput Phys 227(7):3465–3485. doi:10.1016/j.jcp.2007.01.037

Tien Bui D, Pradhan B, Lofman O, Revhaug I, Dick OB (2012a) Landslide susceptibility assessment in the Hoa Binh province of Vietnam: A comparison of the Levenberg–Marquardt and Bayesian regularized neural networks. Geomorphology 171–172:12–29. doi:10.1016/j.geomorph.2012.04.023

Tien Bui D, Pradhan B, Lofman O, Revhaug I, Dick OB (2012b) Landslide susceptibility mapping at Hoa Binh province (Vietnam) using an adaptive neuro-fuzzy inference system and GIS. Comput Geosci 45:199–211. doi:10.1016/j.cageo.2011.10.031

Tien Bui D, Pradhan B, Lofman O, Revhaug I, Dick OB (2012c) Spatial prediction of landslide hazards in Hoa Binh province (Vietnam): a comparative assessment of the efficacy of evidential belief functions and fuzzy logic models. CATENA 96:28–40. doi:10.1016/j.catena.2012.04.001

Tien Bui D, Pradhan B, Lofman O, Revhaug I, Dick ØB (2013) Regional prediction of landslide hazard using probability analysis of intense rainfall in the Hoa Binh province, Vietnam. Nat Hazards 66(2):707–730. doi:10.1007/s11069-012-0510-0

Williams JR (1975) Sediment—yield prediction with universal equation using runoff energy factor. Proceedings of the sediment—Yield Workshop, USDA Sedimentation Laboratory

Wischmeier WH, Smith DD (1978) Predicting rainfall erosion losses—a guide to conservation planning. Agriculture Handbook (537)

Yin KL, Yan TZ (1988) Statistical prediction models for slope instability of metamorphosed rocks. Landslides. In: Proceedings of the Fifth International Symposium on Landslides, vol 2. pp 1269–1272

# Sand Dunes and Mangroves for Disaster Risk Reduction and Climate Change Adaptation in the Coastal Zone of Quang Nam Province, Vietnam

**U. Nehren, Hoang Ho Dac Thai, N.D. Trung, Claudia Raedig and S. Alfonso**

**Abstract** Ecosystem-based measures to reduce the risk of coastal hazards and adapt to climate change have attracted increasing attention in science, policy, and planning. In the coastal zone of Quang Nam province in Central Vietnam, natural ecosystems such as coastal dunes and mangroves can serve as natural buffers against typhoons, storm surges, waves, and even small tsunamis and protect the shoreline from coastal erosion and sea-level rise. Apart from these protective ecosystem services, intact dune and mangrove ecosystems perform various other regulating services, such as carbon storage and sequestration and groundwater protection, as well as a variety of provisioning and cultural services. Moreover they support a high biological diversity, which forms the basis for secure livelihoods of coastal communities. However, important ecosystem functions and services have already been lost or diminished due to overexploitation and a lack of integrated coastal management approaches. In this chapter, we (1) provide an overview of the current distribution and status of dune and mangrove ecosystems in the coastal zone of Quang Nam province, (2) analyze the actual and potential ecosystem services of and threats to these ecosystems in selected study sites, (3) assess the awareness and preparedness of coastal communities to coastal hazards, and (4) discuss the potential for conservation, restoration, and sustainable use of coastal dunes and mangroves.

**Keywords** Coastal dune systems (CDS) · Mangroves · Ecosystem-based Disaster Risk Reduction (Eco-DRR) · Ecosystem-based Adaptation (EbA) · Central Vietnam

U. Nehren (✉) · H. Ho Dac Thai · N.D. Trung · C. Raedig · S. Alfonso
Institute for Technology and Resources Management in the Tropics and Subtropics (ITT), TH Köln, University of Applied Sciences, Betzdorfer Straße 2, Cologne 50679, Germany
e-mail: udo.nehren@th-koeln.de



A. Nauditt and L. Ribbe (eds.), *Land Use and Climate Change Interactions in Central Vietnam*, Water Resources Development and Management,
DOI 10.1007/978-981-10-2624-9_13

## Introduction

The coast of Central Vietnam is prone to natural hazards; in particular typhoons and storm surges are frequent occurrences in the monsoon season. According to GFDRR & The World Bank (2010), on average about ten tropical typhoons hit the Vietnamese coast every year, often causing severe damages. It is estimated that between 1953 and 1991 more than 60 % of all disaster events in Vietnam originated from typhoons, and 78 % of killed and missing people in that period were typhoon victims (Fritz and Blount 2007). In general, the risk of a large typhoon is considerably higher in North and Central Vietnam compared to the southern parts of the country, but at the same time southern Vietnam with the Mekong Delta is particularly prone to storm surges and coastal flooding (Takagi et al. 2014).

Compared to typhoons, the tsunami risk seems to be considerably lower along the Vietnamese coast. However, based on the few available data, a final conclusion on future tsunami risk is not possible. According to Vu (2014), there is no reliable evidence for historical large and damaging tsunamis, only for smaller events that did not cause significant damage. On the other hand, historical documents indicate five significant tsunami events between 1877 and 1978 of which two might have been triggered by earthquakes (1877, 1982), and one each by a submarine volcanic eruption (1923), a submarine landslide (late nineteenth/early twentieth century), and strong winds (1978). One of these tsunamis hit the coast of North Central Vietnam in the province of Dien Chau. It caused inundation damages up to one kilometer inland (Vu and Nguyen 2008). Based on computer models, Vu and Nguyen (2008) assume that significant tsunamis could be generated by a high magnitude earthquake of at least 7.0 along the fault of the Central Vietnam shelf, or by an earthquake with a magnitude larger than 8.0 at the Manila Trench. Even though the tsunami risk along the Vietnamese coast is only moderate, Takagi et al. (2014) suggest that preparedness against coastal hazards should consider also tsunamis apart from typhoons and storm surges.

Ecosystem-based Disaster Risk Reduction (Eco-DRR) and Ecosystem-basedAdaptation (EbA) measures have increasingly been recognized and addressed in coastal management, as it has been realized that coastal ecosystems can provide effective protection against tropical cyclones, storm surges, waves, coastal flooding, and other coastal hazards, where a combined protection of various ecosystems such as coral reefs, seagrass beds, and sand dunes or mangrove forests are particularly effective (Badola et al. 2005; Granek and Ruttenberg 2007; Batker et al. 2010). Moreover, healthy ecosystems can reduce coastal erosion by stabilizing sediments and weakening wave energy (Prasetya 2007; Barbier et al. 2011). To a certain extent, they can even mitigate climate change impacts such as sea-level rise and salinization of coastal aquifers (Carter 1991; Heslenfeld et al. 2004; Saye and Pye 2007). At the same time, particularly in tropical and subtropical countries coastal ecosystems are under increasing pressure due to urbanization processes and tourism development (Martínez et al. 2004), land use intensification (French 2001; Nehren et al. 2016), sand mining (Sridhar and Bhagya 2007; Miththapala 2008;

Takagi et al. 2014), and the development of aquaculture systems (Phan and Nguyen 2006). Depending on the severity of impacts going along with these activities, coastal ecosystems show varying degrees of degradation. The degradation status in turn has a direct influence on the provision of ecosystem services including the protection against coastal hazards.

Today, the coast of Vietnam is among the most densely populated regions in South-East Asia and a large part of Vietnams Gross National Product is produced through fishery and aquaculture industries, marine transport and tourism (Nguyen and Shaw 2010). Along with these activities, there is an ongoing degradation of nature-near coastal ecosystems, such as mangroves, wetlands and dune forests, which had already been partly destroyed and degraded in historical times, in particular during the Vietnam War (Ross 1974; Sterling et al. 2006). The clearing or degradation of the natural buffer zones results in higher exposure and vulnerabilities of coastal communities to disasters and a loss of biodiversity and associated ecosystem services. According to WWF (2013), the Southern Vietnam Lowland Dry Forest ecoregion is among the most degraded in Asia with an estimated 90 % of the natural habitats already been cleared and high land use pressure on the remaining coastal ecosystems.

Within the next years, the pressures on coastal ecosystems will likely increase in some coastal regions due to intensified on- and offshore oil and gas exploitation. The possibilities to develop new oil and gas deposits are currently being investigated by the Vietnamese government (Vietnam Oil and Gas Report Q1 2013). Climate change impacts, in particular sea-level rise, will likely put additional pressure on the coastal zone. According to Dasgupta et al. (2009), Vietnam is among the five most affected countries to 1-M sea-level rise with major effects on livelihoods, ecosystems, water and soil resources, and economy.

Within the frame of the LUCCi project, we took a closer look on two important coastal ecosystems: coastal dune systems (CDS) and mangroves. Both ecosystems offer important ecosystem services related to hazard mitigation and coastal protection and at the same time support local livelihoods. However, in the coastal zone of the Vu Gia Thu Bon watershed, the original vegetation cover of CDS has often been replaced by tree plantations and aquaculture systems and most former mangrove areas have already been deforested or are severely degraded. Several publications show that intact coastal ecosystems and vegetation structures, such as mangrove forests and coastal dunes, provide protection from storms (McIvor et al. 2012) and waves (Mazda et al. 1997; Tanaka et al. 2007; Laso Bayas et al. 2011), while their degradation reduces the adaptive capacity of societies to deal with disaster risk (Alliance Development Works 2012). Apart from the ecological status, several research papers point out that the effectiveness for coastal protection depends on the physical conditions of an ecosystem; in the case of CDS in particular on their height, width, shape, and continuity (Gómez-Pina 2002; Dahm et al. 2005; Takle et al. 2007; Thao et al. 2014) and in the case of mangroves on their width and structural properties, such as canopy diameter and tree height (Tri et al. 1998; Nghia 2004; Das and Vincent 2009; Yanagisawa et al. 2010).

Conserving, sustainably managing and restoring dune and mangrove ecosystems can therefore be a cost-effective alternative to hard infrastructure measures to mitigate the risk of climate-related disasters. In addition to the protective functions, these ecosystems provide a variety of co-benefits, which could be used by the coastal communities to secure their livelihoods in the long run. So far there is only little research on the potential of CDS for coastal protection in Vietnam, while mangroves have been researched particularly in the delta regions of the Mekong and the Red River (Quartel et al. 2007; Van Santen et al. 2007; Boateng 2012; Christensen et al. 2014; Schmitt and Albers 2014; Thanh 2014; Trung and Tri 2014; Anthony et al. 2015; Besset et al. 2015; Phan et al. 2015). For Central Vietnam, there is a lack of reliable data on the ecosystem services and protective functions for disaster risk reduction and climate change adaptation for both CDS and mangrove ecosystems. This in turn prevents from implementing suitable conservation and restoration measures that aim at combined risk reduction, climate change adaptation, and biodiversity conservation. In our research in the coastal zone of Quang Nam province, we address these research gaps by (1) providing an overview of the current distribution and status of dune and mangrove ecosystems, (2) analyzing the actual and potential ecosystem services of and threats to these ecosystems, (3) assessing the awareness and preparedness of coastal communities to coastal hazards, and (4) discussing the potential for conservation, restoration, and sustainable use of coastal dunes and mangroves.

## Methods

In a first step, dune and mangrove ecosystems in the coastal zone of Quang Nam province were identified based on Landsat images. The identified areas were visited to get an overview of the ecological status and human impact on these ecosystems and select study sites for in-depth research. Based on this rapid field assessment, Nui Thanh district with the commune of Tam Hai was selected for in-depth research, as it is the only commune where both mangrove remnants and coastal dunes are present. Expert interviews with representatives from local authorities were conducted to get first-hand information on the status and uses of coastal ecosystems, the trend of land use change, as well as protection and restoration policies. The visited authorities include the Department of Natural Resources (DONRE), the Department of Agriculture and the Rural Development (DARD) of Nui Thanh district with the Flood and Storm Control Board, the People's Committee of Tam Hai commune as well as the heads of the villages. Expert interviews were conducted with Mr. Tran Dinh Son, Deputy Director of Nui Thanh DARD, who is also the Vice Chairman of the Green Shield mangrove restoration project (funded by USAID), and Mr. Nguyen Tan Hung, Deputy Director of the Agriculture and Rural Development Division, Tam Hai Commune People's Committee.

Tam Hai commune is an island commune located in southern end of Nui Thanh district and Quang Nam province (Figs. 1 and 2). It comprises 1568 ha and is divided into seven villages, namely: Thuan An, Dong Tuan, Long Thanh Dong, Binh Trung, Xuan My, Long Thanh Tay, and Tan Lap. According to the

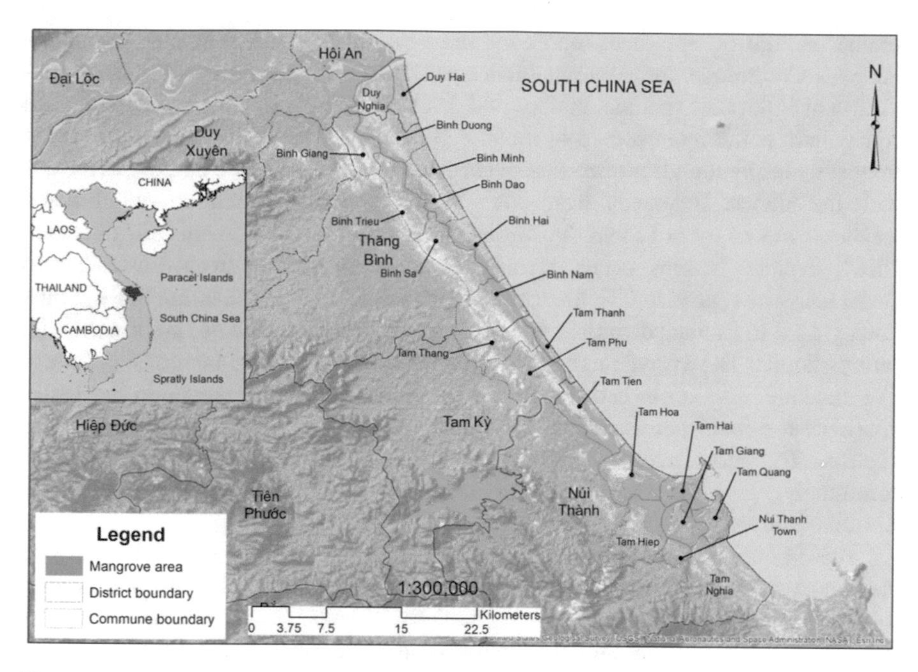


**Fig. 1** Overview map of the study area in Nui Thanh district. Mangroves are only found in small fragments in the An Hoa lagoon, sheltered from the sea

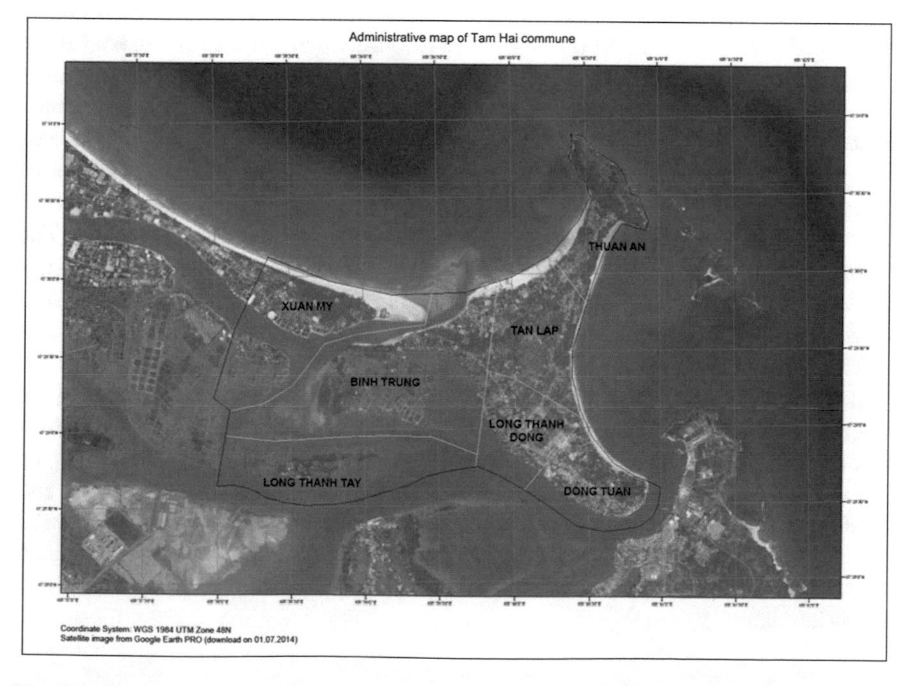


**Fig. 2** Location of Tam Hai commune. Rectangular grid cells indicate aquaculture activities (shrimp ponds)

socioeconomic development report for the year of 2013 of Tam Hai Commune People's Committee, the commune has around 2200 households with a total population of about 8300 people. In about 80 % of the households, offshore fishery and aquaculture is the main source of income (Tam Hai CPC 2013). The commune is frequently hit by thunderstorms that occur from late March to the end of November with the highest frequency from May to October. The maximum wind speed reaches levels of up to 11 and the storms are accompanied by heavy rainfall (Pham 2009). Tropical storms occur during the monsoon season from September to December with a peak in October. Some storms weaken over the ocean and are then downgraded to tropical depressions, while those hitting the land are associated with serious floods. The average number of tropical storms reaching the coast is four to five annually. The storm and flooding events have negative impacts on the livelihoods of the local communities, especially to offshore fishery and shrimp production. The most severe disasters in historical times according to the local community were the storms Klara (1964), Eve (1999), Xangsane (2006), and Ketsana (2009).

In 2014, a household survey based on a semi-structured questionnaire was conducted in Tam Hai commune. The questionnaire included questions in four main categories: (a) characteristics of rural livelihoods, (b) awareness of disaster risk, (c) disaster impacts and preparedness, (d) ecosystems and their protective services. The household survey was carried out in one village with mangrove forest (Binh Trung) and another with sand dune forest (Xuan My). In each village, 20 household heads were interviewed.

Moreover, a Participator Rural Appraisal (PRA) was carried out based on the guidelines provided by FAO (2006). The PRA tools that were selected are (a) the *village history tool* to reconstruct the development of the village and remarkable historical events, (b) the *participatory village mapping tool* to identify the distribution and status of mangroves and sand dunes in the study area and evaluate their potential protection function, (c) *transect walks* to create schematic profiles of dunes and mangroves, and (d) an *analysis of the seasonal calendar* to get an overview of the production activities of the local communities around the year. The respondents of each PRA tool were selected based on the primary principles according to the guideline. The number of responders varied from four to eight in the different tools, with the participation of both male and female, different ages and different occupations.

Based on the experience of the surveys in Xuan My village, another survey with a higher number of households ($n$ = 120) and an extended catalogue of questions was carried out in 2015 in the commune of Duy Hai in Duy Xuyen district (Fig. 1). The commune is located within the delta of the Vu Gia Thu Bon river system and mainly affected by typhoons, coastal flooding and droughts. The questions related to (a) observed changes of the coastal environment, (b) impacts of climate-related disasters, (c) disaster preparedness, and (d) the potential for ecosystem-based measures.

# Results

## *Distribution and Status of Coastal Dune and Mangrove Ecosystems*

### Coastal Dune Systems (CDS)

In Quang Nam province, CDS are found parallel to the coastline from the Vu Gia Tu Bon delta to the southern border of the province. The dune system is about 50 km long, reaches maximum heights of about 20 m, and stretches up to 4.8 km inland. The dunes are strongly influenced by human activities with a high degree of fragmentation due to a dense network of roads and paths that connect the many small settlements that are dotted across the dunes. Wide parts of the CDS are covered by shrimp ponds (Fig. 3a) and the natural vegetation has been either completely cleared or replaced by monocultures, in particular *Casuarina* plantations, and smaller patches with *Acacia* species that have been introduced since the 1990s. However, there is also a mosaic of remnants of stands of the indigenous tree species *Melaleuca cajeputi* that covers about 10 ha within Duy Xuyen district. Moreover, native tree and shrub species such as *Carallia brachiata, Lithocarpus concentricus, Cinnamomum burmannii, Lindera myrrha, and Eurycoma longifolia* are scattered through the coastal dunes.

In Xuan My village in Nui Thanh district, an active dune system of about 25 ha has been studied more in detail. It shows a zonation with embryo dunes that are widely covered by *Ipomoea pes-caprae*, fore dunes with *Casuarina equisetifolia* and different marram grass species, as well as yellow dunes dominated by *C. equisetifolia* and *Acacia auriculiformis* (Fig. 4). *C. equisetifolia* and other *Casuarina* species are growing extensively in the fore dunes and yellow dunes. They were planted in the second half of 1990s, mostly in 1997 for the purpose of

**Fig. 3** **a** Construction of shrimp ponds in a dune area. **b** Community-based mangrove restoration in Tam Hai commune

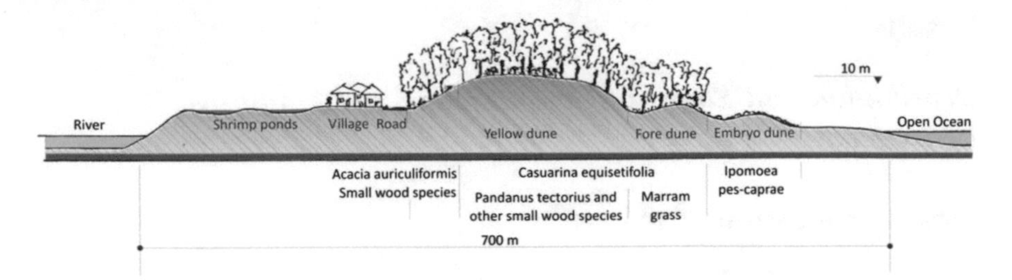


**Fig. 4** Schematic profile of a coastal dune system in Xuan My Village

protecting shrimp farms from sand transported by ocean winds. According to Mr. Nguyen Tan Hung, officer from the DARD in Tam Hai, another reason for the establishment of *Casuarina* plantations was the degradation of the native forest due to overexploitation for charcoal production and extraction of timber as construction material. Toward the inland there is a vegetation belt dominated by *Acacia auriculiformis*, a species that has been introduced by the local communities for the purpose of timber production, and smaller patches of native small wood species. Among them are *Pandanus tectorius*, a species that has been widely used for fire wood and charcoal production, as well as *Severinia monophylla.* Grass species include *Fimbristylis sericea*, *Cynodon dactylon* var. *dactylon*, *Dactyloctenium aegyptiacum*, *Digitaria petelotii*, *Eremochloa ciliaris*, *Fimbristylis lasiophylla*, and *Paspalum vaginatum*. Moreover, some introduced cactus species such as *Cereus peruvianus* are found.

Despite the widespread use of gas cookers in recent years, trees are still being used as an extra source for fire woods. Moreover, both timber and sand are extracted from the CDS as construction materials. In the 1990s, sand was also exploited in industrial scale and exported to other villages. However, these activities were prohibited in 1995 by the local authorities. After the extensive plantation of *Casuarina* in 1997, the use of sand for construction had been halted. Up to now, only few small areas with planted *Acacia* trees are used for this purpose. Sand is also used for the construction of shrimp farms. Furthermore, the households cultivating shrimps install the seawater pumping pipe lines through the dunes. An emerging threat to the sand dunes of Xuan My village is the resettlement of households coming from other villages as well as the needs of housing for workers in the large-scale shrimp production farms. Some new houses were already built in the dune area, despite the risk of storms and sand drift.

## Mangrove Ecosystems

Under natural conditions mangrove forests in Quang Nam province are delimited to relatively small estuaries and lagoons protected against the storms and the rough sea. Extended mangrove ecosystems, such as in the large delta regions of the northern and southern parts of the country are not found in Central Vietnam, and

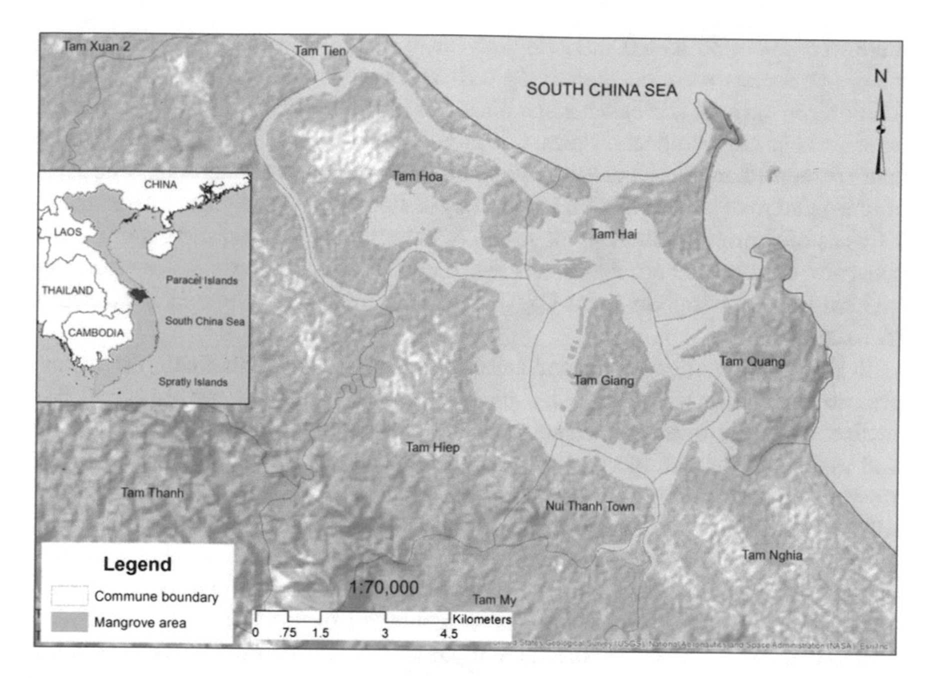


**Fig. 5** Location of mangrove remnants in An Hoa lagoon

due to historical overexploitation even these small mangrove areas have been widely lost or degraded. Only in some river estuaries and lagoons without the direct impact of tidal waves and low human interference, few mangrove patches are found. This is the case in the southern part of Quang Nam province in the district of Nui Thanh (Fig. 5). However, the few remnants found there are small, highly fragmented, degraded, and under high land use pressure. The main causes for mangrove degradation in this region are the extension of aquaculture systems, road construction, pollution, and the expansion of residential areas. Since 2015, some small-scale restoration measures are conducted by local communities (Fig. 3b).

Tam Hai in the district of Nui Thanh is an estuary commune delimited by the ocean in the North and the East and Truong Giang River in the South, where small areas of mangroves forest are located near to the river mouth and around some small islands. The People's Committee of Tam Hai commune stated that the commune had about 22.7 ha of wetland area in 2008, of which 17.8 ha were mangroves forest (Pham 2009). In the village of Binh Trung as part of Tam Hai commune, the authors tried to reconstruct the historical development of the mangrove area based on interviews with community members and participatory mapping techniques. The community members stated that there were about 40 ha of mangrove forest in this village before 1975. However, during the Vietnam-US war, about a quarter of the mangrove area was destroyed by bombs and rockets, so that there were about 30 ha left in 1975, the year of the unification of the country. The mangrove forests were comparatively well managed by the government until the Asian tiger shrimp (*Penaeus monodon*) was introduced in late 1980s and the large-scale shrimp production plan was approved in 1990 (Pham 2009). The mass conversion of mangrove forests to tiger shrimp farms led to an alarmingly decrease of the mangrove area from 30 ha to only 8 ha within two years. After nearly 20 years of shrimp production, the local community had realized the importance of mangrove forests to their livelihoods and they started restoring smaller areas with external financial support from USAID, so that the mangrove area increased to 16 ha in 2010.

In Tam Hai commune in general and Binh Trung village in particular, mangrove trees were traditionally used as the main source for fire wood and charcoal production. With the mass conversion of mangrove area to shrimp farms in 1990 as well as the introduction of other cooking material such as muddy coal and gas, this use became less important. According to the household survey in Binh Trung village, the second important use of mangroves is for fishery. Local people, mainly women, usually go to the mangroves to pick up shellfishes such as clams, oysters, and other kinds of fish and crab. This is the only use of mangrove forest which is still being conducted by the local people in present times. Other traditional uses, such as extracting timber for construction purposes, collecting medicinal plants, and preparing organic fertilizers are no longer practiced.

Figure 6 shows the mangrove area of Binh Trung village in 2014. At that time only few small patches with three mangrove species were present: *Rhizophora apiculata*, *Avicennia germinans*, and *Sonneratia caseolaris*. The stands are dissected by a large shrimp farm and settlement area. Toward the north there is a larger patch of *Casuarina* forest. The small patches of *S. caseolaris* are highly degraded; tree heights reach between 7 and 10 m compared to average heights of 15–20 m under nature-near conditions. *S. caseolaris* is a native mangrove species which originally grow in this area. However, the largest part of its distribution area has been destroyed for the construction of the shrimp farm. *A. germinans* dominates the sedimentation area in the estuary, also known as the indicator species for this habitat. *A. germinans* trees in Binh Trung reach about 5 m in height, compared to a maximum height of 10 m. This species is important due to its function of erosion control, tidal wave impact reduction, water filtration, salinity reduction, and barrier for other mangrove species to grow and develop. An area of about eight hectares of the original mangrove zone was still in good shape, located in the southwestern part of Binh Trung village. *R. apiculata* is found in semi-flooded areas and along the river in the estuary. *R. apiculata* is the species which had been selected for the first mangrove restoration activities in the village. The first restoration project was initiated in 2005. After severe damages caused by the storm Ketsana more than 50,000 trees of *R. appiculata* were planted in 2013 and 2014, which is equivalent to about 8 ha (see Fig. 6).

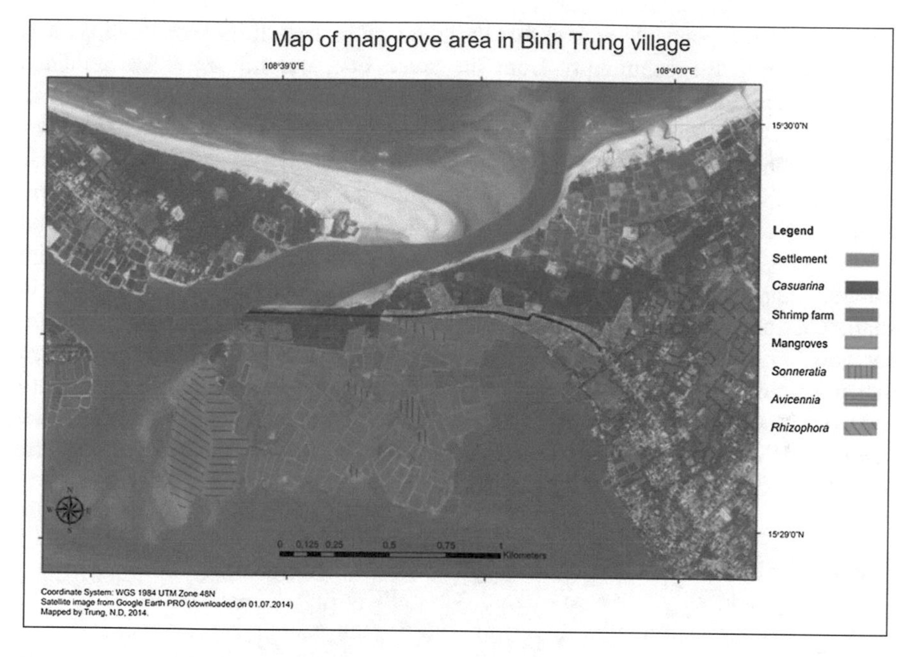


**Fig. 6** Distribution of mangroves in Binh Trung village in 2014

## *Perception and Awareness of the Hazard Protection Function of Dune and Mangrove Ecosystems*

### Coastal Dune Systems (CDS)

The plantation of *Casuarina* trees in late 1990s protected the community from sand drift (called "flying sand" and "jumping sand" in Vietnamese). However, it also led to the stereotype of local people to consider only the plantation cover of the sand dune area as the protection material. Sand dunes without its protecting vegetation cover have been—and are still considered as the reason of sand drift which created many negative impacts. These negative impacts have often been weighted as more severe than the positive effects including the buffering function against storms and waves.

This is also reflected by a survey of 20 households in Xuan My village that was conducted in 2014. It showed that storms and sand drift are considered the two most important natural hazards. The households were also asked in how far dykes, early warning systems, and Eco-DRR measures in form of tree buffers offer protection. With regard to storms, all respondents considered a combination of the three measures as the most effective and the majority stated that both the engineered structure and the ecosystem-based solution can offer high to very high protection against storms that can be even improved with an early warning system. However, five of the 20 respondents argued that the dyke offers only little protection, as it is only protecting the community from the river side, but no sea dyke has been constructed in this village to protect the community from the ocean side. In contrast, only two respondents stated that ecosystem-based measures offer only little protection against storms.

With respect to sand drift, all respondents argued that an early warning system does not protect the communities at all and the majority of 14 respondents stated that a dyke also offers no effective protection. Another three interviewees each answered that a dyke may offer little or medium protection, respectively. This is in sharp contrast to the protective functions of ecosystem-based measures that are considered effective against sand drift. The majority of 14 interviewees answered that protection forest on sand dunes has a high potential to prevent sand drift, and another six saw the potential even as very high, while only three see a medium potential. With respect to coastal hazards and their impacts on local livelihoods, an interesting statement is given by Mr. Le Van Nam (Box 1).

**Box 1: Statement of a shrimp farmer in Xuan My village**
I am Le Van Nam, 54 years old. My family is living on a 0.5 ha shrimp production farm since 1992. In our village, the people depend mostly on offshore fishery and shrimp production. The natural disasters have impacted our livelihoods very seriously in both human and financial aspects. The most dangerous disaster is storm, which happened usually from October to December. Sand drift (flying sand) is also a serious problem occurring all throughout the whole year. It is even more serious before the occurrence of a storm. We are living between the river dyke system in the south along the Truong Giang River and the *Casuarina* forest in the north along the coastline. The dyke and forest help us quite well to reduce the impacts of disasters in this area. The *Casuarina* trees are very effective to block the sand transported by wind from the ocean and also effective in reducing the magnitude of storm surges. This forest area should be well managed and protected in the future to continue its important role to protect the village behind.

Based on the experience of the community survey in Xuan My village, a larger survey ($n$ = 120) with an extended catalogue of questions was carried out in 2015 in the commune of Duy Hai in Duy Xuyen district (see Chapter “Vu Gia Thu Bon River Basin Information System (VGTB RBIS)—Managing Data for Assessing Land Use and Climate Change Interactions in Central Vietnam”). The majority of the interviewed household heads stated that they have observed more extreme events (67 %) and a shorter rainy season (63 %) in the last years, and 74 % of the interviewees think that climate-related disasters have occurred more frequently. When asked about risk mitigation and preparedness measures, 25 % of the respondents answered that they plant *Casuarina* trees to maintain water resources, but at the same time they extract timber for firewood. However, 42 % said that they do nothing, because they feel powerless against the forces of nature (33 % did not answer the question). People were also asked about ecosystem-based measures. The majority was aware that forests reduce the risk of landslides and sand drifting. However, only 30 % were aware that dunes and dune forests also reduce the risk of storms, coastal flooding, and droughts.

In conclusion, the two surveys show that coastal communities are aware of some beneficial protective ecosystem services of dune forests, in particular those related to sand drift, but that that many others are not fully understood or at least not considered. Moreover, *Casuarina* and *Acacia* plantations are regarded as beneficial, because they offer a direct value in form of timber use apart from the regulating services.

## Mangrove Ecosystems

Tropical storms and coastal flooding are considered the most dangerous natural hazards in Tam Hai commune. According to Nui Thanh Natural Resources Department (2013), in recent years tropical storms and depressions have already occurred in May or June. Those events are even more dangerous to local people, because they are less prepared in the peak time of their different production activities. The local community is well aware of the role that mangrove forests play to mitigate the impacts of tropical storms and storm surges. In Binh Trung village, all interviewees ($n = 20$) stated that they know about the coastal protection services of mangroves. However, they also said that due to the loss and degradation of the local mangrove forest, the protection function is considerably reduced. For this reason, there was no or only very little protection from the severe storms Eve (1999), Xangsane (2006), Ketsana (2009), and Nari (2013). The impacts were noticeably high in the storms Eve, where parts of Binh Trung village eroded into the sea, and Xangsane, where 10 people lost their lives and many more were reported missing. In 2009, the storm Ketsana swept away nearly 50 ha of shrimp farms and destroyed two offshore fishing ships.

In Binh Trung village, community members were asked about the protection potential of mangrove forests as an ecosystem-based measure in comparison to dykes and early warning systems. With respect to both storm and flood hazards, dykes were considered by far more effective than mangroves. The vast majority of the respondents believe that dykes offer a high to very high protection against both storms (85 %) and floods (80 %), while the protection potential of mangroves is seen as low or very low (for storms = 90 %; for floods = 100 %). Regarding early warning systems, the interviewees stated that they have a high or very high potential against storm impacts (95 %) and floods (100 %). For instance, in the two most recent storms in 2013, the early warning system operated quite effective, which enabled the local community to get to the shelter houses and protect their properties in due time, help reducing the losses and damages in both human and economic aspects. An opinion from an interviewee is summarized in Box 2.

**Box 2: Statement of a community member with respect to mangroves and coastal protection**
I am Le Thi Dinh and I am a participant in the Green Shield project. I know mangrove forests are very important to help us, the people who living in the coastal area, to mitigate the negative impacts of storms and floods. Many forests in Do Son (Hai Phong province) and Xuan Thuy (Nam Dinh province) are helping the people there very well during such strong tropical storms which occurred few years ago. I always hope that the mangroves in Binh Trung village could work like that. Unfortunately, a large area of mangroves was destroyed for shrimp production, and now we have to replant the mangrove trees. The trees are still young, about 3–5 years, and cannot be able to reduce the impacts of storms. The width of mangrove band is also not enough to protect the settlement area behind. We are now putting many efforts to replant and protect the mangrove forest in our village, with the hope that it could somehow work as the "green shield" for our community in the near future.

In conclusion, people in Binh Trung village are aware of the protection function of mangroves, but due the high degree of degradation they feel that engineered measures such as dykes and nonstructural measures such as an early warning system can protect them more effectively.

## Discussion on the Potential for Conservation, Restoration, and Sustainable Use of Mangrove and Coastal Dune Ecosystems

### *Coastal Dune Systems (CDS)*

As shown in Section "Distribution and Status of Coastal Dune and Mangrove Ecosystems", CDS in Quang Nam province are highly degraded due to the replacement of the native vegetation cover, fragmentation by roads and paths, soil sealing, local sand extraction, and others. These degradation patterns can be widely observed along the coast of Central Vietnam, for instance in the province of Thua Thien-Hue, as shown by Nehren et al. (2016). The government has reacted to this negative development by defining a category of protection forest on sandy soils to prevent from sand encroachment, movement, and sliding (Fig. 7). However, this protection status does not consider the geoecological complexity of dune systems and the various ecosystem services they provide. In fact it manifests the distinction between substrate (sand) and vegetation cover, where the sand is seen as a threat and the vegetation cover as a protective structure. Interestingly, this observation has been

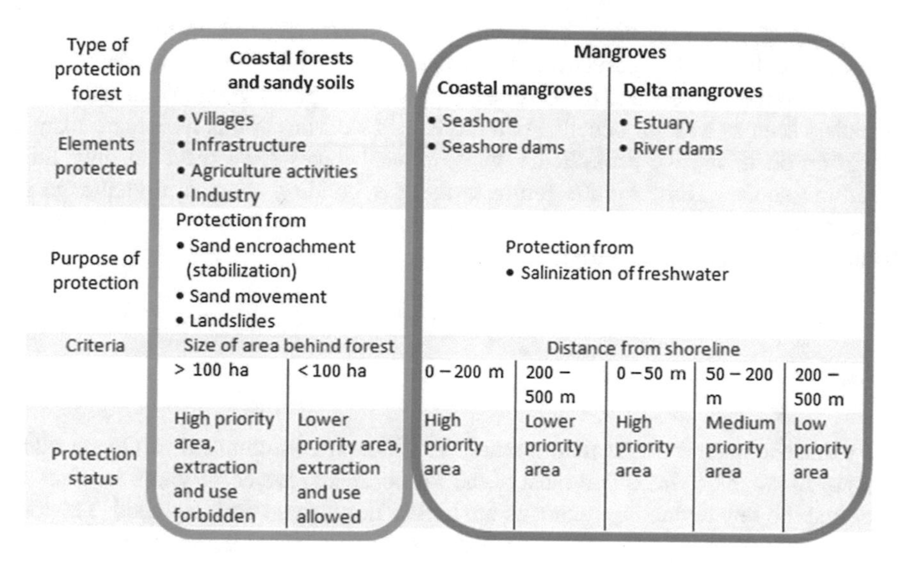


**Fig. 7** Definition of coastal protection forest in Vietnam based on Decision 61/2005/QDBNN (2005)

also made in other countries around the globe. In Chile, for instance, dune sand was seen as a major threat to the development of the coastal zone of the country. Albert (1900, p. 6ff) dedicated a whole book to this topic and talked about "dead sands" and "the invasion of the sands", and Gundian (1947, p. 11) even used the term "cancer of the soil" related to erosion problems, as dune sand was covering fertile soils in several Chilean provinces and also caused sedimentation problems in the harbors. However, for the case of Thua Thien-Hue and two dune systems in Chile and Indonesia, Nehren et al. (2016) have shown that CDS provide numerous important ecosystem services that result from the dune as a geoecosystem including substrate and vegetation cover, such as water storage and purification, protection from coastal erosion, or touristic and recreational services. Moreover, also the biodiversity of coastal dunes depends on the whole system and not only on the vegetation cover. This should be considered in legal frameworks and management plans.

The field research in Xuan My village underlines that the concept of sand dunes as a geoecosystem, which provides various ecosystem services, is new to the local people. Therefore awareness-raising on this concept is very important and highly suggested by the respondents. In the household survey in Xuan My village ($n = 20$), 90 % of the people stated that there is an urgent need on this issue. Moreover, 60 % said that dune forests should be managed by the community, while only 20 % saw a need for a protected area and only 15 % for a restoration project. This shows that local communities are much more aware of the various functions and services mangroves offer than those of coastal dune forests. However, this is not surprising, because many coastal dwellers directly depend on mangroves, for example as breeding grounds for fish and shrimps, while the direct use of dune forests is rather limited and the *Casuarina* plantations protect them from sand drift, which is seen as a major benefit. Also the NGOs working in this region are mainly focusing on mangrove ecosystems, while coastal dunes have received only little attention so far. Thus, for the future awareness building remains a challenge to enable the local communities to take into account the various direct and indirect values and services CDS provide.

## Mangrove Ecosystems

As a main outcome of the field research in different communities of Quang Nam province, it can be stated that most of the former mangrove ecosystems are already lost and the few remaining remnants are highly fragmented and degraded. The loss of wide parts of mangrove ecosystem dates back to the Indochina wars. Phan (1991) stated that almost 40 % of the total mangrove forest area in Vietnam was destroyed by the napalm and herbicide bombardment (Agent Orange) in the Vietnam-US war (1963–1975), so that the total area decreased from about 400,000 to about 250,000 ha. However, the large extended mangrove areas are located in the North (with the Red river delta) and the south (with the Mekong delta) of the country. In the coastal area of Central Vietnam, mangroves are limited to relatively small deltas and lagoons (Phan and Hoàng 1993a, b), which is also the case for Quang Nam province.

Phan (1991) divided the mangroves of Vietnam in four main zones: (1) the Northeast coastal area from Ngoc cape to Do Son beach, (2) the northern delta region from Do Son to Lach Truong cape, (3) the central coastal area from Lach Truong to Vung Tau cape, and (4) the southern coastal area from Vung Tau cape to Nai cape. Region 3 with Quang Nam province has by far the longest coastline, but only a very small share of mangrove forests. According to Phan and Hoang (1993a, b) in 1983 the whole region comprised 14,300 ha of mangrove forests, which were 5.7 % of the total mangrove area of Vietnam at that time. From 1983 to 2001, this area was further reduced to only 2000 ha, or 1.28 % of the total mangrove areas of the country (155.3 ha in 2001). The main driver for this loss was the introduction of the Asian tiger shrimp and the mass conversion of different land uses to shrimp production farms in the early 1990s, which affected the mangroves of the whole country.

Considering the various ecosystem services that mangroves offer, including the provision of wood and nonwood forest products, coastal protection, carbon storage and sequestration, conservation of biological diversity, and the provision of habitat, spawning grounds and nutrients for a variety of fish and shellfish (FAO 2005), the large-scale destruction and degradation can be considered an ecological disaster, but more than this, also a tragedy for the coastal communities that depend on the various ecosystem services mangroves provide. To counteract the negative impacts, in recent times, the Vietnamese government has strengthened its efforts to conserve, restore, and sustainably manage mangrove forests. However, in Quang Nam province there are so far only small areas that have been restored mainly based on communal initiatives and there is a lack of an integrated strategy for the sustainable use and conservation of mangrove ecosystems. Moreover, legal frameworks have included delta mangroves for coastal protection, in particular against salinization of freshwater (Fig. 7). It is highly recommended also for legal frameworks to extend the protection functions of mangroves to other hazards, in particular as bioshields against storms and waves, and geoprocesses, in particular coastal erosion.

In the long history of Binh Trung village, the mangrove forests were managed by the local authorities. During the Vietnam-US war, mangroves and coastal areas were used for military defense purposes. After the reunification of the country, due to the mismanagement of local government, mangrove forests were overexploited for fire wood and charcoal production and other uses. Low awareness and capacity of the entire authority system led to the approval of the conversion of mangroves to shrimp production farms in late 1980s and early 1990s. At the time, this study was conducted in Binh Trung village, the mangrove forest was still managed by the Agriculture and Rural Development Division under the supervision of People's Committee. However, the local people are now proposing another management scheme which is based on the community. A "Forest Guard Squad" was founded with six members from Binh Trung village, including a representative from Commune People's Committee, the head of the village, a local public security officer and the participants of the mangroves restoration project. There were two female members in the "Forest Guard Squad" to ensure the gender equality in forest management.

In the household survey in Binh Trung village ($n$ = 20), people expressed their expectations and made recommendations on how the mangrove forest should be managed in the village. The majority of 90 % stated that the mangroves should be managed by the community and 65 % suggested the creation of a protected area. Moreover, 80 % called for more investments for restoration projects and 75 % think that there should be more activities on awareness rising and capacity building. However, the head of the village expected NGOs or donors of development projects to support restoration projects in both financial and technical aspects.

## Conclusions and Recommendations

Along the coastline of Quang Nam province, as in wide parts of Central Vietnam, mangrove and coastal dune ecosystems are highly degraded or already entirely lost due to historical land use and resource exploitation processes. Today, there are various initiatives by the national and provincial governments as well as local authorities, communities, and NGOs to conserve, sustainably manage, and restore coastal ecosystems and thereby secure the livelihood of coastal dwellers and counteract the impacts of coastal hazards and climate change. At the same time, there is high land use pressure in the densely populated and economically important coastal zone.

With the introduction of ecosystem-based approaches to climate change adaptation and disaster risk reduction (EbA, Eco-DRR) that have been recognized in international conferences and agreements such as the UN climate summit in Paris (COP 21) and the Sendai Framework for Disaster Risk Reduction 2015–2030, there is a rising awareness among policymakers on the importance of coastal ecosystems in the context of sustainable development. This is also the case in Vietnam, where various ecosystem-based strategies and measures have been developed.

In our case study in Quang Nam province, it became evident that a rethinking process toward "living with nature" instead of "fighting the forces of nature" has been started at the policy level but also in coastal communities, but that there is still a long way to go to achieve the goal of coastal resilience. We identified several fields of action, which are summarized below:

**Research**: So far mangroves ecosystem were pretty much in the focus of the scientific community, while other coastal ecosystems such as coastal dunes, but also seagrasses and coral reefs are underresearched. With respect to all coastal ecosystems, there is a need to assess their services and develop models for multi-hazard and multi-risk assessment. Moreover, the monetary values of these ecosystems need to be evaluated as a basis for a possible implementation of Payment for Ecosystem Services (PES) schemes. Another important field of research is the reconstruction of the natural vegetation to conserve, manage, and restore coastal ecosystems in a near-natural manner instead of introducing non-endemic species such as *Casuriana* trees or vetiver grass (*Chrysopogon zizanioides*). In this way, coastal protection and biodiversity conservation can go hand in hand (Fig. 8a).

**Fig. 8** **a** Endemic tree and shrub species such as *Wikstroemia indica* should therefore be considered for conservation and restoration activities; **b** Students of Hue University learn about suitable tree species and soil substrates for restoration and reforestation measures

**Environmental education**: Environmental education should be fostered in all levels from school kids to university graduates and also in the communities in form of adult education (Fig. 8b). A main task should be the improvement of the system understanding of CDS and mangrove ecosystems, and the awareness of the benefits of healthy ecosystems particularly in terms of regulating ecosystem services.

**Policy and governance levels**: The long-term costs and benefits of different use options of coastal resources and ecosystem should be given stronger consideration in political decisions. It should also be thought about the possibility of implementing incentive systems for smallholders to sustainably manage their lands, for instance in form of PES schemes. Moreover, legal frameworks should be further developed toward integrated approaches rather than the protection of defined types of vegetation cover.

**Management approaches**: The concept of Integrated Coastal Zone Management (ICZM) should be stronger considered in the planning and implementation of policies. Furthermore, it should be thought about geoecological protection zones with monitoring and control mechanisms to protect CDS and mangroves and restore their natural vegetation. The expanding shrimp production sector which already has led to the loss of most mangrove areas is now creating new threats to dune systems. The mass installation of seawater pumping pipes and construction of houses for workers in the studied dune area negatively affects the structure and dynamics of the CDS. It is also reported that there have been shrimp farms constructed in protected dune areas. To prevent from these negative impacts, a stricter set of regulations should be implemented. A community-based forest management plan is also needed, where the local community could supervise the appropriate uses of the CDS and report violations to the local authorities. With respect to mangroves, their vital role for coastal protection seems to be mostly understood by the local communities. Initiatives and projects on mangrove restoration have received high interest in the case of the studied community. However, those projects have been implemented in a small village with the participation of about 200 participants, and are still at the pilot level. There is a need to disseminate the lessons learned to other villages.

## References


Albert F (1900) Las dunas—o sean las arenas volantes, voladeros, arenas muertos, invasion de las arenas, playas i médanos del Centro de Chile. Santiago de Chile, Imprenta Cervantes, Bandera 46

Alliance Development Works (2012) World Risk Report 2012

Anthony EJ, Brunier G, Besset M et al (2015) Linking rapid erosion of the Mekong River delta to human activities. Sci Rep 5:14745

Badola R, Hussain SA (2005) Valuing ecosystem functions: an empirical study on the storm protection function of Bhitarkanika mangrove ecosystem, India. Environ Conserv 32:85–92

Barbier EB, Hacker SD, Kennedy C et al (2011) The value of estuarine and coastal ecosystem services. Ecol Monogr 81(2):169–193

Batker DP, de la Torre I, Costanza R et al (2010) Gaining ground-wetlands, hurricanes and the economy: the value of restoring the Mississippi River Delta. Earth Economics, Tacoma

Besset M, Brunier G, Anthony EJ (2015) Recent morphodynamic evolution of the coastline of Mekong river Delta: towards an increased vulnerability. Geophysical Research Abstracts vol 17, EGU2015-5427-1, EGU General Assembly 2015

Boateng I (2012) GIS assessment of coastal vulnerability to climate change and coastal adaption planning in Vietnam. J Coast Conserv. 16:25–36

Carter RWG (1991) Near-future sea level impacts on coastal dune landscapes. Landscape Ecol 6:29–39

Christensen SM, Tarp P, Hjortso CN (2014) Mangrove forest management planning in coastal buffer and conservation zones, Vietnam: a multimethodological approach incorporating multiple stakeholders. Ocean & Coast Manag 51:712–726

Dahm J, Jenks G, Bergin D (2005) Community-based dune management for the mitigation of coastal hazards and climate change effects: a guide for local authorities. https://www.boprc.govt.nz/media/32260/ClimateChange-0505-CoastalhazardsandclimateReport.pdf

Das S, Vincent JR (2009) Mangroves protected villages and reduced death toll during Indian super cyclone. Nat Acad Sci USA 106(18):7357–7360

Dasgupta S, Laplante B, Meisner C et al (2009) The impact of sea level rise on developing countries: a comparative analysis. Clim Change 93:379–388

FAO (2005) Global forest resources assessment 2005: thematic study on mangroves, Vietnam country profile. Forestry Department, Food and Agriculture Organization of the UN

FAO (2006) Participatory Rural Appraisal (PRA) Manual. Food and Agriculture Organization of the UN

French PW (2001) Coastal defences: processes, problems and solutions. Routledge, London

Fritz HM, Blount C (2007) Thematic paper: Role of forests and trees in protecting coastal areas against cyclones; chapter 2: protection from cyclones. In: Braatz S, Fortuna S, Broadhead J, Leslie R (eds) Coastal protection in the aftermath of the Indian Ocean tsunami: what role for forests and trees? Proceedings of the regional technical workshop, Khao Lak, 28-31 August 2006, RAP Publication (FAO), No 207/07

GFDRR, The World Bank (2010) Weathering the storm: Options for disaster risk financing in Vietnam

Gómez-Pina G (2002) Sand dune management problems and techniques, Spain. J Coastal Res 36:325–332

Granek EF, Ruttenberg BI (2007) Protective capacity of mangroves during tropical storms: a case study from 'Wilma' and 'Gamma' in Belize. Mar Ecol Prog Ser 343:101–105

Gundian VB (1947) Erosion—Cancer del Suelo. Imprenta Universitaria, Santiago de Chile

Heslenfeld P, Jungerius PD, Klijn JA (2004) European coastal dunes: ecological values, threats, opportunities and policy development. In: Martínez ML, Psuty NP (eds) Coastal dunes: ecology and conservation. Springer, Berlin, Heidelberg

Laso Bayas JC, Marohn C, Dercon G et al (2011) Influence of coastal vegetation on the 2004 tsunami wave impact in west Aceh. Proc Natl Acad Sci USA 108:18612–18617

Martínez ML, Maun MA, Psuty NP (2004) The fragility and conservation of the World's coastal dunes: geomorphological, ecological and socioeconomic perspectives. In: Martínez ML, Psuty NP (eds) Coastal dunes. Springer, Berlin, Heidelberg

Mazda Y, Magi M, Kogo M, Hong PN (1997) Mangrove on coastal protection from waves in the Tong King Delta, Vietnam. Mangroves Salt Marshes 1:127–135

McIvor A, Spencer T, Möller I, Spalding M (2012) Storm surge reduction by mangroves. Natural Coastal Protection Series

Miththapala S (2008) Seagrasses and sand dunes. Coastal ecosystems series, vol 3. Ecosystems and Livelihoods Group Asia, IUCN, Colombo, Sri Lanka

Nehren U, Hoang HDT, Marfai A et al (2016) Assessing ecosystem services and degradation status of coastal dune systems for Eco-DRR and EbA: Case studies from Vietnam, Indonesia, and Chile. In: Renaud F, Sudmeier-Rieux K, Estrella M, Nehren U (eds) Ecosystem-based disaster risk reduction and adaptation in practice. Series: Advances in natural and technological hazards research, Springer, pp 401–434

Nghia NH (2004) Mangrove forest conservation and development planning in Nghe An—Vietnam. In: International symposium on geoinformatics for spatial infrastructure development in earth and allied sciences

Nguyen H, Shaw R (2010) Climate change adaptation and disaster risk reduction in Vietnam. In: Shaw R et al (eds) Climate change adaptation and disaster risk reduction: an Asian perspective (community, environment and disaster risk management, vol 5) Emerald Group Publishing Limited, chapter 18, pp 373–391

Nui Thanh Natural Resources Department (2013) Nui Thanh Statistical Yearbook. Nui Thanh District People Council

Pham DL (2009) The assessment of mangrove forest exploitation for shrimp farming to develop recommendations for appropriate uses in Tam Hai Commune, Nui Thanh District, Quang Nam Province. Duc Tri College, Da Nang

Phan NH (1991) The ecology of Mangroves in Vietnam. Doctoral dissertation

Phan NH, Hoang TS (1993a) Mangroves of Vietnam. IUCN, Bangkok 173 pp

Phan NH, Hoang TS (1993b) Mangroves of Vietnam. IUCN, Bangkok

Phan TGT, Nguyen VH (2006) Cost-benefit analysis for coastal sand shrimp farming in Vietnam. Nong-Lam University, Faculty of Economics, Vietnam

Phan LK, van Thiel de Vries JSM, Stive MJF (2015) Coastal Mangrove Squeeze in the Mekong Delta. J Coast Res 31(2):233–243

Prasetya GS (2007) The role of coastal forest and trees in combating coastal erosion. In: Braatz S, Fortuna S, Broadhead J, Leslie R (eds) Coastal protection in the aftermath of the Indian Ocean Tsunami: what role for forests and trees?. FAO, Bangkok

Quartel S, Kroon A, Augustinus PGEF et al (2007) Wave attenuation in coastal mangroves in the Red River Delta, Vietnam. J Asian Earth Sci 29(4):576–584

Ross P (1974) The effects of herbicides on the mangrove of South Vietnam. National Research Council, Washington DC

Saye SE, Pye K (2007) Implications of sea level rise for coastal dune habitat conservation in Wales, UK. J Coast Conserv 11:31–52

Schmitt K, Albers T (2014) Area coastal protection and the use of bamboo breakwaters in the Mekong Delta. In: Thao ND et al (eds) Coastal disasters and climate change in Vietnam: engineering and planning perspectives, pp 107–132

Sridhar KR, Bhagya B (2007) Coastal sand dune vegetation: a potential source of food, fodder and pharmaceuticals. Livest Res Rural Dev 19(6). http://www.lrrd.org/lrrd19/6/srid19084.htm. Accessed 17 Apr 2016

Sterling JE, Hurley MM, Minh DL (2006) Vietnam: a natural history. Yale University Press, pp 1–21

Takagi H, Esteban M, Ngyuen DT (2014) Introduction: coastal disasters and climate change in Vietnam. In: Thao ND et al (eds) Coastal disasters and climate change in Vietnam—engineering and planning perspectives. Elsevier, London

Takle ES, Chen T-C, Wu X (2007) Protection from wind and salt. In: Braatz S, Fortuna S, Broadhead J, Leslie R (eds) Coastal protection in the aftermath of the Indian Ocean Tsunami: What role for forests and trees?. FAO, Bangkok

Tanaka N, Sasaki Y, Mowjood MIM et al (2007) Coastal vegetation structures and their functions in tsunami protection: experience of the recent Indian Ocean tsunami. Landscape Ecol Eng 3:33–45

Thanh ND (2014) Climate change in the coastal regions of Vietnam. In: Thao ND et al (eds) Coastal disasters and climate change in Vietnam: engineering and planning perspectives, pp 175–198

Thao ND, Takagi H, Esteban M (eds) (2014) Coastal disasters and climate change in Vietnam. Elsevier, London, Waltham

THC (2013) Tam Hai Commune People's Committee socio-economic development report 2013

Tri NH, Adger WN, Kelly PM (1998) Natural resource management in mitigating climate impacts: the example of mangrove restoration in Vietnam. Global Environ Change Hum Policy Dimensions 8(1):49–61

Trung NHT, Tri VPD (2014) Possible impacts of seawater intrusion and strategies for water management in coastal areas in the Vietnamese Mekong delta in the context of climate change. In: Thao ND et al (eds) Coastal disasters and climate change in Vietnam: engineering and planning perspectives, pp 219–232

Van Santen P, Augustinus PGEF, Janssen-Stelder BM et al (2007) Sedimentation in an estuarine mangrove system. J Asian Earth Sci 29:566–575

Vietnam Oil and Gas Report Q1 (2013) Wall Street Journal, April 4, 2013

Vu TC (2014) Tsunami hazard in Vietnam. In: Thao ND et al (eds) Coastal disasters and climate change in Vietnam: engineering and planning perspectives, pp 277–302

Vu TC, Nguyen DX (2008) Tsunami risk along Vietnamese coast. J Water Resour Environ Eng 23:24–33

WWF (2013) Mainstreaming ecosystem-based adaptation in Vietnam: policy note. World Wild Fund for Nature

Yanagisawa H, Koshimura S, Miyagi T, Imamura F (2010) Tsunami damage reduction performance of a mangrove forest in Banda Aceh, Indonesia inferred from field data and a numerical model. J Geophys Res 115

# Hydrological Drought Risk Assessment in an Anthropogenically Impacted Tropical Catchment, Central Vietnam

**Alexandra Nauditt, A.B.M. Firoz, Viet Quoc Trinh, Manfred Fink, Harro Stolpe and Lars Ribbe**

**Abstract** Seasonal meteorological and hydrological droughts are a recurrent phenomenon in water-abundant tropical countries and are expected to become more frequent in the future. Unusual water shortage in recent years has severely affected societies living in South East Asia in general and Vietnam in particular. Preparedness, however, is absent and site-appropriate water management measures and strategies are not available. While drought-related research and water management in recent years have been addressed in water scarce sub-tropical regions, the US and Europe, more limited research has been focused on drought risk in tropical catchments. In this study, the drought characteristics of a large tropical catchment in Central Vietnam, the Vu Gia Thu Bon, were analysed in an integrated assessment framework. Daily precipitation and runoff time series for the VGTB catchment were analysed applying statistical methods to compare historical meteorological and hydrological drought, in addition to low flow frequency and seasonality. The role of tropical catchment characteristics, storage and climate variability in seasonal drought evolution were analysed by applying statistical analyses and the spatially distributed J2000 rainfall-runoff model. To assess

A. Nauditt (✉) · A.B.M. Firoz · V.Q. Trinh · L. Ribbe
Institute for Technology and Resources Management in the Tropics and Subtropics, Technical University of Applied Sciences, Cologne, Germany
e-mail: alexandra.nauditt@th-koeln.de

A.B.M. Firoz
e-mail: abm.firoz@th-koeln.de

V.Q. Trinh
e-mail: trinhquocviet1981@googlemail.com

L. Ribbe
e-mail: lars.ribbe@th-koeln.de

V.Q. Trinh · H. Stolpe
Environmental Engineering and Ecology, University of Bochum, Bochum, Germany

M. Fink
Department of Geography, Friedrich-Schiller University Jena, Jena, Germany
e-mail: manfred.fink@uni-jena.de



A. Nauditt and L. Ribbe (eds.), *Land Use and Climate Change Interactions in Central Vietnam*, Water Resources Development and Management,
DOI 10.1007/978-981-10-2624-9_14

anthropogenic impacts on hydrological drought, human interventions in the hydrological system due to hydropower development were quantified with the HEC-ResSim Reservoir operation model and the implications of low flows for salt water intrusion in the delta were simulated by the hydrodynamic Mike 11 model. It can be concluded that such an integrated model-data analysis which accounts for both landscape controls and anthropogenic impacts on the local hydrological system is a useful approach for drought management in tropical countries.

## Introduction

National economies worldwide are increasingly threatened by droughts affecting agriculture, hydropower, navigation, forestry, urban water supply, public health and the environment. Recent severe droughts in California, central Chile or southeast Australia demonstrate that even well-developed regions have limited preparedness for extreme drought conditions. It is equally apparent that serious droughts in water abundant tropical regions such as Southeast Brazil, Colombia, Central America and South East Asia are occurring with increasing frequency in recent years culminating markedly worldwide during the severe El Niño year of 2015–2016.

Drought monitoring worldwide is mainly based on large scale observations (global, national or regional) looking at parameters as precipitation and soil moisture typically based on historical and real time satellite observations (Van Loon et al. 2016). Hydrological drought and water scarcity, however, is usually experienced most acutely at the smaller catchment scale and in addition to lack of precipitation, or other hydroclimatic factors, anthropogenic factors such as over abstractions of water affect the hydrological system and exacerbate droughts (Van Loon and Van Lanen 2013; Van Loon et al. 2016). Consequently there is a strong need for research related to drought assessment and management which integrates anthropogenic impacts and seasonal water demand with wider assessment of hydroclimatic controls. Drought hazard monitoring and management has been most widely applied to arid and semiarid regions of the developed world (NDMC 2012; UCL 2012; UN-ISDR 2005; Wilhite et al. 2000), while droughts in tropical regions have been relatively neglected in science, water management and international disaster management strategies (UNDP 2011; UN-ISDR 2009). Here most attention has been paid to flood risk management although droughts are the most frequent and devastating disasters in terms of economic losses occurring in South East Asia (Adamson and Bird 2010; Geng et al. 2016; Terink et al. 2011; Navuth 2007).

As a result, historical drought frequency, risk and types are still poorly understood for South East Asian river basins. Such spatial and temporal drought characterization, however, is urgently needed as drought risk management strategies have to be based on a robust scientific understanding about the role of climate, hydrology and human water demand in drought evolution and propogation.

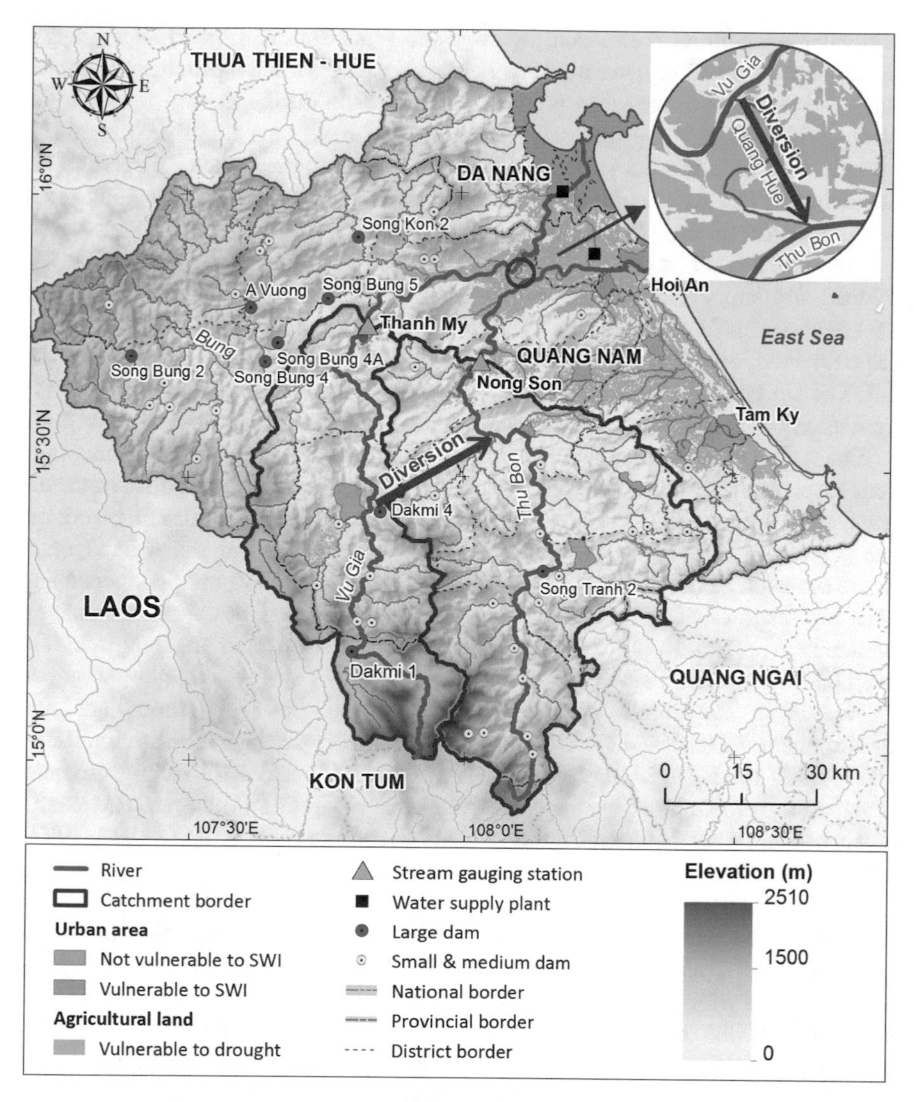


**Fig. 1** The VGTB River basin showing the agricultural and urban areas vulnerable to hydrological drought during the dry season. In the upstream, the current hydropower plants (2013) are shown as well as the diversion (highlighted with *red arrow*) from Vu Gia to Thu Bon at the Dak Mi 4 which started its operation in 2012. Downstream, the Quang Hue diversion is highlighted

## Drought-Related Challenges in the Region

Hydrological droughts and subsequent saltwater intrusion in coastal areas are also a prominent phenomenon in the Vu Gia Thu Bon (VGTB) river basin in Central Vietnam (Fig. 1). Droughts have a strong effect on the local socio-economy; in particular agricultural production, hydropower generation and urban water supply during the extended dry season from January to August. Rice production is most impacted as the two rice crops per year are cultivated during the dry season (Viet et al. 2016). Besides the increasing frequency of heat waves and prolonged dry periods with little rainfall (IMHEN 2012), human activities, such as hydropower development and over-exploitation of fresh water in the region, are putting additional pressure on available water resources.

The VGTB basin is composed of two main sub-catchments, the smaller Vu Gia (North) and larger Thu Bon (Southern part) which have their confluence around 36 km upstream from the East Sea, forming a major delta where 120,000 ha of land are cultivated, essentially with rice paddies and where major settlements are located. The Vu Gia river supplies water to the City of Da Nang and large parts of the agricultural area in the North and is gauged at Than My discharge station (Fig. 1).

The region experiences the longest dry season in Vietnam with 8 months of the year receiving only 20–35% of the annual rainfall, frequently resulting in severe water shortages and too little stream water to push back saltwater intruding from the sea (Viet 2014). Furthermore, relatively low annual average rainfall with 570 mm in the coastal area (at the Da Nang station), contrasts with 1250 mm more upstream (Tra My station at 150 m of elevation). This, along with high atmospheric temperatures averaging to 30 °C and high evapotranspiration rates frequently lead to dry conditions in the delta region.

According to reports of the Provincial Ministries for Agriculture and Environment, DONRE and DARD, droughts in the dry seasons of the years 1976–1977, 1982–1983, 1993–1994, 1997–1998, 2005, 2009, and 2012–2016 were most severe in the provinces of Quang Nam and Da Nang (IMHEN 2008; RCHM 2013). Some of these extreme droughts fell into El Niño years such as 1976–1977, 1997–1998, 2002–2003, and 2015–2016.

However, anthropogenic alterations in recent years have aggravated drought disasters in this rapidly changing region. Since 2007, the construction of hydropower reservoirs and plants has been a major human intervention in the hydrological system with major implications for upstream ecosystems (ICEM 2008). Ten large hydropower projects (>30 MW) with a total installed capacity of 1147 MW are planned for the river basin. Of these, seven plants were finalized within the period from 2008 to 2013, with a total installed capacity of 895 MW. The majority of the potential hydropower sites are located upstream the Vu Gia sub-basin (Fig. 1). Since 2012, when the Dak Mi 4 hydropower plant went into operation, a polemic controversy between the Provinces Da Nang and Quang Nam Province started regarding the impacts of hydropower development on downstream water uses. Although upstream areas form part of Quang Nam Province, the Vu Gia catchment delivers water to Da Nang city and surrounding agricultural areas. The city of Da Nang consequently blamed Quang Nam for developing large-scale hydropower which has negative effects on water availability downstream impacting on water supplies during the dry season. Figure 1 shows the that one of the hydropower plants in the Vu Gia Basin, Dak Mi 4, diverts water of Dak Mi River (a major source of Vu Gia River) to the wetter Thu Bon sub-catchment.

In addition, water abstractions for agriculture, urban water supply and industry have increased during the last decade. Paddy rice in the lowlands is cultivated on 62,000 ha and other annual crops on approximately 42,000 ha. Rice production consumes the largest amount of freshwater resources and the water works in the delta were mainly constructed for paddy irrigation. During the dry seasons of recent years, a minimum discharge of 30.2 $m^3/s$ was required for irrigation and another 2.5 $m^3/s$ for drinking water supply (Viet et al. 2016).

As Da Nang City and the North of the VGTB basin are particularly affected and socio-economically vulnerable to droughts (Fig. 1), this study has the overall objective to take an integrated, holistic perspective to analyse the spatial and temporal distribution of drought risks and their origin for the Northern Vu Gia part of the VGTB river basin. Specific objectives are (1) to characterize typical drought duration and frequency in the region and identify historical hydrological droughts, (2) to analyse sensitivity of low flows to land use change, (3) to assess to which extent low flows in the Vu Gia have been impacted by hydropower development and (4) to quantify the spatial and temporal extent of saltwater intrusion risk in coastal areas after upstream hydropower development.

## Drought Assessment Concept

In Europe, drought research has had an increasing prominence in hydrological research since the late 1990s (Tallaksen and Van Lanen 2004; Van Loon and Van Lanen 2012) and was a focus even earlier in the US especially for arid and semiarid regions (Wilhite and Glantz 1985). As a result, drought types were categorized into meteorological, soil moisture, hydrological, agricultural and socioeconomic droughts, a concept which is widely discussed in, e.g. reviews of Van Loon (2015) and Mishra and Singh (2010). An essential development for the purposes of the current study is the concept of 'drought propagation' (Tallaksen and van Lanen 2004) and the evaluation of a hydrological drought evidenced by negative anomalies in surface and subsurface water indices (Van Loon 2015). The assessment of the individual processes controlling stream flow generation and characterizing the evolution from a meteorological drought to a hydrological drought is subject of a large number of studies (Tallaksen and van Lanen 2004; Van Loon and van Lanen 2012; Van Loon and Lahaa 2015; Stoelzle et al. 2014).

While meteorological droughts are essentially assessed by analyzing precipitation and other climate parameters as well as soil moisture and vegetation anomalies, understanding the development of hydrological droughts requires information on discharge and groundwater levels as well as on catchment water storage. Subsurface storage reflects geology, soil characteristics and land use; these factors interact with hydroclimatic drivers to control the seasonal water balance and the evolution of long term hydrological drought.

Drought assessment methodologies using hydro-meteorological time series can be classified in standardized and threshold approaches (Van Lanen et al. 2013; Van

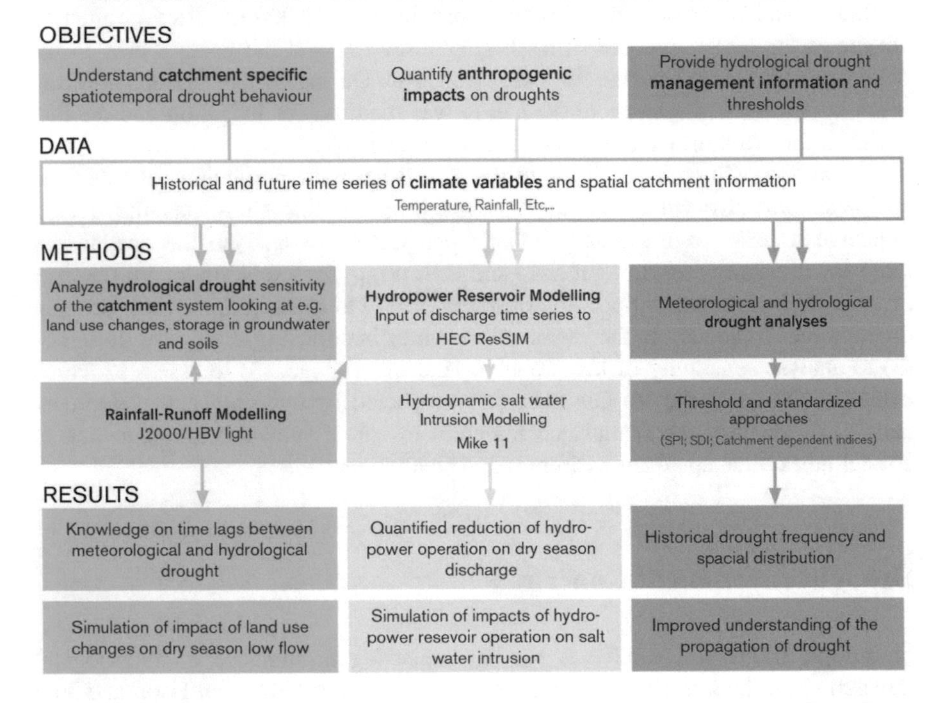


**Fig. 2** Catchment specific drought assessment methodology for the Vu Gia Thu Bon River Basin, Central Vietnam

Loon 2013). The effects of the above-described climate and catchment related controls on the generation of low flows and the identification of drought thresholds in relation to drought propagation is a major research challenge and one where also hydrological modelling can be valuable (Tallaksen and van Lanen 2004; Van Loon and Van Lanen 2012).

Also, in anthropogenically modified catchment systems, a drought additionally needs to be distinguished from human-induced hydrological drought and water scarcity (Van Loon and Van Lanen et al. 2013). The Vu Gia Thu Bon river basin is an increasingly modified hydrological system in its downstream reaches due to hydropower operations on the one hand and intensive irrigated agriculture in the estuary on the other. Therefore a complex assessment and modelling framework is needed to evaluate and quantify the different impacts leading to hydrological drought and salt water intrusion. Figure 2 gives an overview on the objectives, methods and outcomes of this drought analysis.

In summary, the following analyses were combined to obtain an improved understanding of both, catchment and human impacted droughts.

Precipitation and runoff time series from the 16 stations from 1982 to 2013 were corrected and analyzed to assess drought severity and typology. A K means Cluster analysis was applied to the daily discharge time series from 1976 to 2013 to identify dry, medium and wet years for the two dry seasons from January to April and from May to August (MacQueen 1967; Hartigan and Wong 1979). The Standardized Precipitation Index (SPI) and its runoff homologue (SRI) were calculated and compared. SPI and SRI—with the time steps of 1, 3, 6 and 12 months values—were determined using the available rainfall and record from the precipitation gauging stations in the Vu Gia River and the Thanh My discharge station, following the approach of McKee et al. (1993). The rainfall–runoff relationship as well as the sensitivity of stream flow against land use and climate changes was evaluated applying the HRU distributed model J2000 (Fink et al. 2013; Krause 2002) while the impacts of the human-induced hydrological alterations on drought risk were quantified by simulations of the reservoir operation model *HEC ResSim* (Firoz et al. 2016) which had been calibrated for the existing and planned hydropower reservoirs and operational rules.

The salt water intrusion risk during the low flow period in dependence of reservoir operation impacted water inflow at Thanh My and Ai Nghia stations was simulated by the hydrodynamic numerical *Mike 11* model (Viet et al. 2016; Chapter “Land Use Adaption to Climate Change in the Vu Gia—Thu Bon Lowlands: Dry Season, Rainy Season”).

## Historical Drought Characterization

To identify and assess historical hydrological droughts, the dry season was divided in two periods which also correspond to the two main rice growing periods of highest water demand—from January to April and from May to August-using a k-means cluster analysis (MacQueen 1967; Hartigan and wang 1979). This was applied to daily discharge time series from 1976–2013 (Fig. 3).

For the dry season period from May–August, 20 dry years and for the January–April period 14 dry years were clustered. Also for the period from May–August only three wet and 15 medium years were identified while for the January–April period 7 wet and 16 medium years were clustered. Based on this analysis, we found that the years 1983, 1998, 2005 and 2013 were extremely affected by hydrological drought during both dry periods from January to April and from May to August. The figure also shows that the period between 1977 and 1998 was drier than the following 15 years.

***SPI/SRI Analyses***

Severity, seasonality and length of historical dry periods were assessed by applying the Standardized Precipitation Index (SPI) and its runoff homologue Standardized Runoff (SRI), after McKee et al. (1993). The data are fitted to a probability distribution and later transformed into a normal distribution so that the mean SPI/SRI for the location and desired period is zero. Positive SPI/SRI values indicate greater

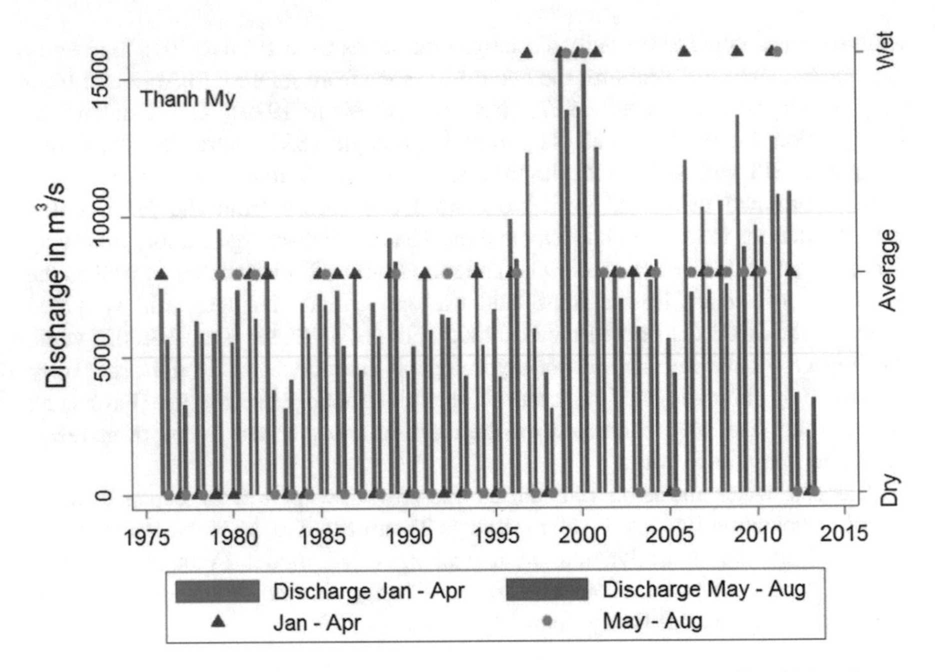


**Fig. 3** Wet, medium and dry clusters for the dry season periods January–April and May–August at Thanh My station

**Table 1** SPI and SRI drought classification (McKee et al. 1993)

| SPI values | SRI values | Condition |
|---|---|---|
| −0.25 to −0.49 | −0.25 to −0.49 | Mild drought |
| −0.5 to −0.99 | −0.5 to −0.99 | Moderate drought |
| −1.0 to −1.44 | −1.0 to −1.44 | Severe drought |
| −1.45 to −1.99 | −1.45 to −1.99 | Very severe drought |
| −2.0 and less | −2.0 and less | Extreme drought |

than median precipitation or runoff and negative values indicate less than median precipitation or runoff (Table 1).

The 1–12 month time steps for SPIs and SRIs were calculated for the entire period (1976–2013) showing similar results as the discharge cluster analysis, indicating strong negative anomalies for the dry seasons of the years 1982–1983, 1998 and 2005.

Comparing SPI and SRI values for the different time steps, on the one hand gave an insight to catchment storage related systematic deviations of SRI from SPI values before hydropower development started in 2008 in the upstream (Fig. 4). On the other, they were used to observe impacts of anthropogenic alterations on streamflow. Strong differences between both values indicate either catchment-related propagation from meteorological to hydrological drought or human alterations of the hydrological system. Figure 4a shows similar values of SPI and SRI

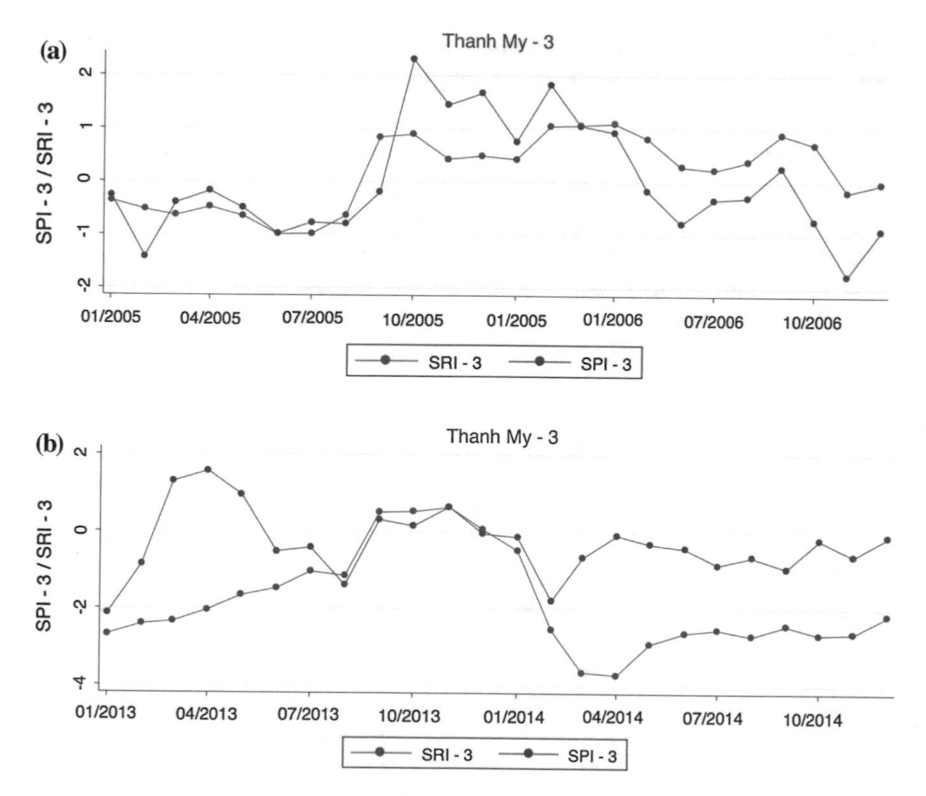


**Fig. 4** Relationship between SPI-3 and SRI-3 in the Vu Gia River basin for the years 2005 (before hydropower operation) and 2013 (after Dak Mi 4 operation). Values below 0 indicate meteorological (SPI) or hydrological drought conditions (SRI)

during the dry season of 2005. During the wet season, however, positive anomalies are higher for the standardized precipitation index while during the following dry season in 2006 discharge keeps more positive values. This might indicate that flood season precipitation was stored in catchment groundwater, vegetation and soils slowly discharging to the stream during the dry season. However, data uncertainties related to catchment wide precipitation and groundwater storage behavior do not allow a final interpretation of such results.

For the year dry seasons of 2013 and 2014, however, results are more obvious. The standardized runoff index (SRI) shows a strong negative anomaly with values below −0.7 during both dry seasons 2013 an 2014 compared to the positive SPI which in previous recorded years tends to be lower than SRI (Fig. 4b). Dak Mi 4 hydropower plant came into operation in early 2012 starting to divert water from Vu Gia to Thu Bon leading to a severe drought in 2013 in both periods.

## Sensitivity of Low Flows Against Land Use Changes and Other Catchment Characteristics

As previously stated, for data scarce tropical catchments there remains a large research gap concerning drought studies and the role of catchment characteristics in drought susceptibility and recovery. A particular issue is that of land use change and the way it affects water partitioning, groundwater recharge and the maintenance of low flows. The impact of land use change and deforestation on floods and water availability in the tropics has a long history (e.g. Calder 1993, 2000; Bruijnzeel 1990; Bradshaw et al. 2007), encouraged by international institutions such as the UN Food and Agricultural Organization (FAO) and the UNESCO International Hydrological Programme when large-scale deforestation became perceived as a major threat to water resources in tropical countries. Although research results indicated that forests might help to protect downstream areas from smaller/moderate floods and flash floods, the effects of land cover change and forest loss on low flows have been more ambiguous.

The rainfall runoff relationship as well as the sensitivity of stream flow to land use and climate change were evaluated applying the hydrological model J2000 which is based on distinct hydrological response units (HRUs) (see Krause 2002; Fink et al. 2013 for further details). J2000 is a distributed, process-based hydrological model used for discharge simulations of meso-and macro-scale catchments (Kralisch and Krause 2006; Kralisch et al. 2007). Its distribution concept is based on three different aggregation levels reflecting the different spatial and temporal dynamics of specific processes which control the variability of: 1. runoff generation processes, 2. runoff concentration and 3. flood routing. The hydrological response units (HRUs) are delineated by GIS overlay techniques of the data-layers elevation, slope, aspect, land use, soil-type, hydrogeological units and subbasins. The model is explained in detail in Krause (2002) which focusses specifically on application in the evaluation of land use change impacts on discharge. It was calibrated for the entire Vu Gia Thu Bon river basin to simulate the hydrological processes of each HRU carrying out data regionalization and correction, calculation of potential and actual evapotranspiration, canopy interception, soil moisture and groundwater processes (Fink et al. 2013; Kralisch et al. 2007). Addressing the individual spatially distributed processes and simulating the flow for each HRU outlet, it is a valuable tool to assess individual catchment and storage processes also during the dry season.

The overall simulation performance showed good results as shown by the efficiency value of 0.77 (Nash and Sutcliffe 1970), a coefficient of determination of 0.84 and an average deviation of 0.8%. The simulations show that the model is able to describe the water balance of the Vu Gia on a daily basis almost exactly capturing the hydrograph during wet and dry periods (Fig. 5a). The sensitivity of discharge and especially dry season low flows against land use changes was tested by converting forest land cover as well as grassland to agricultural land. Results

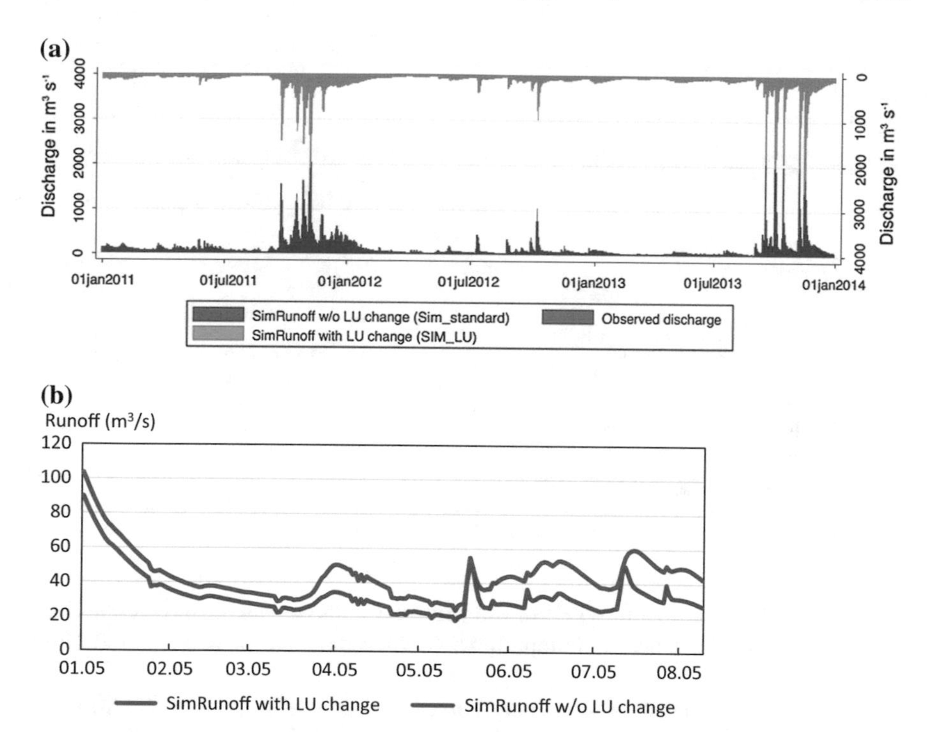


**Fig. 5** **a** J2000 model performance for Vu Gia catchment at Thanh My station for the period between 2011 and 2013. Discharge visibly increases in the simulation of the converted land cover (*orange*) to agricultural area. **b** J2000 simulated discharge with the current forested land cover compared to simulated discharge with land cover converted to agricultural land for dry season May 2005 at Thanh My station

show that discharge would increase by 9% in average during the dry season if land cover would be entirely converted to agricultural land (Fig. 5a, b).

## Impact of Hydropower Development on Downstream Water Availability in the Vu Gia River

The consequences of human-induced hydrological alterations on drought risk were quantified by simulations of the reservoir operation model HEC ResSim (Firoz et al. 2016) which had been calibrated for existing and planned hydropower reservoirs considering the current operational rules. For the purpose of this study the model was set up for the Reservoir Dak Mi 4 which was built on the Vu Gia river diverting its discharge to the Thu Bon and thus has the most dramatic effect on re-distribution of water within the river basin. HEC-ResSim facilitates the development of simulations of single or multiple reservoirs in a hydrological network, based on the

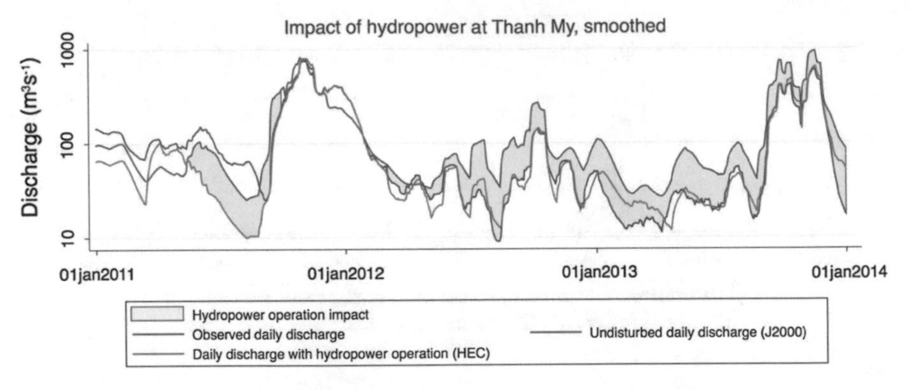


**Fig. 6** J2000 simulations for undisturbed flow and HecReSim hydropower operation model simulations are compared with observed discharge for Thanh My station (*blue line*). Dak Mi 4 Hydropower plant came into operation in 2012. In 2011, observed flow matches well with J2000 simulations and after April 2012 observed discharge matches with HEc Resim Modelling. A ±10 days time step is used for the presentation to avoid flattering lines

available hydrological (inflow) data, the physical reservoir characteristics and the operating rules. The model is comprehensively documented in Klipsch and Hurst (2013).

To assess the impact of the operation of the currently existing hydropower reservoirs on the Vu Gia stream flow, the reservoir operation model HecResim was applied which was established in the scope of the LUCCi project on the basis of the available 33 years of historical hydro-meteorological data record with upstream discharge data provided by the J2000 model (Fink et al. 2013; Firoz et al. 2016). For the purpose of this research, the discharge at Thanh My was simulated with a model setup considering the reservoir operation of the Dak Mi 4 hydropower plant and the corresponding diversion to the Southern catchment Thu Bon (Fig. 1). The simulated daily flow with and without reservoir impact was compared to daily observed flow and J2000 simulated undisturbed flow for Thanh My discharge station (Fig. 6).

Table 2 summarizes the mean monthly change of discharge (1980–2013) under the hypothesis that all current reservoirs (2013) were under operation during this period.

Figure 7 highlights the impact of the Dak Mi 4 reservoir operation on daily flows at Thanh My station for the dry period from 1 April to 30 August, 2013. Undisturbed discharge (blue line) by far exceeds the discharge values impacted by hydropower operation (red line). The dry meteorological conditions are shown by

**Table 2** Mean monthly discharge reduction in $m^3/s$ at Thanh My station with an operating Dak Mi 4 hydropower plant diverting water to the Thu Bon river system as simulated by HecReSim

| Jan | Feb | Mar | Apr | May | Jun | Jul | Aug | Sep | Oct | Nov | Dec |
|---|---|---|---|---|---|---|---|---|---|---|---|
| −30 | −15 | −13 | −17 | −32 | −33 | −26 | −30 | −81 | −125 | −122 | −84 |

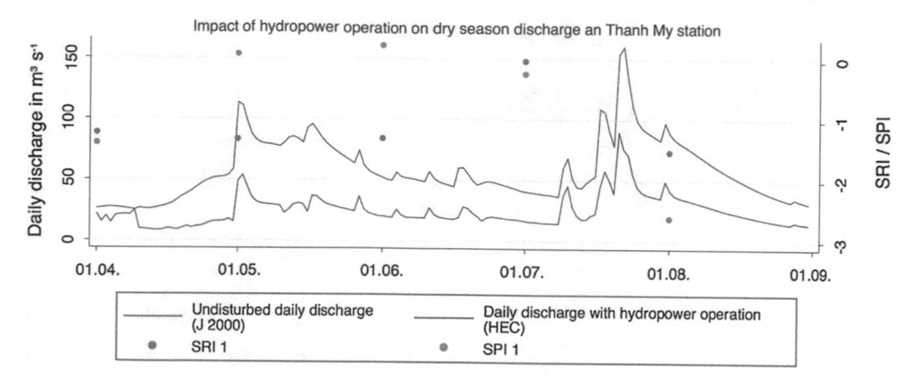


**Fig. 7** Impact of the operating Dak Mi 4 hydropower plant on the Vu Gia River stream flow during the dry season of 2013 at Tanh My station, 1 January to 30 August; Dak Mi 4 diverts water from Vu Gia to Thu Bon

the orange SPI 1 dots below 0 indicating a negative precipitation anomaly. Green dots show the respective negative discharge anomalies which do not coincide with the SPI values.

## Drought and Saltwater Intrusion Risk

Drought disasters in the VGTB river basin estuary are essentially characterized by salinity intrusion affecting agricultural production and drinking water supply. Salinity thresholds for irrigation are 1.0 g/L. As soon as salinity values in river water exceed 1.0 g/L, irrigation pumping stations are turned off.

In the scope of the LUCCi project, a Mike 11 hydrodynamic model was set up to determine the salt water intrusion risk in the delta region under different scenarios (compare Chapter “Land Use Adaption to Climate Change in the Vu Gia—Thu Bon Lowlands: Dry Season, Rainy Season”; Viet 2014). For this analysis, the salt water intrusion hazard risk for the dry reference period in 2005 was calculated for the current situation and compared to the hazard risk after Dak Mi 4 hydropower plant came into operation. The inflow data to the Mike 11 model for this simulation were taken from the Hec Resim daily discharge simulations at Thanh My as described in the previous section (Fig. 8).

In Fig. 8 the simulated saltwater intrusion hazard risk before and after reservoir operation and the consequent water diversion to Thu Bon are shown as well as a diversion further downstream—Quang Hue. Reference season is the relatively dry year 2005. Under the strong interaction of river flow and tide propagation, the length and the boundaries of salt hazards are constantly changing. Therefore, the average time periods per year were chosen to determine the risk level. These risk levels during up to 90 days a year are mainly occurring in the dry season.

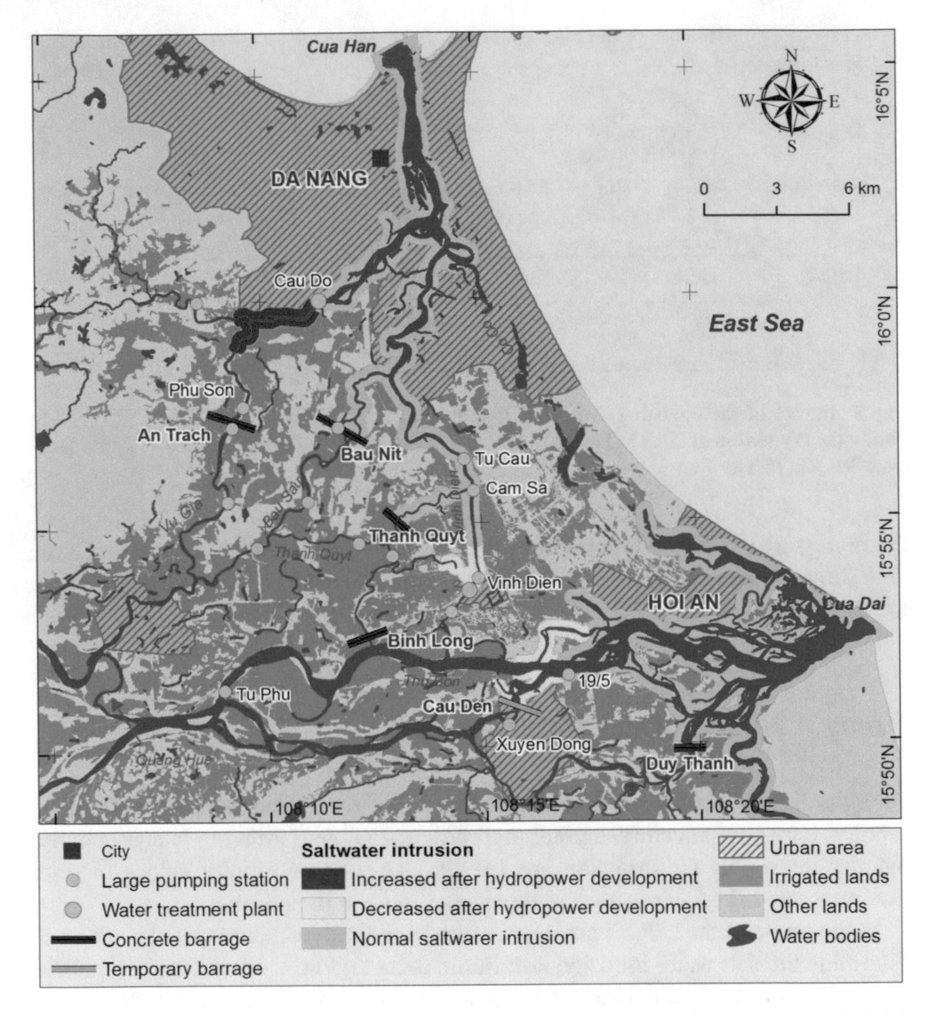


**Fig. 8** Spatial extent of the saltwater intrusion hazard risk in the estuary before and after Dak Mi 4 hydropower came into operation

It shows that the remarkable reduction of Vu Gia flow due to the diversions leads to an increasing saltwater intrusion hazard risk in Vu Gia and Vinh Dien rivers while the salinity hazard risk in the Thu Bon River is decreasing due to the diversion. In the Vu Gia, high salt hazard reaches upstream Cau Do drinking water treatment plant and threatens drinking water supply for approximately one million inhabitants of Danang City. On Vinh Dien River, the high salt hazard appears at the intakes of Cam Sa and Tu Cau Pumping Stations but does not affect the normal operation of Vinh Dien Pumping Station. Increased inflow of Thu Bon after water diversion from Vu Gia pushes back saltwater intrusion in the river and protects the intakes of large pumping stations along the river from high saltwater intrusion.

## Conclusion

The tropical Central Vietnamese Vu Gia—Thu Bon river basin is frequently affected by droughts and saltwater intrusion strongly impairing local socio-economy in particular the agricultural sector (though rice irrigation), drinking water supply and hydropower. Hence there is an urgent need to understand local drought phenomena and how different aspects of the hydrological cycle are related to climate, catchment characteristics as well as to human intervention. However, drought research has so far been less focused on tropical less-developed regions compared to other geographical areas. In this research, common drought research concepts were applied to assess the particular drought behaviour in the Vu Gia Thu Bon (VGTB) river basin using an ensemble of models to simulate low flows as pragmatic tools for drought management. Meteorological and hydrological droughts in the Vu Gia Thu Bon were assessed by using a K-means Cluster Analysis and standardized precipitation and discharge indices applied to the daily precipitation and discharge time series from 1976 to 2013. The rainfall runoff relationship as well as the sensitivity of stream flow to land use and climate changes was assessed with the physically-based distributed model J2000. The impacts of human induced hydrological alterations to drought risk were quantified by simulating operational consequences using the reservoir model HEC-ResSim which had been calibrated for the existing and planned hydropower reservoirs and operational rules. The models capture low flows quite well in the simulations and are valuable tools to detect the impacts of land use changes and reservoir operation. Also the additional spatial and temporal salt water intrusion risk could be quantified by informally linking models with the (1) spatially distributed J2000 simulating discharge for key inflow locations to the reservoirs (HEC) and irrigation scheme (Mike 11), (2) Hec ResSim delivering the reservoir impacted inflow time series to (3) Mike 11 simulating the salt water behaviour in dependency of sea level and inflow from upstream impacted by reservoir operation. In order to improve the decision making process for future drought situations, simulation results, related uncertainties as well as expected changes of the system and correspondent scenarios need to be further discussed with the hydropower generators and the water users in the estuary. Minimum water releases from the reservoirs need to be defined for the dry season adapted to the irrigation cycle and downstream water demand. Early warning based on modelled measured real time discharge data should be established based on the thresholds calculated in the scope of this study.

## References

Adamson P, Bird J (2010) The Mekong: a drought-prone tropical environment? Int J Water Resour Dev 26:579–594

Birkel C, Soulsby C, Tetzlaff D (2012) Modelling the impacts of land-cover change on streamflow dynamics of a tropical rainforest headwater catchment. Hydrol Sci J doi:10.1080/02626667.2012.728707

Bradshaw CJA, Sodhi NS, Peh KS-H, Brook BW (2007) Global evidence that deforestation amplifies flood risk and severity in the developing world. Glob Change Biol 13:2379–2395. doi:10.1111/j.1365-2486.2007.01446.x

Bruijnzeel LA (1990) Hydrology of moist tropical forests and effects of conversion: a state of knowledge review by the faculty of earth sciences, Free University, PO Box 7161, 1007MC Amsterdam, The Netherlands, 224, UNESCO IHP

Calder IR (1993) Hydrologic effects of land-use change. chapter 13. In: Maidment DR (ed) Handbook of hydrology. McGraw-Hill, New York, p. 50

Calder IR (2000) Land use impacts on water resources. Background paper 1. In: FAO electronic workshop on land-water linkages in rural watersheds. 18 September–27 October 2000

Fink M, Fischer C, Führer N, Firoz AMB, Viet TQ, Laux P, Flügel W (2013). Distributive hydrological modeling of a monsoon dominated river system in central Vietnam. International congress on modelling and simulation(20th), pp 1826–1832

Firoz ABM, Nauditt A, Fink M, Ribbe L (2016) Modelling the impact of hydropower development and operation on downstream discharge in a highly dynamic tropical central Vietnamese river basin, submitted to HESS

Geng G, Wu J, Wang Q, Lei T, He B, Li X, Mo X, Luo H, Zhou H, Liu D (2016) Agricultural drought hazard analysis during 1980–2008: a global perspective. Int J Climatol 36

Hartigan J-A, Wong MA (1979) Algorithm AS 136: a k-means clustering algorithm. Appl Stat 100–108

ICEM (2008) Strategic environmental assessment of the Quang Nam Province Hydropower Plan for the Vu Gia-Thu Bon River Basin. Prepared for the ADB, MONRE, MOITT & EVN, Hanoi, Viet Nam, 1–205

IMHEN (2008) Mapping drought and water deficient level in the South Central Coast and Highlands. Hanoi

IMHEN (2012) Climate change, sea level rise scenarios for Vietnam. Ministry of Resources management and Environment, Hanoi

Krause P (2002) Quantifying the impact of land use changes on the water balance of large catchments using the J2000 model. Phy Chem Earth 27:663–673

Kralisch S, Krause P (2006) JAMS—a framework for natural resource model development and application. In: Voinov A, Jakeman A, Rizzoli AE (eds) proceedings of the iEMSs third biannual meeting. Burlington, USA

Kralisch S, Krause P, Fink M, Fischer C, Flügel W-A (2007) Component based environmental modelling using the JAMS framework. In: Kulasiri D, Oxley L (eds) proceedings of the MODSIM 2007 international congress on modelling and simulation. Christchurch, New Zealand

MacQueen JB (1967) Some methods for classification and analysis of multivariate observations. In: proceedings of 5th berkeley symposium on mathematical statistics and probability. I:281–297, Univercity of California Press, Berkeley, California

McKee TB, Doesken NJ, Kleist J (1993) The relationship of drought frequency and duration of time scales. Presented at the eighth conference on applied climatology, Anaheim, CA. American Meteorological Society, pp 179–186

Mishra AK, Singh VP (2010) A review of drought concepts. J Hydrol 391:202–216

Nash JE, Sutcliffe JV (1970) River flow forecasting through conceptual models. Part I—A discussion of principles. J Hydrol 10:282–290

Navuth T (2007) Drought management in the Lower Mekong Basin. MRC, Vientiane

NDMC (2012) Centre for Ecology and Hydrology, Maclean Building, Crowmarsh Gifford, Oxfordshire OX10 8BB, UKNDMC, 2012, National Drought Mitigation Center Website, International Early Warning. http://drought.unl.edu/MonitoringTools/InternationalEarlyWarning.aspx. Assessed 20 July 2012

RCHM (Regional Center for Hydro-meteorology) (2013) Time series of meteo-hydrologic data in 1977–2010 in the Vu Gia Thu Bon River Basin

Stoelzle M, Stahl K, Morhard A, Weiler M (2014) Streamflow sensitivity to drought scenarios in catchments with different geology. Geophys Res Lett 41:6174–6183 doi:10.1002/2014GL061344

Tallaksen LM, van Lanen HAJ (2004) Introduction. In: Tallaksen LM, von Lanen HAJ (eds) Hydrological drought—processes and estimation methods for streamflow and groundwater. Developments in water science, vol 48. Elsevier Science B.V, Amsterdam

Terink W, Immerzeel WW, Droogers P (2011) Drought monitoring and impact assessment in the Mekong River Basin. Report Future Water: 104, Wageningen

UCL (2012) Global Drought Monitor, University College London, Department of Space and Climate Physics, Aon Benfield, UCL Hazard Research Centre. Assessed 22 July 2012

UNDP (2011) Mainstreaming drought risk management—a primer, UNON Printshop, Nairobi United Nations Office at Nairobi (UNON), Publishing Services Section, ISO 14001:2004-certified/March 2011

UN-ISDR (2005) The Hyogo framework of action, Geneva: UN-ISDR. http://www.unisdr.org/2005/wcdr/intergover/official-doc/L-docs/Hyogo-framework-for-action-english.pdf. Assessed 15 July 2012

UN-ISDR (2009) Drought risk reduction, framework and practices: contributing to the implementation of the Hyogo framework of action. UN-ISDR, Geneva

Van Lanen HAJ, Wanders N, Tallaksen LM, Van Loon AF (2013) Hydrological drought across the world: impact of climate and physical catchment structure. Hydrol Earth Syst Sci 17:1715–1732

Van Loon AF, Van Lanen HAJ (2012) A process-based typology of hydrological drought. Hydrol Earth Syst Sci 16(7):1915–1946. doi:10.5194/hess-16-1915-2012

Van Loon AF, Van Lanen HAJ (2013) Making the distinction between water scarcity and drought using an observation-modeling framework. Water Resour Res 49:1483–1502. doi:10.1002/wrcr.20147

Van Loon AF (2013) On the propagation of drought—how climate and catchment characteristics influence hydrological drought development and recovery. PhD thesis, Wageningen University

Van Loon AF (2015) Hydrological drought explained. WIREs Water 2:359–392. doi:10.1002/wat2.1085

Van Loon AF, Laaha G (2015) Hydrological drought severity explained by climate and catchment characteristics. J Hydrol doi:10.1016/j.jhydrol.2014.10.059

Van Loon AF, Gleeson T, Clark J, Van Dijk AIJM, Stahl K, Hannaford J, Di Baldassarre G, Teuling AJ, Tallaksen LM, Uijlenhoet R, Hannah DM, Sheffield J, Svoboda M, Verbeiren B, Wagener T, Rangecroft S, Wanders N, Van Lanen HAJ (2016) Drought in the anthropocene. Nat Geosci 9:89–91. doi:10.1038/ngeo2646

Viet TQ (2014) PhD Thesis: estimating the impact of climate change induced saltwater intrusion on agriculture in estuaries-the case of the Vu Gia Thu Bon river basin, Central Vietnam, University of Bochum/Cologne University of Applied Sciences, ITT

Viet TQ, Stolpe H, Ribbe L, Nauditt A (2016) Impact of sea level rise on saltwater intrusion in the Vu Gia Thu Bon estuary, Central Vietnam, submitted to Continental Shelf Research

Vogt JV, Safriel U, Maltitz GV, Sokona Y, Zougmore R, Bastin G, Hill G (2011) Monitoring and assessment of land degradation and desertification: towards new conceptual and integrated approaches. Land Degrad Dev 22:150–165

Wilhite DA, Glantz MH (1985) "Understanding the drought phenomenon: the role of definitions". Drought Mitigation Center Faculty Publications. Paper 20

Wilhite DA, Sivakumar MVK, Wood DA, Svoboda MD (2000) Early warning systems for drought preparedness and drought management. In: Proceedings of an expert group meeting held in Lisbon, Portugal, 5–7 September, 2000. World Meteorological Organization, Geneva, Switzerland

# Conclusion

**Lars Ribbe, Viet Quoc Trinh and Alexandra Nauditt**

This book summarizes key elements of the research output of a five-year project conducted by a consortium of researchers from Germany, Vietnam, and International Organisations together with stakeholders from the national and regional level in Vietnam. Compared to the state of knowledge at the beginning of the project, significant advances could be achieved. However, the tremendous lack of basic spatial, socioeconomic, and hydro-meteorological data was a strong inhibitor to reach an advanced level of research or the building up of scientifically sound scenarios and other bases for planning. Partially this data gap could be filled by the immediate research results but ultimately a monitoring process with the aim to provide such fundamental data on the human and natural environment is the task of established, continuously running state governed monitoring programs. The LUCCi project hopefully supported the setup of improved monitoring programs of the relevant institutions working closely together with the scientist during research.

The establishment of Science-Policy or Science-Society interfaces is widely discussed in the research-, practitioner-, and funding communities The LUCCi project is with no doubt a positive example of the interaction of researchers with institutions in the study region from the initial steps of research to the dissemination of results. However, the different institutions involved clearly see a demand for a continuous exchange in form of a support of research for key development issues. It is a general question how such interfaces can be generated in a sustainable format,

L. Ribbe (✉) · V.Q. Trinh · A. Nauditt
Institute for Technology and Resources Management in the Tropics and Subtropics, TH Köln, Cologne, Germany
e-mail: lars.ribbe@th-koeln.de

V.Q. Trinh
e-mail: trinhquocviet1981@googlemail.com

A. Nauditt
e-mail: alexandra.nauditt@th-koeln.de



A. Nauditt and L. Ribbe (eds.), *Land Use and Climate Change Interactions in Central Vietnam*, Water Resources Development and Management,
DOI 10.1007/978-981-10-2624-9_15

who takes responsibility to organize them and how to finance them. With the current institutional arrangements, neither the research community nor the stakeholders responsible to plan and implement natural resources policies have the means available to engage in this process.

Next to a clear benefit for development issues within the Vu Gia Thu Bon river basin the research results have a potential value for other regions—in Vietnam or internationally—facing similar challenges. Within the LUCCi consortium, potential transfer basins were identified and one of the current tasks is to define the way forward to implement results from the project also in other regions in Vietnam. The Vietnamese Ministry of Agriculture and Rural Development already took action to apply the research results and tools developed under the Lucci project to provinces in South Central Vietnam with a pilot province being Ninh Thuan. This could be a positive example of a continuation of a science policy dialog involving researchers and institutions at different levels in Vietnam and Germany.

## Transferring Project Results to Other Vietnamese Regions

The potential transfer of the project results to other central Vietnamese river basins and provinces was a major task for the last LUCCi project phase. The experiences and knowledge gained during the first four years of the LUCCi project regarding the institutional setup, the identification of stakeholders responsible for land and water resources management on national, provincial, and district level will allow a systematic process of transferring the results to other river basins and provinces. Based on environmental and socio-economic conditions as well as on current land and water resources management challenges alltogether seven basins were identified as potential transfer basins (Fig. 1).

The developed methods and tools, which can be transferred, comprise assessment, planning and management tools, support for improved monitoring and data storage for hydrometeorology, land resources and water quality, conservation strategies as well as capacity building measures. The developed tools are appropriate for similar environments with low data availability and similar challenges and namely are for example: River Basin Information System, River Basin Status Report Template, modeling and decision support systems, handbook for sustainable land and water resources management strategy. These products will be offered to the leading institutions at National level as MARD and MONRE and to the PPCs of the involved provinces.

The following maps show the key environmental and socio-economic characteristics of those catchments in Central Vietnam which have been selected as potential transfer basins (Table 1).

In recent years drought disasters culminating in El Niño-induced drought and saltwater intrusion of 2015–2016 have been impacting agricultural production in the Mekong Delta and South-Central Vietnam. This worst drought has destroyed an estimated area of 249,944 ha of rice, 18,960 ha of other annual crops, 149,704 ha

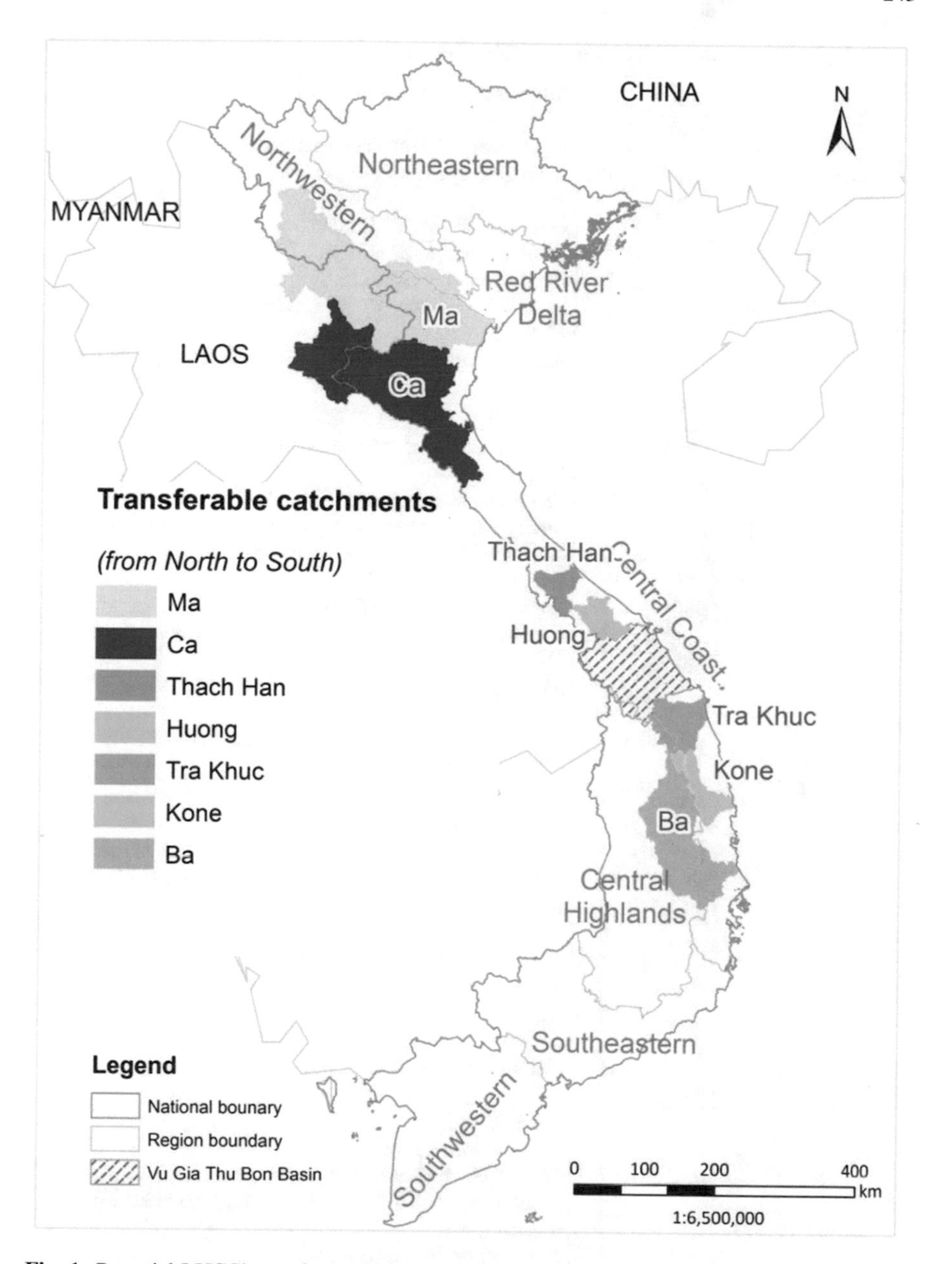


**Fig. 1** Potential LUCCi transfer basins

of industrial crops, and 6,857 ha of aquaculture. Approximately one million tons of rice has been lost due to drought and saltwater intrusion. Thus, next to Central Vietnam the transfer of knowledge from the LUCCi project is very urgent for South-Central Vietnam. The proposed research will be carried out in the driest region of Vietnam: the Dinh River Basin, Ninh Thuan Province. From 2000 to

**Table 1** Basic features of selected transferable river basins

| | |
|---|---|
| 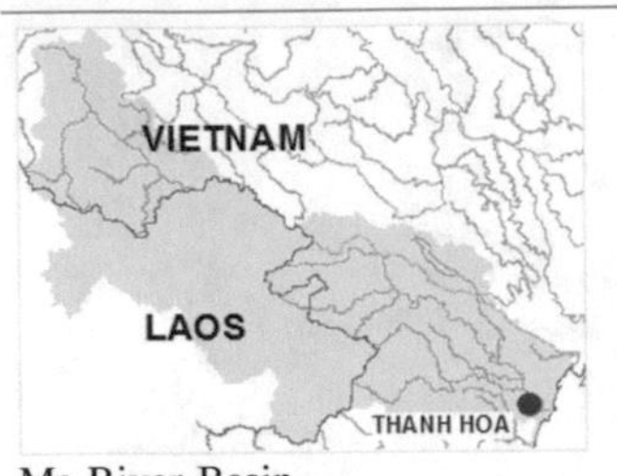  Ma River Basin | *Area*: 28,400 km$^2$, of which 17,600 km$^2$ (62 %) in Vietnam<br>*Representative province*: Thanh Hoa<br>*Main Soil*: Ferralsol (75 %), Fluvisol (12 %) Acrisol (6 %)<br>*Land use*: forest (29 %), annual crops<br>*Water Flow*: 18–22 Bill. m$^{3\ -1yr}$<br>*Population*: 4.6 Mio. (2009)<br>*Economic*: industry 42 %, agriculture 15 %, services 43 % (2010)<br>*Main hydropower reservoir*: Muong Hinh, Canh Pet, Cua Dat<br>*Challenges*: wildfire, drought, floods, salinity intrusion |
| Ca River Basin | 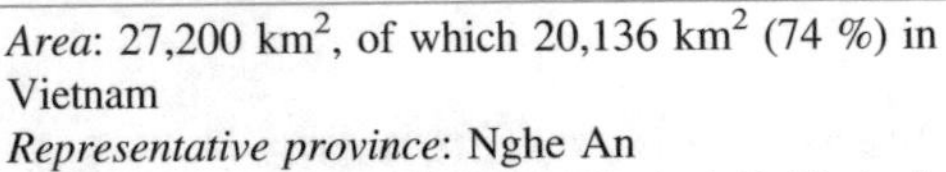 *Area*: 27,200 km$^2$, of which 20,136 km$^2$ (74 %) in Vietnam<br>*Representative province*: Nghe An<br>*Main Soil*: Ferralsol (83.5 %), Fluvisol (8 %) Acrisol<br>*Land use*: forest (41 %), annual crops<br>*Water Flow*: 21–24 Bill. m$^{3\ -1yr}$<br>*Population*: 3.4 Mio. (2009)<br>*Economic*: industry 39 %, agriculture 21 %, services 40 % (2010)<br>*Main hydropower reservoir*: Ban Ve, Ban Mong, Thac Muoi<br>*Challenges*: drought, floods, extreme hot in Föhn effecting days |
|   Thach Han River Basin | *Area*: 2500 km$^2$<br>*Representative province*: Quang Tri<br>*Main Soil*: Arenosols (13 %), Fluvisol (21.3 %) Acrisol<br>*Land use*: forest (40.4 %), agricultural land (15.1 %), unused (35 %)<br>*Water Flow*: ~3.1 Bill. m$^{3\ -1yr}$<br>*Population*: 0.67 Mio. (2009)<br>*Economic*: industry 26 %, agriculture 32 %, services 42 % (2010)<br>*Main hydropower reservoir*: Rao Quan<br>*Challenges*: floods, drought, sand expansion |
|   Huong River Basin | *Area*: 3948 km$^2$<br>*Representative province*: Thua Thien Hue<br>*Main Soil*: Acrisols (68.7 %), Fluvisol (19.5 %) Acrisol<br>*Land use*: forest (45 %), agricultural land (12.3 %), unused (37 %)<br>*Water Flow*: ~5.64 Bill. m$^{3\ -1yr}$<br>*Population*: 0.9 Mio. (2009)<br>*Economic*: industry 42 %, agriculture 12 %, services 46 % (2010)<br>*Main hydropower reservoir*: Binh Dien, Hung Dien<br>*Challenges*: floods, drought, landslide |

(continued)

**Table 1** (continued)

| | |
|---|---|
|   Tra Khuc River Basin | *Area*: 5200 km$^2$<br>*Representative province*: Quang Ngai<br>*Main Soil*: Ferrasols (1.58 %), Fluvisol (18.9 %) Acrisols (73.42 %)<br>*Land use*: forest (29.4 %), agricultural land (14.3 %), unused<br>*Water Flow*: ~6.1 Bill. m$^{3\ -1yr}$<br>*Population*: 1.2 Mio. (2009)<br>*Economic*: industry 55 %, agriculture 25 %, services 30 % (2010)<br>*Main hydropower reservoir*: Dakdrinh<br>*Challenges*: floods, drought, salinity intrusion, pollution |
|   Kone River Basin | *Area*: 3640 km$^2$<br>*Representative province*: Binh Dinh<br>*Main Soil*: Arenosols (2.6 %), Fluvisol (7.6 %) Acrisol (70 %)<br>*Land use*: forest (41 %), agricultural land (10 %), unused (25 %)<br>*Water Flow*: ~4.36 Bill. m$^{3\ -1yr}$<br>*Population*: 1.1 Mio. (2009)<br>*Economic*: industry 27.4 %, agriculture 35 %, services 37.6 % (2010)<br>*Main hydropower reservoir*: Vinh Son<br>*Challenges*: drought, floods, salinity intrusion |
| 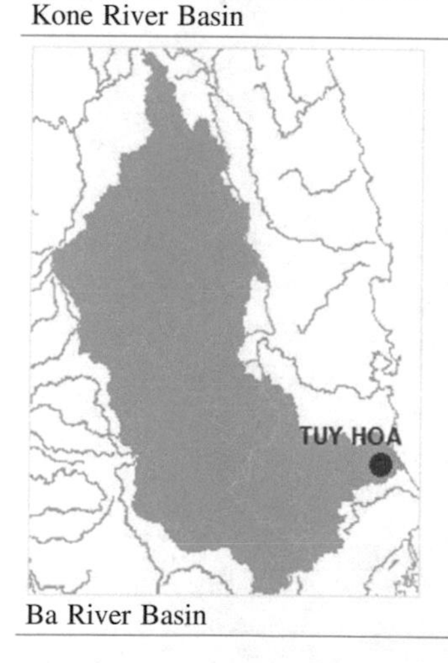  Ba River Basin | *Area*: 13,900 km$^2$<br>*Representative province*: Phu Yen<br>*Main Soil*: Arenosols (13 %), Fluvisol (21.3 %) Acrisol<br>*Land use*: forest (51.2 %), agricultural land (26.3 %), unused (18.6 %)<br>*Water Flow*: ~9.79 Bill. m$^{3\ -1yr}$<br>*Population*: 2.1 Mio. (2009)<br>*Economic*: industry 26 %, agriculture 32 %, services 42 % (2010)<br>*Main hydropower reservoir*: An Khe-Kanal, Krong Hnang, Song Hinh, Ba Ha, Ia Mla<br>*Challenges*: drought, floods |

2015, the basin has been experiencing rainfall deficits, which alongside with socioeconomic development and increasing water demand have provoked severe drought disasters pushing rural population into poverty and hunger. Taking the most recent drought of the year 2015 as an example, the fallowed food production areas in winter–spring crop were 6,100 ha, including 3,214 ha of paddy rice and 2,886 ha of annual crops to cause the loss of 30,000 tons of food. Drought continued to cause severe damage in next summer-autumn cropping season and 5,430 ha of paddy farm and annual crops were fallowed. Diminished flows in dry seasons lead to intensification of saltwater intrusion to threaten irrigation downstream. Also ecosystems are affected and desertification is expanding to 12 % of uncultivated areas of the basin.

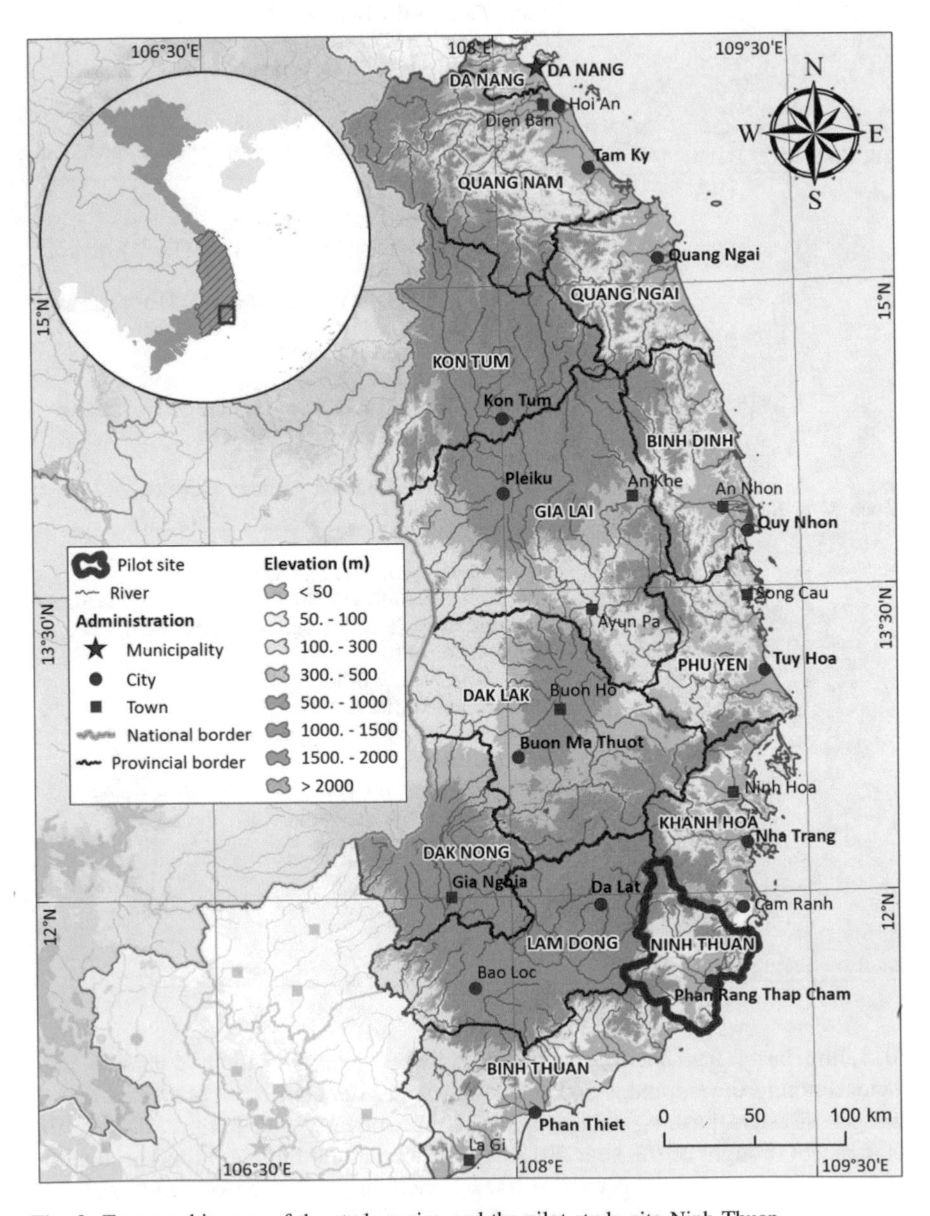


**Fig. 2** Topographic map of the study region and the pilot study site Ninh Thuan

Figure 2 shows the drought prone study region highlighting the pilot basin of Dinh River, Ninh Thuan Province, South Central Vietnam.

To address the challenges of meteorological and hydrological droughts and to identify related research demands, we have invited involved institutions from regional and National level as well as ODA institutions to discuss drought monitoring assessment and management for the Southern Central Region and South East Asia. Furthermore, funding options and research demand have been explored based on a moderated science policy dialog. In conclusion the tools and experiences gained by the LUCCi project were considered as very adequate to be applied in the pilot region of Ninh Thuan. In this context, a German–Vietnamese project has been prepared and was presented to the German Ministry of Education and Research by the end of June.

The LUCCi project yielded fundamental contributions to advance the understanding of socio-ecological systems of the Vu Gia Thu Bon River Basin. The new insights are related to hydrology, floods and droughts, soil and sediment balance, carbon balance, climate change, vegetation and ecosystems; biodiversity; agricultural and hydropower systems. As the research was planned and coordinated from the beginning with local stakeholders and referred to their demands and challenges, results could often be transferred into an improved decision making in the region. Examples include the selection of sites for weir construction to control saltwater intrusion and the creation of a reservoir level information system. The design and implementation of capacity development for the local institutions as well as the establishment of a River Basin Information Center (RBIC) and stakeholder friendly information products contributed to the dissemination of the knowledge into the region and this way hopefully started processes which will create tangible benefits for development in the years to come. However, if this will actually be the case is difficult to say and should be questioned. In general the sustainability of transdisciplinary research projects can be contested if there are no mechanisms in place which foster a science-policy dialogue or support the implementation of project results. However, with regard to the LUCCi project it should be noted positively that the Government of Vietnam consulted members of the research consortium regarding the recent drought problem in South-Central Vietnam which opens up an opportunity to continue the solution oriented research in the region.

# Index

*Note*: Page numbers followed by *f* , *t* and *n* indicate figures, tables and notes respectively





A. Nauditt and L. Ribbe (eds.), *Land Use and Climate Change Interactions in Central Vietnam*, Water Resources Development and Management,
DOI 10.1007/978-981-10-2624-9

Printed in the United States
By Bookmasters